Psychopathy Unmasked

Psychopathy Unmasked

The Rise and Fall of a Dangerous Diagnosis

Rasmus Rosenberg Larsen

The MIT Press
Cambridge, Massachusetts
London, England

The MIT Press
Massachusetts Institute of Technology
77 Massachusetts Avenue, Cambridge, MA 02139
mitpress.mit.edu

The MIT Press would like to thank the anonymous peer reviewers who provided comments on drafts of this book. The generous work of academic experts is essential for establishing the authority and quality of our publications. We acknowledge with gratitude the contributions of these otherwise uncredited readers.

This book was set in Stone Serif and Stone Sans by Westchester Publishing Services. Printed and bound in the United States of America.

Library of Congress Cataloging-in-Publication Data

Names: Larsen, Rasmus Rosenberg, author.
Title: Psychopathy unmasked : the rise and fall of a dangerous diagnosis / Rasmus Rosenberg Larsen.
Description: Cambridge, Massachusetts : The MIT Press, [2025] | Includes bibliographical references and index.
Identifiers: LCCN 2024037464 (print) | LCCN 2024037465 (ebook) | ISBN 9780262552202 (paperback) | ISBN 9780262382854 (epub) | ISBN 9780262382861 (pdf)
Subjects: MESH: Antisocial Personality Disorder | Forensic Psychology
Classification: LCC RA1148 (print) | LCC RA1148 (ebook) | NLM WM 190.5.A2 | DDC 614/.15—dc23/eng/20250113
LC record available at https://lccn.loc.gov/2024037464
LC ebook record available at https://lccn.loc.gov/2024037465

EU product safety and compliance information contact is: mitp-eu-gpsr@mit.edu

·The hallmark of scientific behaviour is a certain scepticism even towards one's most cherished theories.

—Imre Lakatos

Contents

Acknowledgments

This book has a relatively long history, being the output of almost 15 years of academic interest in *Psychopathic Personality Disorder* or "psychopathy." I first began working on the subject as a graduate student at the University at Buffalo, which later evolved into a targeted research program in my role as an assistant professor at the University of Toronto. Without the support from these two institutions, this book would never have come to fruition.

During my years in Buffalo, the Department of Philosophy facilitated an intellectually stimulating environment that encouraged critical thinking and interdisciplinary collaboration, two aspects that have proven crucial for writing a book on psychopathy. I am indebted to the entire faculty at Buffalo, but especially Barry Smith, Kah-Kyung Cho, Neil Williams, and David Braun, who each played an elemental role in maturing me as an academic.

Since 2017, I have been part of the faculty in the Department of Philosophy and the Forensic Science Program at the University of Toronto, where I have had the privilege of teaching a diverse set of courses to an incredibly talented bunch of students. I am thankful for all the conversations I have had with them, and this book is in no small part an attempt to reproduce our discussions about psychopathy. My gratitude also goes out to all my colleagues, and especially Tracy Rogers, Dianna Raffman, Gurpreet Rattan, Martin Pickavé, and Amy Mullin, for their support and guidance over the years.

Psychopathy research dates back centuries, which means that novel contributions never grow out of a vacuum. Over the years, my work has been influenced by many prominent scholars in the field. Of central value to me was the work by Scott Lilienfeld, John Edens, David DeMatteo, Daniel Murrie, Jennifer Skeem, Thomas Widiger, Robert Hare, Stephen Hart, Marcus Boccaccini, Devon Polaschek, Christopher Patrick, Heidi Maibom, and Luca Malatesti. However, the two most influential voices behind this book are

Jarkko Jalava and Stephanie Griffiths. Their research inspired me to critically re-think what I thought I knew about psychopathy, and their mentorship is, in no small part, what led to my ambition of writing this book.

Writing a book that aims to synthesize knowledge from multiple scientific disciplines, as well as legal and clinical practices, brought tremendous and at times agonizing challenges. To ease the process, I was helped by friends and colleagues who generously spent countless hours providing indispensable feedback while I was polishing the ideas and material for this book. These include Lauren Schroeder, Jarkko Jalava, Stephanie Griffiths, Sonya McLaren, David Sackris, David DeMatteo, Shadd Maruna, Stephen Hart, Peter Koch, Philip Deming, Nathan Howard, Kristian Larsen, Harjeet Parmar, Dylan Jones, Zach Richer, Janna Hastings, Barry Smith, René Rosfort, Noor Abbas, Julia Tang, and the four anonymous reviewers from the MIT Press. In their own way, each of these individuals made important suggestions that led to significant improvements, and I quickly lost track of the number of embarrassing blunders they pointed out.

Finally, I would like to thank my family and friends for their help and encouragement throughout the years, especially to those in Buffalo and Toronto who opened their doors and hearts to welcome a Danish immigrant seeking new opportunities in North America. But one person shines brighter than anyone else: my wonderful and brilliant wife, Lauren Schroeder. I met her on a dark and cold October night in 2015, but since then the world has seemed lighter and warmer. She is both the love of my life and my role model as a scientist. Without her, these pages would most certainly never have been written.

Psychopathy research is an interdisciplinary field, spanning dozens of research topics that have morphed into subfields in their own rights, using different sophisticated methods and statistics. So, in shaping the focus of this book, I was forced to make decisions about which topics to prioritize and exclude. Despite these limitations, I hope the final product is decent and that it will stimulate critical thought and conversation about a topic that many experts agree so urgently needs it. Enjoy!

Introduction: The Problem of Psychopathy

Individuals with psychopathic personality, or psychopaths, have a disproportionate impact on the criminal justice system. Psychopaths are twenty to twenty-five times more likely than non-psychopaths to be in prison, four to eight times more likely to violently recidivate compared to non-psychopaths, and are resistant to most forms of treatment . . . Given psychopathy's enormous impact on society in general and on the criminal justice system in particular, there are significant benefits to increasing awareness of the condition.

—Kent A. Kiehl and Morris Hoffman in *The Criminal Psychopath: History, Neuroscience, Treatment, and Economics* (2011)[1]

The history of humankind is fraught with violence and destruction. Although our species occasionally inspires admirable achievements for the betterment of the planet and fellow citizens, it is undeniable that human civilization also casts a trailing shadow of pain and ruin. In the past two centuries, more than 200 million people have died because of armed conflicts raged by national armies and guerilla militias.[2] Many of these victims were innocent civilians— mostly women and children—and many suffered the brutal fate of a genocidal onslaught, preceded by relentless ideological campaigns of persecution and denigration.[3] As of this writing in the year 2024, United Nations estimates that a record-high 110 million people are currently displaced from their homes due to conflict and human-caused devastation.[4] Unfortunately, wars and political disorder are not the sole cause of moral wreckage. Even in times of peace and prosperity, societies still struggle to keep our more unsavory inclinations at bay. According to data from Statista, during the year 2020, the United States—one of the richest countries in the world—had more than 6.4 million property crimes, and violent assault exceeded 1.3 million reported cases. These numbers include around 140,000 rapes, 20,000 murders, and more than 150,000 incidents where children were either physically or sexually abused.[5] Following a 2017 report, the aggregate yearly cost of crime in the United States is estimated to scale as high as $3.41 trillion, roughly 50% of what the US Government budgets every year for the military, public schooling, health care, social security, and so forth.[6]

But *why* is human civilization tainted with such an abundance of violence and destruction? According to experts in forensic psychiatry and psychology, an important part of the explanation can be found in an endemic personality disorder known as *psychopathy* (or *Psychopathic Personality Disorder*). People with this disorder—colloquially referred to as *psychopaths*—have the most peculiar psychological disposition, which for centuries has bewildered leading clinicians and mental health scientists around the world. The hallmark of psychopathic persons is that they do not appear to have any immediate mental dysfunctions, often coming across as utterly normal and perhaps even intelligent, charming, and affable. However, this outward appearance of normalcy is merely a façade that the person has carefully devised to mask an underlying disturbing psychology largely incapable of social functioning. What is hiding under this *mask of sanity*—as the American psychiatrist Hervey Cleckley famously characterized the disorder—is a cold and barren psychological profile, characterized by a profound impairment of moral and emotional capacities, such as lacking conscience and the ability to empathize and feel remorse.

What makes psychopathy such a compelling explanation of human violence is, according to prominent researchers, that the disorder comprises a strikingly antisocial and predatory disposition, which propels psychopathic persons into a tumultuous life shaped by continuous transgressions of the law and social norms—not because they necessarily take a sadistic pleasure in the havoc they cause but mostly because they are deprived of the psychological mechanisms that foster prosocial attitudes and keep aggressive impulses in check. As the lauded researcher Robert Hare writes in the opening pages of his 1993 best-selling book, *Without Conscience: The Disturbing World of The Psychopaths among Us*:

> Psychopaths are social predators who charm, manipulate, and ruthlessly plow their way through life, leaving a broad trail of broken hearts, shattered expectations, and empty wallets. Completely lacking in conscience and in feelings for others, they selfishly take what they want and do as they please, violating social norms and expectations without the slightest sense of guilt or regret.[7]

The Scope of the Problem

Although psychopathy is a relatively rare condition, believed to be present in only 0.6% to 2.3% of the general population,[8] it is broadly recognized as an alarming social problem with overwhelming implications for public health and safety.[9] For one, researchers report that psychopathic persons are far more likely than the average individual to commit a legal offense grave enough to warrant a prison sentence, where some estimates suggest that up to 35% of all incarcerated people in North America are clinically psychopathic. As one study reported, "There is no other variable that is more highly correlated to being in prison than psychopathy," allegedly dwarfing other well-known criminogenic variables such as socioeconomic factors and substance misuse.[10] Secondly, psychopathy is also believed to be disproportionately correlated with serious crimes such as homicide, physical violence, and sexual assault.[11] Some researchers have reported that psychopathic persons are more likely to use weapons and cause more harm during violent assaults,[12] and others have speculated that they might be responsible for up to 50% of all cases where police officers are killed in the line of duty.[13] Thirdly, researchers often stress that convicted offenders with psychopathy are more likely than the average justice-involved person to be released early from prison due to their deceptive skills and sleek charisma, fooling prison staff and parole boards into thinking that they are no longer a risk to society, only to repeat the cycle of violence over and over.[14] Altogether, this information suggests that psychopathic persons place an enormous monetary toll on society, including the cost of communal turmoil, property damage, treatment of victims, and strain on the legal system. A 2023 study by Dylan Gatner and colleagues estimated that the cost of psychopathy might scale $1,591 billion in the United States and $53 billion in Canada, far exceeding the burden of other social challenges such as alcohol and substance misuse, tobacco smoking, and obesity.[15]

On top of these observations, it is often highlighted that a substantial portion of the overall problem of psychopathy might be fundamentally elusive. Although the numbers above indicate that psychopathic persons are easily caught up in a criminal lifestyle and thus tracked in the public data through their endless in-and-outs of prison and probation systems, some researchers argue that many psychopathic persons are comparatively more "successful" and manage to curtail the judicial institutions altogether. These individuals will either develop creative strategies that allow them to dodge law enforcement or they may choose a path that is not strictly criminal yet still socially and morally problematic.[16] According to leading researchers, key sectors of our society have an overrepresentation of these seemingly successful law-abiding psychopathic persons, including corporate executive management and political governance, and their presence is believed to tacitly contribute to the perpetuation of social dysfunction and injustice.[17] For example, some researchers share the belief that the 2007–2008 global financial crisis might have been premeditated by the overrepresentation of psychopathic individuals in the financial sector, including stock exchanges, banks, and regulatory agencies.[18] Others have argued that the more disruptive forms of social unrest—such as local and global armed conflicts—to a considerable degree are the outcome of psychopathic political leaders, who are precariously self-serving, irrationally aggressive, and blatantly oblivious to the moral worth of other peoples.[19] For example, a 2012 study by Scott Lilienfeld and colleagues[20] analyzed the personality of 42 former US presidents and found evidence of clinically significant psychopathic traits in more than one-third of the former US leaders, including recent presidencies such as John F. Kennedy, Franklin Roosevelt, Ronald Reagan, and George W. Bush. On a similar note, a 2022 study by Robert Hare and colleagues[21] found evidence of psychopathy in people convicted of crimes against humanity such as state-sponsored terrorism.

How the Criminal Justice System Is Responding to the Problem

While the problem of psychopathy is typically framed as this alarming criminal justice issue with dizzying implications for public health and safety, there is nonetheless growing optimism about the prospects of implementing effective intervention strategies. During the past three decades, scientists have reportedly made enormous progress in studying psychopathic persons, which in turn has influenced and assisted the legal system in facilitating an increasingly coordinated response to the problem.[22]

This change began to steadily take shape in the early 1990s. Before this time, there were no shared, standardized protocols for addressing the problem of psychopathy head-on, and, as a result, psychopathic criminals were roaming much more freely, essentially operating *under the radar* of our legal system. Many psychologists were calling for a change in practice—such as implementing systematic screening for psychopathy in prison intake protocols—but these calls were largely ignored across legislative and judiciary institutions.[23] Some correctional initiatives were tailored to respond to the psychopathy problem, such as the Social Therapy Unit at the Oak Ridge maximum security facility in Ontario, Canada—operating roughly between the years 1965 to 1983—but programs like this one that was designed to treat and rehabilitate psychopathic persons were rare and often seen as exploratory in nature.[24]

However, judicial practices began to rapidly change in the 1990s, ostensibly due to remarkable progress in the scientific understanding of psychopathy. One central pillar of this progress is the development of state-of-the-art clinical tools that make it possible to reliably assess for psychopathy, which not only made it much easier to scientifically study the disorder but also gave the legal system a viable way to implement systematic screening of the disorder in justice-involved persons.[25] Throughout the twenty-first century, we have thus seen a remarkable and continuing increase in using psychopathy assessments across judicial institutions in North America, often implemented as a key component in the routine psychiatric and psychological evaluations.[26] It is estimated that millions of such psychopathy assessments have been carried out since the 1990s, making it one of the most utilized instruments in forensic evaluation and management.[27] Where psychopathic persons had before been able to slide under the radar of the legal system, there is now an active and coordinated effort—supported by substantial fiscal and human resources—to detect who the psychopathic criminals of our society are.

The biggest advantage of implementing a systematic use of psychopathy assessments is that it lays the foundation for providing a targeted judicial response to the problem of psychopathy, insofar that these individuals are now formally and explicitly distinguished from the less problematic (non-psychopathic) offenders. Accordingly, once a person is assessed with psychopathy, this is likely to have significant consequences on their journey through the legal system. This entails, but is not limited to, an increased likelihood of receiving a non-mitigated sentence, refused bail, placement in maximum-security institutions, exclusion from rehabilitation programs, being given a *dangerous offender* status, denial of parole, juvenile transfer to adult courts, increased probation supervision, and so forth.[28] Researchers estimate that on an annual basis, hundreds of thousands of psychopathic offenders in North America are legally impacted in these restrictive ways, either as a direct or indirect result of a psychopathy assessment.[29]

In light of such historic developments, psychopathy research—and the legal practices it facilitates—has been widely celebrated as one of the most important innovations in contemporary forensic psychiatry and psychology, recognized for its instrumental role in intervening with the social problem of psychopathy.[30] It is an incredible story about a personality disorder that few people in the legal system took seriously before the 1990s, let alone had any idea of how to deal with in a targeted fashion. But with the work of a small group of persistent forensic scientists, these pioneers managed to fundamentally influence and to some extent shape the policies around how our society deals with psychopathic persons once they violate juvenile and criminal laws.

Psychopathy Assessments as Evidence-Based Practice

To legal systems tasked with deterring criminal activity and keeping society safe, the general justification for implementing psychopathy assessments might seem obvious and perhaps even self-explainable: since psychopathic persons comprise a relatively small, yet criminally prolific subgroup—they are social predators, the few rotten apples that ruins the barrel of society—there is a clear motive to allocate extra resources toward deterring, intervening, and restricting these extraordinary persons. As is frequently emphasized by leading researchers, if we find a way to intervene with psychopathic persons' antisocial trajectory, crime rates and social dysfunction will be drastically reduced.[31]

However, an important element in qualifying the legal practices that has emerged around managing psychopathic persons is that these practices must be rooted in or backed by some form of robust knowledge, preferably scientific evidence. It is not enough to simply believe, claim, or state that psychopathic persons are remorseless social predators, and thus deal with them accordingly. Crucially, a common aspiration across legal systems in North America is that judicial decisions of this kind must not be guided by arbitrary beliefs but instead anchored in measurable and generalizable facts: an orientation often referred to as *evidence-based practices*.[32] In essence, before our legal system can justify the enhancement of restrictive measures of psychopathic individuals, there must be clear empirical evidence to show that those who are assessed with psychopathy are, in fact, extraordinary offenders that warrant extraordinary legal means.

Within this framework, the evidence-based justification for using psychopathy assessments in the legal system has generally rested on five basic claims, which altogether comprise an understanding of psychopathy that has enjoyed relatively broad consensus among forensic researchers and practitioners since the 1990s. That is, these five claims are not supposed to be mere conjectures about psychopathy, but they have historically been seen as empirically demonstrable facts about the individuals who are assessed with psychopathy:

1. *Psychopathic persons are extraordinarily dangerous.* Perhaps the most obvious and central incentive for implementing psychopathy assessments is that psychopathic individuals are believed to be responsible for the majority of the serious crimes that plague society, including homicidal, instrumental, and sexual violence. Psychopathic persons are thus believed to be quintessential examples of social predators who care little about laws and social norms, let alone the people that happen to stand in their way. According to researchers, when psychopathic persons are compared to non-psychopathic criminals in empirical studies, what they find is a remarkable difference in the danger they pose to public health and safety.[33]

2. *Psychopathy is a chronically untreatable disorder.* Psychopathic persons are not just extraordinarily dangerous, but their condition is also believed to be chronic and untreatable. So, when a psychopathic person is enrolled in rehabilitation programs or offered psychiatric treatment for their personality and social dysfunction, it is commonly believed that these efforts are fundamentally futile. This observation stands in clear contrast to ordinary offenders who typically benefit from these programs, aiding them in breaking patterns of criminal behavior and other psychological problems. In addition, many researchers also highlight evidence suggesting that persons with psychopathy get *worse* when treated—that is, they show adverse reactions to clinical care—which manifests in a higher risk of criminal recidivism after treatment completion.[34]

3. *Psychopathic persons are deprived of moral capacities.* A frequently highlighted observation about psychopathy is its association with a complete disregard for moral and social values that has historically been described as a fundamental lack of conscience, empathy, and remorse. According to researchers, when psychopathic persons are tested in experimental settings that involve probing their capabilities to empathize with other people and engage in moral reasoning, their peculiar deficiencies are easily detected when compared to ordinary people.[35]

4. *Psychopathy is "biological."* Mental health diagnostical frameworks include dozens of well-established psychiatric disorders, such as major depressive

disorder, autism spectrum disorder, and schizophrenia spectrum. Though, our knowledge about most psychiatric disorders rarely goes beyond overt clinical observations, such as mapping the prototypical signs and symptoms associated with a diagnosis. Only in rare cases do researchers manage to empirically document biological correlates that further enhance our understanding of a disorder. In the case of psychopathy, however, researchers argue that there is growing evidence to suggest that the disorder is associated with discrete brain abnormalities.[36]

5. *Psychopathy is theoretically corroborated.* A fundamental part of how science operates is the process of theory testing. Scientists posit theoretical accounts about entities they study, and then work toward gathering evidence that either corroborate or falsify these theories. In psychopathy studies, there has historically been broad consensus among researchers that theories about psychopathy have been empirically corroborated, meaning that researchers can rightfully claim to have a scientifically sound understanding of what makes psychopathic persons psychologically different from ordinary (healthy) individuals. More specifically, it is believed that psychopathy boils down to an impairment of basic mechanisms involved in emotional processing and impulse control.[37]

Altogether, these five beliefs neatly summarize how psychopathic persons are broadly seen as an overly prolific subgroup of criminals who remain an extreme, uncorrectable, and chronic threat to public health and safety. Since the mid-1990s, researchers have emphatically iterated that there is robust scientific evidence to back up this understanding of psychopathy as a personality disorder with immense social implications. For example, this understanding of psychopathy reliably figures in the broadly cited review literature published by leading researchers in prominent academic journals,[38] in edited anthologies by the respected academic presses,[39] and in the most popular books written by top researchers.[40] It is largely on this scientific background that the judicial control of psychopathic justice-involved persons is frequently referred to as an evidence-based approach to criminal justice reform and practice.

The Need for a Critical Conversation

Up until this point, what has been outlined is the conventional, mainstream depiction of how psychopathy research has made immense progress through the accumulation of empirical evidence, and how this knowledge has been instrumental in generating coordinated policies and practices designed to facilitate a targeted control of psychopathic persons passing through our legal system. The narrative behind these practices goes something like this: there are obvious moral, social, and economic incentives to identify who the true psychopathic persons in our society are, and the judicial system is then responding appropriately by ensuring that these individuals are subjected to enhanced judicial oversight. In short, the implementation of psychopathy assessments in the legal system is simply making our society a safer place.

However, since this narrative is fundamentally contingent on the empirical evidence behind the five basic claims about psychopathy, it follows that the narrative is inherently fallible: Psychopathy assessments are only making society safer if what we say about psychopathic persons is also factually true. For example, try to imagine a scenario where the five basic views about psychopathy turns out to be empirically false: a scenario where the people we currently assess with psychopathy (1) are no more dangerous than the average offender, (2) are equally treatable and able to rehabilitate, (3) suffer from no apparent impairment of the moral faculties, (4) have no measurable links to biological correlates, and finally (5) that the theories scientists posit about psychopathy are abandoned due to falsifying evidence. In this scenario, psychopathy assessments would not be making our society safer as the targeted individuals exert no extraordinary threat to public health and safety. But since we are still singling out these individuals with the psychopathy label—and systematically restricting and detaining them *as if* they are extremely dangerous social predators—in this version of the story, a psychopathy assessment would instead function as a vessel for unjustified legal discrimination.

This little thought experiment demonstrates how high the stakes really are when it comes to implementing and using a personality disorder like psychopathy to inform legal decisions: either the practice is making society a safer place by delivering crucial information about extreme offenders *or* it puts justice-involved persons at risk of systemic, large-scale discrimination. To legal systems with a long-established tradition of using psychopathy assessments, it is of absolute importance to ensure that they are on the right side of this sliding scale.

So, is our legal system on the right side?

Psychopathy: A Story about Large-Scale Discriminatory Practices

This book aims to provide an answer to this timely question. The answer that will be developed over the following chapters is one that flips the narrative against the mainstream depiction of psychopathy as one of the most important innovations in forensic psychiatry and psychology. At the heart of this critical analysis is the uncomfortable and largely overlooked fact that the five basic views about psychopathy are no longer and probably never were scientifically tenable.

When psychopathy research gained traction in the 1990s–2000s, there was a lot of premature enthusiasm about early experimental results coming from within the research paradigm: results that eventually shaped and cemented the five mainstream views among contemporary and future generations of forensic researchers and practitioners. However, as the experimental work evolved over the past decades—alongside methodological and technological improvements in behavioral and mental health science—it is becoming increasingly clear that not only were these early results highly dubious (in large part due to questionable research practices) but these experiments have continuously failed to replicate in subsequent studies. Moreover, we now seem to have strong reasons to conclude that the five mainstream beliefs about psychopathy are either straightforwardly false or strongly misleading. Not only is this emerging insight seriously challenging the idea of psychopathy as a personality disorder but it also fundamentally undermines the justification for using psychopathy assessments in legal practices.

Today, we are faced with the uncomfortable reality that psychopathy assessments have metamorphized from being an alleged high mark of forensic psychological innovation to being a contender for a much less admirable status: a vehicle and facilitator of legal discrimination of seemingly colossal magnitudes.

Overview of the Book

Psychopathy Unmasked is written with the aim of navigating the reader through the history of psychopathy research and how it was prematurely applied in legal contexts. In this sense, the book seeks to provide an all-encompassing critical analysis of psychopathy research and the judicial and professional practices that have been influenced by or emerged from the research paradigm. It will do so by systematically surveying the experimental work published since the advent of the contemporary research paradigm in the 1990s.

Chapter 1 provides some necessary background information about psychopathy and its most utilized clinical assessment tools, and it surveys how these assessment tools are used in clinical practices and specifically by legal practitioners to inform judicial and correctional decisions.

Chapter 2 explores the evidence behind the claim that psychopathic persons are dangerous social predators, demonstrating that there is little meaningful difference between psychopathic and non-psychopathic criminals in terms of their risk profile. In the 1990s, there was much hype around a few studies that showed large differences between these groups; however, since then, hundreds of published studies have repeatedly shown that these group differences are not only much smaller than initially thought but researchers have also highlighted important caveats in the data that (in obvious ways) makes the "social predator" narrative empirically indefensible.

Chapter 3 explores the claim that psychopathy is "untreatable," revealing growing evidence that suggests a divergent and much more optimistic conclusion about rehabilitation of individuals assessed with the disorder. The pessimistic attitude that dominated scientific conversations in the 1990s was largely based on one single study, which has subsequently been rejected by researchers due to its use of highly unreliable data. The evidence that is coming from contemporary studies overwhelmingly suggests that psychopathic persons benefit from treatment and rehabilitation programs in similar ways as non-psychopathic persons.

Chapter 4 discusses the claim that psychopathic persons lack empathy and are incapable of moral reasoning. Although this view about psychopathy is as old as the history of the research paradigm itself—and is still frequently invoked by contemporary researchers—it has never been supported by any credible scientific evidence. When psychopathic persons are tested in moral psychological experiments, researchers consistently fail to detect any meaningful differences in performance compared to non-psychopathic persons.

Chapter 5 explores the claim that psychopathy is associated with discrete neurobiological correlates, that psychopathy is linked to functional and structural brain abnormalities. This question has been pursued since the advent of brain imaging technologies in the 1950s, but more systematically since the 2000s when these technologies matured. However, contrary to widespread beliefs among psychiatrists and psychologists, these hundreds of experiments have yet to document any clear evidence of psychopathy being linked to such biological correlates.

Chapter 6 surveys the evidence behind the two leading theories of psychopathy, that the disorder is linked to and explained by emotion deficits and/or impulse-control impairments. While theory-testing is always an ongoing process, the evidence that has emerged so far appears to robustly falsify these two theories as viable explanations of the disorder. This includes evidence from research that measure biomarkers as well as behavioral task-based experiments; research that continues to show that there are no meaningful differences between psychopathic and non-psychopathic persons.

Chapter 7 takes a step back from surveying the empirical research to discuss the ethics of psychopathy assessments and how they are used to inform legal decisions. It is argued that the professional ethical obligations of psychiatrists and psychologists precludes them from using or providing psychopathy assessments in forensic contexts. More specifically, the chapter explores how psychopathy assessments may be in direct or partial violations of the ethical codes and standards provided by the American Medical Association and the American Psychological Association.

Chapter 8 explores potential explanations why researchers have failed to find any clear empirical support for the five mainstream claims about psychopathy. If psychopathic persons truly exist—if there truly are individuals who match what researchers have traditionally said about the disorder— why is it that scientists cannot document their existence in psychological experiments? While the chapter explores several answers to this question, it focuses on an explanation that has so far received little attention by mainstream researchers: that perhaps the best way to explain the lack of scientific progress is because the disorder is not *real*. More specifically, the chapter considers the possibility that the idea of psychopathy may be an instance of what scientists colloquially refer to as a *zombie idea*: an idea that is intuitively appealing or "intellectually infectious" but nevertheless an idea that is "dead," that has no corresponding referent in reality.

The book concludes by advocating for an immediate moratorium on the clinical and legal use of psychopathy assessments and discusses future directions for psychopathy research.

1   The Rise of Psychopathy Research and Its Legal Influence

Psychopathy refers to a personality disposition to charm, manipulate, and ruth-
lessly exploit other persons. Psychopathic persons are lacking in conscience and
feeling for others; they selfishly take what they want and do as they please without
the slightest sense of guilt and regret. Psychopathy is among the oldest and arguably the most heavily researched, well-validated, and well-established personality
disorder.

—Robert Hare, Craig Neumann, and Thomas Widiger in *Psychopathy* (2012)[1]

Psychopathic personality disorder, or psychopathy, was one of the first personality disorders to enter the mental health lingo back in the eighteenth
century, placing it among the most debated and researched conditions in
the history of mental health science.[2] However, while psychopathy research
dates back to the early days of modern psychiatry, it was only much later—
in the early 1990s—that psychopathy became widely acknowledged among
the clinical and scientific mainstream establishment, and since this period,
the disorder has been increasingly used in the criminal justice system to
inform a variety of judicial and correctional decisions.

In this chapter, I shall track some of these historical developments while
carefully outlining how, and to what extent, psychopathy is used in legal
settings. This includes outlining what psychopathy "is" or what the concept refers to, how clinicians screen and assess people for psychopathy, and
the concrete ways in which these clinical assessments influence decisions
in criminal justice systems across North America.

Before we continue, I need to address one central disclaimer. Throughout
this book, I will refer to psychopathy as a "personality disorder" and I will
occasionally describe it as a type of "diagnosis" that psychiatrists can issue

to a person. However, describing psychopathy in this way is contentious, as there is historic and ongoing debate in the broader medical community about the place of psychopathy in codified diagnostic manuals like the *Diagnostic and Statistical Manual of Mental Disorders* (DSM) and the *International Classification of Diseases* (ICD).[3] This conversation has been further complicated in recent years as the DSM and ICD frameworks—and mental health professionals in general—are moving away from using "categorical" labels, increasingly defining personality disorders as multidimensional syndromes without clear categorical boundaries.[4] So, when I have opted to still refer to psychopathy as a categorical type of personality disorder, I am doing so because this reflects how the research community has traditionally described psychopathy, and, moreover, it similarly reflects how psychopathy is frequently (but not exclusively) used and talked about in the legal system as a categorical diagnostic entity. Importantly, regardless of what position one takes on these issues, it should have no or only minimal influence on how to interpret the research I survey throughout this book. In contexts where I believe it could have substantial influence, I have done my best to make this explicit in the manuscript.

Psychopathy Research: A Brief History

The main focus of contemporary psychopathy research is the study of a specific patient stereotype—referred to as *psychopathic* persons, or the informal and somewhat derogatory term *psychopaths*—that forensic psychiatrists and psychologists frequently encounter in their daily practices. According to leading experts, what makes psychopathic individuals different from ordinary people is their peculiar combination of showing no apparent signs of cognitive dysfunction—at times coming across as utterly normal—while still exerting extraordinary and persistent problems with adapting to social and interpersonal expectations. Above all, clinicians report that psychopathic persons are both recklessly impulsive and destructive to their surroundings and seemingly incapable of caring for and empathizing with other human beings.

The oldest scholarly account of psychopathy has been attributed to the American polymath Benjamin Rush (1745–1813) in a speech he delivered to a crowd of intellectuals on February 27, 1786.[5] In this presentation, Rush outlines a theory about a novel psychiatric condition—which he labeled *anomia* and, later, *moral derangement*—that brings about an incapacity to make proper moral judgments while all other mental faculties remain fully intact. According to Rush, this condition of being morally deprived but intellectually capable gives rise to the most peculiar and ghastly behaviors. As an illustration, Rush speaks about a young man named Servin, whom Rush portrays as a prodigy of both vice and knowledge, alluding to what he sees as the paradoxical state where profound amorality occurs in an otherwise intelligent person. Rush describes the young Servin as having the most astonishing intellect, as fluent in Greek and Hebrew, and as having an impressive flair for social tact and high-brow pursuits. But besides this affable side of his character, Servin is also "treacherous, cruel, cowardly, deceitful, a liar, a cheat, a drunkard and a glutton, a sharper in play, immersed in every species of vice, a blasphemer, an atheist. In a word—in him might be found all vices that are contrary to nature."[6]

Rush's basic idea that human moral dispositions could be disordered was broadly recognized as an innovative claim by eighteenth-century physicians.[7] Shortly after the February meeting in 1786, Rush's manuscript was printed and distributed across North America and Europe, with a second edition printed already within the same year.[8] While it is difficult to ascertain how much Rush's idea influenced his contemporaries, it is notable that shortly after Rush's speech, European physicians began to write about similar clinical observations and ideas. In the year 1800, the renowned French physician Philippe Pinel (1745–1826) published an influential treatise cataloging mental disorders based on personal experiences from his lifelong work in Parisian asylums.[9] This treatise includes a diagnosis called *mania without delirium*, a disorder where the intellectual faculties are intact but the person exhibits behavioral perversions. Around 1910, one of Pinel's top students, Jean-Étienne Dominique Esquirol (1772–1840), wrote about a category of disorders called *monomanias*, capturing the same idea of a mental condition where only one (i.e., mono) discrete part of the human mind was disordered, such as a person's moral dispositions.[10] In 1835, the prominent British scientist James C. Prichard (1786–1848) coined the term *moral insanity*, which for a long period became the preferred term for the type of disorder Rush had initially outlined.[11]

By the early twentieth century, Rush's original conception of the disorder appears to have gained considerable traction in the psychiatric community, and it is in this period that we begin to see researchers and clinicians relabeling the disorder as *psychopathy*, which seems to have derived from the diagnostic vocabulary of German psychiatrists, especially Julius Koch's concept of *psychopathic inferiorities* (1891–1893) and Emil Kraepelin's *psychopathic personalities* (1909–1915).[12] It is also the scholarly contributions from this period, and in particular from the post–World War II era, that are typically mentioned as having a truly pioneering influence on contemporary researchers—an era that includes contributions from known figures in the history of psychiatry like Ben Karpman (1886–1962),[13] David Henderson (1884–1965),[14] Silvano Areti (1914–1981),[15] Hervey Cleckley (1903–1984),[16] William McCord (1930–1992) and Joan McCord (1930–2004),[17] David Lykken (1928–2006),[18] and Lee Nelken Robins (1922–2009).[19]

However, despite this interest among prominent scholars, for most of the 1900s psychopathy was largely seen as an opaque and fringe psychiatric diagnosis that few mental health professionals took seriously. For example, as late as in 1976, one of the leading advocates of psychopathy research, Hervey Cleckley, regretfully decried in the preface to the fifth edition of his book *The Mask of Sanity*—maybe the most iconic and academically influential book on psychopathy—that the disorder is broadly "ignored" by clinicians and that "no measure at all is taken" to deal with the problem of psychopathy; Cleckley went as far as comparing this neglect to a "universal conspiracy of evasion."[20] And in a personal correspondence with Robert Hare, one of today's most influential researchers, Cleckley refers to himself as "a voice crying in the wilderness," summarizing his disappointment about the fading impact and unrealized potential of psychopathy studies.[21]

Cleckley's depiction of the bleak status of the field in the 1970s is also reflected in how the medical psychiatric establishment in North America had hesitated and eventually refrained from including psychopathy in their official diagnostic manuals—at least not as it was conceptualized by Cleckley and his closest colleagues.[22] For instance, when the American Psychiatric Association released the third edition of the DSM in 1980—their "official" diagnostic manual—psychopathy was still not included as a separate diagnosis, thus remaining an essentially unrecognized disorder. It was not before the publication of the fourth edition of the DSM in 1994 that psychopathy earned its first clear reference by name, where the term psychopathy is mentioned as a synonym for the diagnosis of antisocial personality disorder. But even with this acknowledgment, many researchers who had advocated for the inclusion of psychopathy in the DSM expressed dissatisfaction about how antisocial personality disorder was mostly defined with reference to maladaptive and antagonistic behaviors, thus downplaying the clinical relevance of many of the personality traits that Cleckley and others had long argued truly characterized the disorder (e.g., unemotional).[23] As mentioned earlier, the place of psychopathy in the DSM has remained a debated topic to this day, though in the latest fifth edition of the DSM—published in 2013—psychopathy is described both as a synonym for and a variant or subtype of antisocial personality disorder (i.e., psychopathy is still not seen as a diagnosis categorically distinct from antisocial personality disorder).[24]

Aside from the negligible role psychopathy played in the medical community before the 1990s, there is also little indication that the diagnosis was used in the criminal justice system, at least not in a systematic way. For instance, a 2006 study by David DeMatteo and John Edens[25] surveyed the case law records in the United States dating back to 1990 and found that the use of psychopathy assessments first began to show up in court decisions sometime around the mid-1990s, and not before the 2000s was it considered broadly influential.[26] However, this is not to say that psychopathy was completely ignored by forensic practitioners. There were some organized efforts, particularly in Scandinavia and in Canada during the post–World War II era, to establish forensic mental health programs that specialized in the treatment and detention of criminal psychopathic persons.[27] Although, in hindsight, these programs were mostly exploratory and never made it into mainstream practices (more on this in chapter 3).

According to some scholars, one major reason why psychopathy was effectively ignored by the medical establishment far into the 1970s is that the research community had enduring disagreements about how to theoretically account for the disorder and how to characterize a stereotypical psychopathic person. When reading publications from the early days of Benjamin Rush, Philippe Pinel, James C. Prichard, and others, it is tempting to conclude that these scholars agreed on their basic definition of psychopathy, but when you dig a little deeper into their work you quickly discover that their accounts are overly vague and that they seem to have fundamentally different answers to central questions about psychopathy. Are psychopathic persons predominately criminal or just antisocial? Are they lacking conscience or just careless? Is psychopathy characterized by impulsivity and anxiousness, or is it a condition of emotional apathy and disinterestedness?

The research history is rife with disagreement about these central questions. Consider, for instance, a 1974 paper titled "Psychopathic Personality:

A Most Elusive Category,"[28] written by the British psychiatrist Aubrey Lewis (1900–1975), the long-time chair of the prestigious Institute of Psychiatry in London. In this piece, Lewis lambasts his discipline's many failed attempts to reach a consensus on how to describe psychopathy, concluding that it was the worst defined condition in all of psychiatric diagnostics—a sentiment that was echoed across the upper echelons of experimental psychology. Indeed, during the post-WWII years, it had become custom to refer to psychopathy as a "wastebasket" diagnosis: in short, a label psychiatrists used for patients where they simply did not know what else to call them.[29] But consider also the searing criticism by Hans Eysenck (1916–1997), one of the most influential psychologists during the 1950s–1990s, who is quoted during a conference in 1975 describing psychopathy as a "white elephant" that "ought to be dumped" by the academic and medical community (the term white elephant is a metaphor that refers to a costly and useless item that is difficult to dispose of).[30]

There was not just disagreement among experts about theory and stereotypical traits of psychopathy, but there appears to have been a problem with broad and fundamental confusion even among experts. For example, in 1975, a large conference was held in Les Arcs, France, bringing together scholars from different countries and disciplines aiming to discuss the current and future status of psychopathy. Robert Hare was one of two conference organizers (the other being Daisy Schalling), and he has frequently depicted— rather somberly—how the conference was a "microcosm of the real academic and clinical world, a world that in 1975 produced much armchair speculation and uninformed debate, but relatively little empirical research, about the nature of psychopathy."[31] When the conference proceedings were published in a 1978 anthology, it contained a hodgepodge of contributions that, under closer scrutiny, can reasonably be questioned as to whether the various authors were at all writing about the same patient stereotype.[32] While Hare has described the anthology in critical yet diplomatic terms, acknowledging that "many of its chapters had only an indirect or tenuous relation to psychopathy,"[33] one could note that when recognized experts in a field do not appear to be capable of the most basic communication—that is, agreeing on what they are talking about—it is probably because confusion is abundant.[34]

In addition to these semantic and conceptual disagreements, some researchers went even further with their criticism, interpreting the lack of basic progress in the field as an invitation to doubt whether psychopathy was at all a real personality disorder.[35] For example, one of the strongest versions of this criticism appears as late as 1988 by the lauded senior forensic psychologist Ronald Blackburn, who argued that psychopathy is simply a "mythical entity" that "should be discarded," rendering it little more than a morally stigmatizing label "masquerading as a clinical diagnosis."[36] What Blackburn emphasized is that mainstream definitions of psychopathy from the 1980s are more or less conceptualized around listing an eclectic mix of antisocial behaviors, which effectively means that most criminals would meet the criteria for the diagnosis. So, is psychopathy simply another word—perhaps a stigmatizing adjective—for describing a criminal?

Despite its frivolous status in the 1970s and 1980s, psychopathy research has since gone through dramatic changes. Today, psychopathy is frequently acknowledged in peer-reviewed journals and anthologies as one of the most researched and best validated personality disorders,[37] and the research paradigm makes up a rapidly growing and influential field shaped by interdisciplinary contributions from psychiatry, psychology, sociology, neurobiology, genetics, and criminology.[38] During the past three decades, the field has seen a steady increase in the number of participating researchers, which has resulted in a virtual explosion in research output. For example, according to the American Psychological Association's research database (*PsychINFO*), the number of publications containing the keywords psychopathy, psychopaths, and psychopathic, has been on an exponential upward trend since the 1990s, rising from 685 publications in the years 1980–1989 to a whopping total of 5,411 in the decade of 2010–2019: a near eight-fold increase in scholarly productivity (see figure 1.1).

This increase in research output is also mirrored in developments inside the court system. In 2014, David DeMatteo and colleagues published a study[39] reviewing US case law records and found that between the years 2000 and 2011 there had (similarly) been an eight-fold increase in the yearly number of court cases using psychopathy assessments to inform decisions. A study from 2010 by Jodi Viljoen and colleagues[40] made an analysis of adolescent case law records and found comparable increases from 1980 to 2008, noting that psychopathy assessment in juvenile cases is "becoming increasingly common."[41] In Canada, studies have documented that psychopathy is used routinely to inform decisions about *Dangerous Offender* and *Long-Term Offender* status hearings. For example, in 2010, Caleb Lloyd and colleagues[42] found that in the period between 2001 and 2007, a total

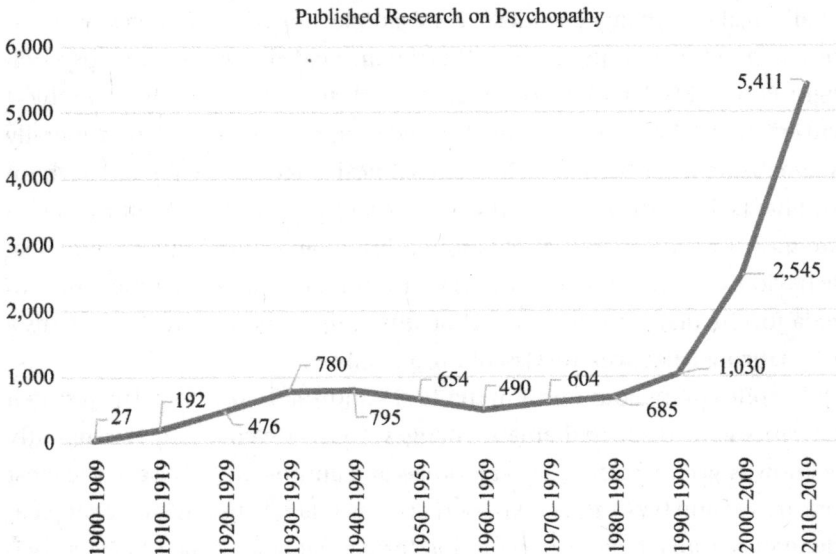

Figure 1.1
Psychological research on psychopathy. This graph depicts the number of academic publications (per decade) containing one of the following keywords: psychopathy, psychopathic, or psychopath. Source: PsychINFO.

of 61.8% of all transcribed hearings included a reference to psychopathy. And a 2015 study by John Edens and colleagues[43] found that psychopathy assessments were regularly used to inform decisions about civil commitment in sex offender cases. These numbers are recognized in commentaries by leading experts—for instance, Robert Hare stated in 2016 that the use of psychopathy assessments is "increasing at a rapid rate in the criminal justice system,"[44] and David DeMatteo and Mark Olver noted in 2022 that the use of psychopathy in "legal contexts has been increasing over the past 15+ years, and there is no reason to think that its use will decline."[45]

Also reflecting the dramatic change of the status of the field is that some of the most respected academic presses are now consistently publishing extensive anthologies about psychopathy—some of them already in their second edition—compared to the days before the 1990s where publications on psychopathy tended to be distributed by less prestigious publishers. For example, Hervey Cleckley's fifth and final edition of *The Mask of Sanity* from 1976—today considered a modern classic by psychopathy researchers—was published by a relatively small imprint in St. Louis, Missouri named C. V. Mosby.[46] Nowadays, psychopathy research is released by top international distributors, including titles such as *Handbook on Psychopathy and Law* by Oxford University Press,[47] *The Clinical and Forensic Assessment of Psychopathy* by Routledge,[48] *Handbook of Psychopathy* by The Guilford Press,[49] *Routledge International Handbook of Psychopathy and Crime* by Routledge, and *The Wiley International Handbook on Psychopathic Disorders and the Law* by Wiley-Blackwell.[50]

In addition, virtually all of today's popular forensic psychology textbooks include elaborate sections or full chapters on psychopathy. For example, perhaps the most frequently used textbook from the United States by Curt Bartol and Anne Bartol[51] includes an extensive section titled "The Criminal Psychopath." The Canada-focused textbook by Joanna Pozzulo and colleagues[52] includes an entire chapter on psychopathy. Further, the topic of psychopathy was included in 2019 in the globally popular book series *A Very Short Introduction*, edited and published by Oxford University Press.[53] There is also a growing number of books written by researchers targeting university students, such as David Lykken's *The Antisocial Personalities*,[54] Adrian Raine and José Sanmartín's *Violence and Psychopathy*,[55] James Blair and colleagues' *The Psychopath: Emotion and The Brain*,[56] Andrea Glenn and Adrian Raine's *Psychopathy: An Introduction to Biological Findings and Their Implications*,[57] Matt DeLisi's *Psychopathy as Unified Theory of Crime*,[58] and Jacqueline Helfgott's *No Remorse: Psychopathy and Criminal Justice*.[59] Add to this an expanding pop-science book market for a broader audience written by senior scientists. This market include works like Robert Hare's global best-seller from 1993, *Without Conscience*,[60] Robert Hare and Paul Babiak's *Snakes in Suits*,[61] Jon Ronson's *The Psychopath Test*,[62] Kevin Dutton's *The Wisdom of Psychopaths*,[63] James Fallon's *The Psychopath Inside*,[64] Martha Stout's *The Sociopath Next Door*,[65] and Kent Kiehl's *The Psychopath Whisperer*.[66]

In contrast to the times prior to the 1990s, it is an observable fact of today's intellectual climate that if you are educating and training to become a forensic mental health researcher or practitioner—or perhaps even a criminologist or legal attorney—you are much more likely to be presented with research on psychopathy than not. As stated by Robert Hare: "The concept of psychopathy clearly is unusually important in the legal and judicial world."[67]

The Clinical Assessment of Psychopathy

One of the most central advancements in psychopathy studies during the past four decades is the development of state-of-the-art assessment tools designed to assist researchers and practitioners in detecting or "measuring" psychopathy. These assessment tools are a necessary first step in conducting scientific research but, equally so, an elemental tool in implementing a meaningful criminal justice response to the problem of psychopathy. Before our legal institutions can be expected to deal with psychopathic persons in a targeted and systematic way, they need a reliable method to identify these individuals. Like doctors tasked with curing a disease, proper assessment and diagnosis necessarily comes before treatment.

In this way, psychopathy assessment tools are, strictly speaking, designed to assist clinicians in ascertaining whether—or to what degree—a person *has* psychopathy (i.e., whether they have clinically significant levels of psychopathic traits). Several types of psychopathy assessments have been developed by different research teams, such as the *Comprehensive Assessment of Psychopathic Personality*,[68] *The Levenson Self-Report Psychopathy Scale*,[69] *The Youth Psychopathy Inventory*,[70] *The Psychopathic Personality Inventory-Revised*,[71] and *The Triarchic Psychopathy Measure*.[72] However, among these many approaches to assessing psychopathy, by far the most popular tool in clinical and forensic settings is the *Hare Psychopathy Checklist-Revised* (PCL-R).[73]

The PCL-R is developed by the Canadian researcher Robert Hare and his team, its first edition published in 1980[74] and later revised in 1991[75] (hence, the term *revised* added to *Psychopathy Checklist*), whereafter it has since been edited into its latest and second edition from 2003.[76] The development of the PCL-R was strongly influenced by the work of Hervey Cleckley, and it is predominately his conception[77] of psychopathy that the PCL-R purports to measure (though with some differences[78]). In addition, there is considerable evidence to suggest that the PCL-R measures a conception of psychopathy that lies on a continuum with (i.e., a severe variant of) the DSM's diagnostic category of antisocial personality disorder (although this remains a debated topic).[79]

The PCL-R consists of a 20-item checklist intended to capture the personality traits and behavioral tendencies that characterize a stereotypical (criminal) psychopathic person, including traits such as lack of empathy, shallow affect, pathological lying, juvenile delinquency, and criminal versatility (see figure 1.2). The item checklist is described as fitting a two-factor and four-facet constellation (from *factor analysis*[80]), where *factor 1* includes interpersonal and affective items and *factor 2* includes lifestyle and antisocial items. While all of the 20 items are believed to represent the full-blown manifestation of psychopathy relevant to forensic practices, the PCL-R manual underlines that factor 1 items are more representative of psychopathy than factor 2 items.[81] This view fits the common understanding of psychopathy being a *personality* disorder insofar as factor 1 items largely represent the core aspects of a clinically psychopathic personality.[82]

The use of factor analysis to categorize the items into factors/facets may sound like some technical lingo, but at its most basic level, it is simply an illustration of how each of the 20 items tend to cluster together during assessment procedures. For example, if a clinician determines that a patient has item 7 (shallow affect), there is going to be a high likelihood that the clinician will determine that the patient has the remaining seven items in factor 1 and even higher likelihood of the other three items in facet 2 (where item 7 belongs). Although there is no need to go into further detail about what can (and cannot) be inferred from a factor analysis, one should note that when a clinical assessment tool has a low number of factors/facets—like in the case of the PCL-R—this is usually seen as a sign of quality or what psychologists call "construct validity."[83] So, it is not unusual for researchers and practitioners to justify their choice of the PCL-R over other tools on the basis of it being promoted as the most validated measure of psychopathy (though there is ongoing debate about how valid the PCL-R is,[84] a discussion that also surveys how well it measures psychopathy on a continuum with antisocial personality disorder or as a potentially separate condition[85]).

A PCL-R assessment must only be conducted by a trained and certified clinician who will proceed to evaluate to what degree a person matches or resembles each of these 20 items (all items are described in more detail in the PCL-R technical manual). In practice, this is done by performing an in-person interview of the clinical subject (or patient) and reviewing their historical records (e.g., forensic or medical), probing for information that may support or rule out the presence of each of the checklist items. The items are then given a score of 0, 1, or 2 points relative to how closely they resemble or *match* the subject's personality and behavior, thus yielding a total score ranging from 0 to 40. The PCL-R is designed on the assumption that psychopathy is a unidimensional disorder—meaning that it is assumed

Hare Psychopathy Checklist–Revised

1. Glibness/superficial charm
2. Grandiose sense of self worth
3. Need for stimulation/proneness to boredom
4. Pathological lying
5. Conning/manipulative
6. Lack of remorse or guilt
7. Shallow affect
8. Callous/lack of empathy
9. Parasitic lifestyle
10. Poor behavioral controls
11. Promiscuous sexual behavior*
12. Early behavioral problems
13. Lack of realistic, long-term goals
14. Impulsivity
15. Irresponsibility
16. Failure to accept responsibility for own actions
17. Many short-term marital relationships*
18. Juvenile delinquency
19. Revocation of conditional release
20. Criminal versatility

Factor 1

Facet 1	Facet 2
Interpersonal	*Affective*
Glib/superficial	Lack of remorse/guilt
Grandiose self-worth	Shallow affect
Pathological lying	Callous/lack of empathy
Conning/manipulative	Fail to accept responsibility

Factor 2

Facet 3	Facet 4
Lifestyle	*Antisocial*
Need for stimulation	Poor behavioral controls
Impulsivity	Early behavioral problems
Irresponsibility	Juvenile delinquency
Parasitic lifestyle	Revocation of cond. release
Lack of realistic goals	Criminal versatility

Figure 1.2

The *Hare Psychopathy Checklist-Revised.* To the left is the full 20-item checklist. To the right is the two-factor and four-facet constellations. Notice that only 18 items appear in the factor analysis (on the right), excluding item numbers 11 ("promiscuous sexual behavior") and 17 ("many short-term marital relationships").

to be a question about to *what degree* a person has psychopathy—and final scores are then interpreted as an expression of dimensionality ranging from low to high psychopathy. However, in forensic settings, a score of 30 and above has over the years become the conventional threshold at which a person is designated with clinical levels of psychopathy (in research settings this threshold is sometimes lowered to 25, though this is not to be confused with a clinical decision). Thus, clinicians might state that a person has clinical levels of psychopathy (score 30), where scores between 30 and 40 might be used to indicate how severe the condition is.

Setting a conventional threshold for clinical psychopathy might at first seem a little arbitrary, but another way to look at this is that the PCL-R score of 40 is an expression of full-blown stereotypical psychopathy, where practitioners believe that slightly lower scores on the PCL-R are still signaling something that is clinically salient—that is, clinically speaking disorderly. But, in some instances, such as scores below the 30-point threshold, it may no longer be clinically justifiable nor meaningful to think of the person as psychopathic (i.e., indicating personality disorder). This perspective is similar to how most mental disorders are thought of as being degree-specific entities—for instance, how one person can be *highly* clinically depressed and another only *moderately* clinically depressed, and a third person might show signs that resemble depressive states, though not to be confused with and therefore importantly different from a clinical depression (i.e., if you feel sad on a Sunday morning, we just say that you are "sad"; we do *not* say you have some slight degree of clinical depression).[86]

In continuation of the work on the PCL-R, Robert Hare and his colleagues have developed two different subversions of the PCL-R, namely a *Youth Version*[87] (PCL:YV) and a *Screening Version*[88] (PCL:SV). The PCL:YV is designed to measure psychopathy in juvenile justice-involved persons, and PCL:SV is developed to perform a comparatively more expedient clinical "screening" for psychopathy in forensic contexts (Hare's team has also developed a self-report measure of psychopathy,[89] but I will not review it here as it is rarely applied in forensic settings). As depicted in figure 1.3, both the PCL:YV and PCL:SV have clear resemblance to the PCL-R in terms of the traits and behaviors they measure, and the assessment procedures are also virtually identical to those of the PCL-R. That is, the assessment is based on information derived from interviews and historic records, and the scoring scheme

Hare Psychopathy Checklist: Youth Version	Hare Psychopathy Checklist: Screening Version
1. Impression management	1. Superficial
2. Grandiose sense of self worth	2. Grandiose
3. Stimulation seeking	3. Deceitful
4. Pathological lying	4. Lacks remorse
5. Manipulation for personal gain	5. Lacks empathy
6. Lack of remorse	6. Doesn't accept responsibility
7. Shallow affect	7. Impulsive
8. Callous/lack of empathy	8. Poor behavioral controls
9. Parasitic orientation	9. Lacks goals
10. Poor anger control	10. Irresponsible
11. Impersonal sexual behavior	11. Adolescent antisocial behavior
12. Early behavioral problems	12. Adult antisocial behavior
13. Lacks goals	
14. Impulsivity	
15. Irresponsibility	
16. Failure to accept responsibility	
17. Unstable interpersonal relationships	
18. Serious criminal behavior	
19. Serious violations of conditional release	
20. Criminal versatility	

Figure 1.3

The two sub-versions of the Hare Psychopathy Checklist. To the left is the 20-item PCL:YV (score range 0–40 points). To the right is the 12-item PCL:SV (score range 0–24 points). The items in the PCL:YV and PCL:SV can be divided into two factors virtually identical to the PCL-R scale.

is set to reflect the dimensionality of psychopathy such that a PCL:YV score ranges from 0 to 40 points and a PCL:SV score ranges from 0 to 24 points.

It is important to point out that the three PCL scales have historically enjoyed enormous popularity among both researchers and practitioners (at times referred to as the *gold standard* in psychopathy assessment[90]), so much so that when experts use the term "psychopathy" or "psychopathic," they are often implicitly referring to an individual who has scored above the clinical threshold on one of the PCL scales (as opposed to any of the other and less popular psychopathy assessment tools mentioned above). There are presumably many reasons why the PCL scales became so widely adopted, but it is worth noting that the PCL scales were, until very recently, the only certified tools available in North America that had been developed specifically for clinician- and interview-based assessments of psychopathy. Almost all other mainstream assessment tools are primarily developed for *research* purposes and/or designed as a self-report measure (which has various limitations in legal settings[91]).

Therefore, throughout this book, I will follow this prevailing convention in the field of using the term "psychopathic persons" to refer to individuals who have been clinically assessed with psychopathy as defined by the PCL scales and, in particular, the PCL-R version (unless otherwise specified). Adhering to this convention is also consistent with how most forensically relevant research on psychopathy is conducted on samples consisting of *PCL psychopathic persons*—that is, individuals who score above 30 on the PCL-R and PCL:YV, or score 18 on the PCL:SV.

The Forensic Use of Psychopathy Assessments

In forensic settings, PCL assessments are usually conducted as part of a broader strategy in the legal system of evaluating people of interest—such as defendants or convicted persons—a process often referred to as a *forensic evaluation.*[92] These procedures might be court-mandated to inform the trier of fact, performed as part of a parole or review board hearing, or conducted during correctional intake protocols. The incentives and purposes behind these evaluations are often multifaceted—and can vary considerably across institutions and jurisdictions—but it is usually motivated by a need to better understand and analyze the person of interest, such as ascertaining their violence risk levels, rehabilitation responsivity, substance abuse problems, mental health needs, or suicide risks.[93]

How exactly a PCL assessment is implemented and utilized during these procedures depends heavily on the specific context (more on this below), but in general it comes down to using the outcome of an assessment as a piece of information that can provide guidance during a variety of decision-making scenarios (e.g., during sentencing or parole decisions). More specifically, the assessment of clinical levels of psychopathy—for instance, when a person scores above 30 on the PCL-R—is interpreted as conveying something generalizing (i.e., nomological) about the person being assessed, based on what is known about—that is, what researchers have learned from studying psychopathic individuals.[94] This is in principle similar to the way a doctor uses a medical diagnosis to make a prognostic inference about a patient. For example, if a person is assessed with having a viral infection (e.g., the SARS-CoV-2 virus that causes COVID-19), a doctor will then use this information to guide an open-ended number of medical decisions, including testing for symptoms related to COVID-19, what medication to administer, whether to isolate the patient, or preparing for severe illness (while this analogy is instructive, it is worth noting that a PCL assessment is not always conducted in a medical/psychiatric context, as it can be conducted by persons without a medical degree and is mostly used to inform nonmedical decisions as outlined below).

Before we explore how exactly a psychopathy assessment is used to guide forensic decisions, it is important to understand the sort of prognostic or predictive information that is commonly associated with psychopathy. The exact information practitioners derive from a PCL assessment, and how they use it to make inferences about psychopathic persons, will or rather should (ideally) depend on a scientific, research-based consensus about the disorder. In other words, forensic practitioners aim to make inferences that correspond to or are conformant with *known facts* about the disorder, again, similar to the way a doctor will (or should) base prognostic decisions about what is known about a specific viral infection and the disease it potentially causes. This way of orienting decision-making according to empirical evidence is broadly referred to as evidence-based practices.[95]

According to the PCL manuals, as well as the researchers who advocate for their forensic use, there are at least five fundamental and interrelated assertions about psychopathic persons that are broadly believed to be supported by scientific evidence. These are:

1. *Psychopathic persons are extremely dangerous:* Often described as social predators, psychopathic persons are both criminally prolific and commit more serious forms of violent crimes compared to ordinary non-psychopathic justice-involved persons.[96]

2. *Psychopathy is chronically untreatable:* Often described as untreatable, psychopathic persons are seen as unresponsive to common psychological intervention and rehabilitation programs, thus comprising a chronic condition incapable of change.[97]

3. *Psychopathic persons are deprived of morality:* Often described as being morally blind or without conscience, psychopathic individuals personify an offender-type incapable of moral reasoning and empathy, thus comprising an extreme behavioral style wholly unconstrained by the law, social norms, and moral values.[98]

4. *Psychopathy is biological:* Often described as being linked to abnormal neurobiological correlates, psychopathy is seen as a brain disorder, which might explain why psychopathic persons are less responsive to common criminogenic factors such as socioeconomic status, education, and age.[99]

5. *Psychopathy is theoretically corroborated:* As it is often described as a well-validated disorder, researchers hold that empirical evidence supports our theoretical understanding of psychopathy, and this scientific account therefore makes up a powerful explanation of their criminal and antisocial lifestyle.[100]

These five basic assertions about psychopathy do not represent an exhaustive list of what is commonly inferred about the disorder, but they are broadly seen as the primary reasons why PCL psychopathy assessments are believed to be relevant and informative to forensic decisions. Or, to put it differently, if none of these five claims were scientifically corroborated—if they were not backed by empirical evidence—it would be unclear why our legal system should use psychopathy assessments to guide forensic decisions, let alone why practitioners should bother spending precious time and resources conducting these assessments in the first place. Why? Because there would be no obvious reason to think that there is any (nomological) inferential value associated with the assessment. It would be akin to a teacher using a specific assessment tool to test for dyslexia in their students but where those who test positive on the assessment still have normal reading and writing skills. Such a test would be both uninformative and a waste of time.

The exact ways in which a PCL assessment ends up informing forensic decisions are rather complex, and it can vary a great deal from jurisdiction to jurisdiction and from case to case.[101] However, the general process usually goes something like this: After a PCL assessment has been conducted, the score will (in most instances) be written into a person's forensic records, a digital document akin to a doctor's journal. As this information is stored, it can then be interpreted and used by an open-ended mix of stakeholders in the legal system, who may use the test results to inform a similarly complex mix of formal and informal decisions. For example, a judge might use the information for legal pre-sentencing and post-conviction sentencing purposes; a parole officer might interpret it as saying something about the parolee's character; and correctional staff may use the information to guide decisions about inmate supervision, and so forth. Another level of complexity to this procedure has to do with the difficulty associated with ascertaining to what degree specific types of information influence specific decisions. For instance, if a parole board consisting of multiple members rules to deny an applicant's request for parole, how do we determine whether, or to what degree, this decision was based on the person's high PCL-R score, as opposed to some other information, say, a record about substance misuse problems? Maybe both pieces of information were important? Or maybe the one piece of information eclipses the relevance of the other? Sometimes a parole board might issue an elaborate formal justification that throws light on the question, while at other times the justification might be brief and less transparent.

Setting these complexities aside, there is a growing body of research documenting what is believed to be some of the most common ways PCL assessments inform and influence forensic decisions. Here I briefly describe some of the most well-documented applications:

Pre-sentencing and sentencing: A judge may interpret a high PCL score as aggravating information when considering pre-sentence detention and post-conviction sentencing outcome. For example, it might contribute to removing the possibility of posting bail or be used as justification for not considering mitigating factors.[102]

Correctional placement: When a person is sentenced to confinement, decisions must be made about their placement needs. Such a decision is usually informed by risk evaluations. Here, persons with a high PCL score are commonly consider "high-risk," thus potentially supporting placement in high/maximum security facilities.[103]

Capital punishment: Psychopathy has been used as an aggravating factor in capital sentencing cases, where courts have interpreted a high PCL score as a chronic risk factor for institutional infraction and maladjustment, which usually weigh in favor of a capital sentence, as opposed to life without parole.[104]

Dangerous and long-term offender status: Many jurisdictions have authority to enhance sentencing and supervision if a person is considered an extraordinary risk to society. This designation usually implies indeterminate or longer sentencing, limitation of rights to parole, and/or extended period of supervision after serving a sentence. It has been found that a high PCL score is frequently invoked in support of such designation.[105]

Sexually violent predator status: This categorization is designated for individuals who have committed a sexual crime and have a personality disorder that place them in chronic risk of sexually reoffending. The status may involve indefinite civil commitment and subsequent supervision. Some jurisdictions acknowledge a high PCL score as supportive of this status.[106]

Juvenile transfer: These decisions involve cases where a young person (usually under the age of 17 or "age of majority") has committed an offense that would be considered a crime if it was committed by an adult. Such a person may be transferred to an adult court and prosecuted as an adult. Juvenile transfers are ordinarily commissioned on the basis of a high-risk profile and low rehabilitation prospects—two aspects that are implicated by a high PCL score.[107]

Treatment and rehabilitation: Justice-involved persons may be offered an opportunity to participate in behavioral treatment and rehabilitation programs during or after incarceration, which are ordinarily tailored to minimize a person's risk of criminal recidivism. There is evidence that a high PCL score has been seen as an "exclusion criterion" that bars a person from these programs, ostensibly because psychopathy is seen as a chronic condition making rehabilitation efforts a futile waste of resources.[108]

Parole: Parole board decisions are primarily determined on whether the applicant poses a significant and continuing risk to society. Since psychopathic persons are widely assumed to be at high risk of re-offending, a PCL assessment may then confer such aggravating information during evaluation. Related information that may be inferred from a PCL assessment is a person's lack of empathy/remorse and benefit from rehabilitation programs, which all may disadvantage the parole applicant.[109]

Correctional and probation supervision and management: Correctional staff are tasked with supervising inmates, minimizing infractions, and facilitating well-being. Probation officers are tasked with supervising a person and ensuring conditions of the probation are met. To achieve these goals, correctional and probation staff will align their service with the risk profile of the individual person under supervision, where a high PCL score will indicate high risk of infraction and probation violation. Such information may contribute to enhanced supervision.[110]

It should be stressed that the issues listed above are unlikely to be decided solely on the basis of a PCL assessment, as decisions usually involve evaluating many different sources of information. However, we know from empirical studies (cited above) that a PCL assessment tends to be seen as a high-quality piece of information that provides all sorts of nuanced insights. For instance, consider how a PCL-R assessment might impact a capital punishment decision in the United States. During these hearings, a judge will normally be tasked with deciding between capital punishment (i.e., death sentence) vs life in prison without the possibility of parole. One aspect that weighs in favor of capital punishment is when the person of interest is believed to exert an elevated and continuing risk to fellow inmates and correctional staff. In this context, a PCL-R assessment can, in various ways, be interpreted as supporting a capital sentence, insofar as psychopathic persons are assumed to (a) constitute a high risk of future institutional infraction and maladjustment (i.e., extremely dangerous), (b) be incapable of changing and assimilating to institutional routines (i.e., chronic untreatable), and (c) be rooted in biological and incurable causes (i.e., chronic risk profile). So, a judge may interpret a psychopathy assessment as an all-encompassing piece of information that directly supports a capital sentencing and/or trumps information that would otherwise support a decision about life without parole.[111]

Overall, what all these different examples have in common is that a high PCL assessment is uniformly interpreted as aggravating information, which in turn promotes a variety of *restrictive* measures—measures that, all things being equal, a person would be less likely to undergo if they were not assessed with psychopathy. Or to put it differently: if two individuals commit the exact same type of crime, under the exact same circumstances but the one person has received a very low PCL-R score (e.g., less than 20) and the other has a very high score (e.g., more than 30), there is an increased probability that the psychopathic individual will be dealt with in a relatively much more precautionary and restrictive way. For instance, the psychopathic person will have a higher likelihood of receiving a non-mitigated sentence, being denied parole, or getting a dangerous offender status. So, when researchers celebrate psychopathy as "the most important and useful psychological construct yet discovered for criminal justice policies,"[112] a substantial part of this celebration has to do with the way PCL assessment is believed to have facilitated a reduction in the impact psychopathic persons have on public health and safety, namely by enabling a targeted management and detainment of such individuals in (at least) all the ways sketched above.

The Scope of Psychopathy Assessments

Up until this point, we have established that psychopathy research rose to prominence during the 1990s, and its forensic use steeply increased in the 2000s and onward. However, to get an idea of the scope of these uses, we might ask whether PCL psychopathy assessments are used only in a small subset of decisions, or whether they are used much more broadly? Is it common for it to be used in court and across judicial contexts? How many people are assessed for psychopathy on a yearly basis? Indeed, it is one thing for a clinical tool to be occasionally used to inform decisions, but it is quite another thing if it is being used on a systematic, routine basis.

One potential avenue for addressing this question is to look at studies that analyze court records. For example, earlier I mentioned two different studies by David DeMatteo and colleagues,[113] which examined case law records in the United States between 1990 and 2011, documenting how psychopathy assessments were virtually nonexistent in the court system in the early 1990s but that an eight-fold increase occurred between 2000 and 2011.[114] However, studies like these have important limitations. For one, they only analyze *court cases* and say nothing about the multitude of other legal decisions that are informed by forensic evaluations (e.g., probation and correctional practices). Secondly, case law studies are based on data pulled from written case summaries uploaded to publicly accessible databases (e.g., *Lexis QuickLaw, CanLii, WestLaw*, and others), but these resources do not include all court cases, and the ones that are included only provide highly selective information about a case. So, even if a PCL psychopathy assessment was used in a specific court case, this information might not show up in the case summary.

Another avenue for answering the question about scope could be to analyze whether forensic practitioners view psychopathy assessments as important, and if they do, this might indicate something about how likely they are to use these assessments in their daily work. Overall, one could simply, and reasonably, assume that if the use of psychopathy assessments is increasing with rapid speed—as continuously stated by forensic experts with "insider" knowledge[115]—this must mean that practitioners see them as being worth their time and effort, and therefore view them as relatively important. A PCL-R assessment is certainly time- and resource-consuming (e.g., it requires a personal interview by a trained and certified forensic mental health clinician), so in a legal system already operating with scarce and stretched resources, there is a strong incentive to gut redundant and uninformative practices. It is therefore fair to assume that if PCL-R assessments are being used at all in such an environment, it must be because decision-makers find them valuable (otherwise they would probably have been abandoned altogether and replaced by other assessment tools).

Another good indicator of the influence of PCL psychopathy assessments is the evidence from survey research. Consider a 2017 study by Marcus Boccaccini and colleagues[116] that questioned 95 evaluators in sexual violent predator (SVP) hearings. They found that 60% of the respondents used the PCL-R to inform *all* or *most* of their cases, indicating that psychopathy assessments are an important part and parcel of these decisions (comparable results were also found by Rebecca Jackson and Derek Hess in 2007[117]). The authors further speculate that these relatively high numbers might be a result of the legislation in 20 states requiring evaluators to assess for potential personality disorders, and some states (for example, Texas) have even made PCL-R assessments mandatory. It is also noticeable that SVP legislation historically highlights psychopathy as a potential *cause* for sexual predatory behavior,[118] and practitioners might therefore view psychopathy assessments as something akin to best practice supported by the legislation itself.

Similarly, and as mentioned earlier, a 2010 study by Caleb Lloyd and colleagues[119] found that psychopathy assessments are used to inform the majority (61.8%) of dangerous offender and long-term offender cases in Canada. To this end, it is notable that the government of Canada has published a *Handbook* designed to guide such decisions, which explicitly recognizes the PCL-R as an appropriate tool. Further, this *Handbook* lists seven reasons that are recognized to support judicial restraint under the legislation, one being that the person has "been diagnosed as a psychopath" and another being that the person "lacks empathy, remorse, or insight into their problems" (i.e., aspects already included in a PCL assessment).[120]

Other evidence of relevance is a 2012 study by John Edens and Jennifer Cox,[121] which distributed a questionnaire among 100 attorneys working for defense teams in capital cases, asking about their personal experiences with psychopathy assessments. On average, respondents estimated that prosecutors used the term *psychopath* to describe defendants in 31% of all cases. Participants were also asked about the prosecution's reasons for using a psychopathy assessment, where 68% said it was used to provide *future dangerousness testimony*. Interestingly, 73% answered that the disorder was used to challenge or rebut the defense's evidence of mental illness in the defendant during the penalty phase, and 60% highlighted similar uses during the guilt-innocence phase. That is, insofar as severe mental illness might reduce criminal responsibility or mitigate sentencing, it appears that a psychopathy assessment might balance such mitigating tendencies. When asked about their perception about the impact of the PCL-R on court decisions, 32% said it had *minimal impact*, 21% said it had *considerable impact*, and 32% said it had *extensive impact* (the rest were undecided). In short, what Edens and Cox's study shows is that the label of psychopathy is widespread in capital cases, and the majority of attorneys agree that the impact of the diagnosis is considerable or extensive.

Correspondingly, there are several studies that appear to support the notion that psychopathy assessments have a wide-ranging impact on court decisions. A study by Daniel Murrie and colleagues[122] surveyed 326 judges in the United States on the effect of psychopathy assessments on their sentencing decisions in mock juvenile cases and found that it had significant impact on a variety of decisions, such as higher support for psychological treatment or less support of deferred prosecution.[123] Consider also a 2021 study by Melissa Jonnson and Jodi Viljoen,[124] who surveyed 170 judges in North America on their perception of the usefulness of assessment tools like the PCL scales and found that judges agreed with the statement that these tools are helpful for making decisions about placement, offender's supervision level, and program/service referrals.[125]

In addition, there are studies that investigate how jurors perceive psychopathic defendants and how a psychopathy assessment might influence their decisions. A study from 2013 by John Edens and colleagues[126] surveyed 285 community members attending jury duty in United States on their perception about a defendant's psychopathic personality traits (loosely based on the PCL-R) in a mock capital case and found that jurors strongly associate such personality traits with enhanced risk levels.[127] A study from 2014 by Shannon Smith and colleagues[128] asked 400 community members attending jury duty in United States how they would characterize a person who was diagnosed as a "psychopath" and what sort of behaviors they associated with the diagnosis. The most common and strongest associations were traits like *unempathetic, manipulative, domineering, deceitful,* and *lack of remorse.* Participants also associated psychopathy with a higher likelihood of committing crimes.[129]

Overall, there seems to be robust evidence to suggest that both judges and jurors associate psychopathy with qualities that support decisions about restrictive measures, which in turn should predict that prosecutors would find it important to pursue the use of psychopathy assessments.[130]

Aside from tracking the importance of psychopathy assessments in the court system, there is evidence that these assessments are a central element in routine forensic evaluations. As mentioned earlier, forensic evaluations often center around making an all-encompassing assessment of a person's needs and risks. A large 2014 study by Jay Singh and colleagues[131] surveyed 2,135 practitioners from 44 countries about their preferred instrument to calculate risk during the past 12 months, and they found that the PCL scales were the most frequent and prioritized among North American practitioners (35% mostly used the PCL-R and 9% used the PCL:SV). A study from 2010 by Jodi Viljoen and colleagues[132] surveyed 199 North American forensic practitioners and found that 64.8% use the PCL-R and PCL:SV to assess risk. Of those who use the PCL scales for risk assessment purposes, 54% reported that they *always* use it, 13% said *frequently,* and 12% said *sometimes.* Similarly, in a 2019

survey of 110 Canadian forensic psychologists, David Hill and Sabrina Demetrioff[133] found that the PCL scales were the most preferred tools in forensic
evaluations, 67% reporting using the PCL-R and PCL:SV and 9% reporting
using the PCL:YV. Practitioners using the PCL scales reported that they made
an average of 40.66 assessments during the past five years (i.e., roughly eight
assessments per year). An international study from 2014 by Tess Neal and
Thomas Grisso[134] surveyed 434 forensic evaluators in court cases (51.6%
from North America) about their practices in their two most recent cases.
They found that the PCL-R was the most frequently used tool for violence
risk assessments (used in 35.6% of cases), it had comparable frequencies in
sex offender risk assessments (35.2%) and civil commitment cases (24.3%),
and it was reported in 16.4% of cases to aid sentencing decisions.[135]

While all of these studies appear to point in one and the same direction—
namely that PCL psychopathy assessments are seen as important and have
become standard practice in legal systems across North America—it is still
difficult to give a reliable estimate about how many assessments are conducted each year, how many of these report clinical levels of psychopathy
(e.g., PCL-R score > 30), and how exactly each one of these assessments
contribute or lead to a further legal constraint of the individual. However,
it might be possible to make some rough conjectures.

The mere fact that PCL assessments have become a standard tool during
forensic evaluations suggests—on its own—that hundreds of thousands of
such assessments are conducted each year, insofar that forensic evaluations
are now a widespread practice throughout the entire criminal justice system
in both the United States and Canada.[136] Quite literally, once an offender
enters the legal system (e.g., juvenile courts, criminal court, correctional, probation, and so on) there is a high likelihood that this person will undergo a
forensic evaluation, which is almost certain to have a risk assessment component to it. And if we trust the data from various surveys,[137] in more than
half of such assessments the practitioner will be using one of the PCL scales.
For comparison, in the United States, the criminal justice system currently
has approximately 5.5 million people under supervision (prisons, jail, parole,
probation, and so on),[138] many of whom go through one or more forensic
evaluations on a yearly basis.

Consider also that two decades ago in 2002, Stephen Hart—a codevel-
oper of the PCL scales—estimated that 60,000 to 80,000 PCL assessments
were conducted each year in United States.[139] If we pair this information

with the eight-fold increase seen in the use of psychopathy assessments in court between 2000 and 2011, what we get is a number that figures between 480,000 and 640,000 per year. Unfortunately, we currently have no good data about whether this exponential development continued after 2011, but if it did, it suggests that there are more than one million PCL assessments conducted each year. Whatever the actual number is, many of these assessments are likely to score under the conventional clinical threshold (e.g., 30 on the PCL-R), but if we adhere to the estimate that up to 35% of incarcerated individuals score above the clinical threshold,[140] this will mean that, on average, up to 35% of these assessments will be "positive" or above 30 on the PCL-R, yielding a total figure of at least 350,000 clinically salient psychopathy assessments each year (these would not necessarily be assessments of unique individuals, as one person might get several assessments in one year). Such numbers are nothing but informed conjectures, but it is conjectures like these that support Robert Hare's statement that PCL assessments have become "unusually important in the legal and judicial world."[141]

In summary, since the 1990s, PCL psychopathy assessments have developed into a wide-reaching practice, signaling how the legal system is now orchestrating a coordinated judicial response to the problem of psychopathy. For every person who is assessed with psychopathy—and dealt with accordingly—society is assumed to have momentarily intervened with the potential havoc and harms caused by psychopathic persons.

In the following chapters, we will take a closer look at each of the separate five claims researchers make about psychopathy, the claims that are central to facilitating the judicial practices of utilizing psychopathy assessments. We will investigate whether these claims truly are supported by the empirical research.

2 How Dangerous Are Psychopathic Persons? Surprising Results from Decades of Research

> The association between psychopathy and criminal violence is so resounding that it is as if psychopathy was a sinister blueprint to create a prototypical antisocial person.
>
> —Matt DeLisi and Bryanna Fox in *Psychopathy Is Integral to Understanding Homicide and Violence* (2022)[1]

In the previous chapter, we surveyed central developments in the history of psychopathy research, particularly how the disorder advanced from a state of scientific and clinical obscurity prior to the 1980s to swiftly becoming recognized as one of the most important forensic concepts in the 1990s. What exactly drove this historic change is a complex story, but a central part of it has to do with the development of the Hare Psychopathy Checklist (PCL) scales. Not only did the PCL scales innovate how research was conducted but they were and continue to be seen as the preferred way to screen and assess for psychopathy in the legal system; more specifically, it is broadly seen as the most reliable method to distinguish psychopathic from nonpsychopathic persons.

Although there are different reasons why the PCL scales are so frequently used to inform legal decisions, perhaps the most common one is their alleged ability to assist decision-makers in identifying individuals that are of extraordinary risk of antisocial and criminal behavior. In their daily work, forensic practitioners have access to a variety of clinical tools that are specifically designed to ascertain a person's forensic risk levels (called "risk assessment" tools). And while the PCL scales were not explicitly designed for risk assessment purposes, it is nevertheless for such purposes that the PCL scales are most typically utilized. In this context, the scales might be used together with other risk assessment tools, but the main motivation for using the PCL scales is that they can yield some form of unique information that is not contained in or measured by other risk assessment tools.

One such uniquely relevant form of information is tied to the conception about psychopathic persons being extraordinarily dangerous, as a group of individuals widely assumed to be of greater and more persistent forensic risk compared to the average, non-psychopathic person. For example, the main architect of the PCL scales, Robert Hare, has characterized psychopathic persons as "intraspecies predators"[2] who are "always deadly,"[3] and forensic experts often underscore that even though psychopathy only affects around 1% of the general population, these few individuals are still responsible for the majority of violent crimes.[4] As stated in a 2015 article by Dennis Reidy and colleagues, psychopathic persons "are at least five times more likely to recidivate violently" and "commit twice as many violent crimes" compared to non-psychopathic offenders.[5] In an attempt to fiscally quantify these dangers, a study by Dylan Gatner and colleagues from 2023 estimated that the devastation caused by psychopathic persons might cost society more than $1.5 trillion in North America alone.[6] It is presumably because of research like this that decision-makers in the legal system acknowledge the incentive to assess for psychopathy—and potentially consider enhanced supervision and detainment of psychopathic individuals—whereas the PCL scales have then become the preferred method to accommodate these needs.

In this chapter, I will survey the scientific evidence behind the assumption that individuals assessed with PCL psychopathy are extraordinarily dangerous individuals and, correlatively, that the scales therefore yield unique information (i.e., one of the main motivations for utilizing the PCL scales for risk assessment purposes). This may sound like a strangely backward task, as if we are now going to inspect a question that we already know the answer to. For instance, some people may think that it is obvious that psychopathic persons are extraordinarily dangerous, because this aspect is such a central part of how psychopathy is conceptually defined: if a person is not dangerous, they will, or rather, should not be described as clinically psychopathic.

However, the question about whether PCL psychopathic persons are extraordinarily dangerous—or other questions such as whether these individuals are untreatable or lack empathy—is not an abstract, conceptual question about how to define the meaning of the term *psychopathy*. Of course, when a forensic clinician uses the term psychopathy to describe a person, it usually implies that the clinician believes that their client is dangerous. And we may turn this on its head and say that if a clinician does not personally believe that an individual is dangerous, it would be very unusual if they assessed that person with psychopathy. However, despite these observations, it still does not automatically follow that just because a person has been assessed with psychopathy—for example, by scoring above 30 on the PCL-R—that they are, in fact, extraordinarily dangerous.

There are many reasons why that is, but one basic reason is that the claim about dangerousness is a statement about *future* behaviors. When researchers indicate that psychopathic persons are extraordinarily dangerous, we must interpret this as a prognostic claim that declares something about what to expect from these individuals—that is, something about their future behavioral proclivities. This is largely similar to when a medical doctor gives a prognosis during the onset of a disease, which may include all sorts of predictions about what the patient should expect—for instance, information about common symptoms or mortality rates. And the crucial thing to notice here is that prognostic claims—insofar that they are predictions—are empirically testable; we can always measure (in hindsight) the accuracy of a prognosis. For example, if a medical doctor tells one of their patients that they are predicted to "have between 2–3 years to live," then we can always measure after two to three years if that prognosis was accurate. In the case of psychopathy, scientists can similarly test whether it is true that those individuals who are assessed with PCL psychopathy reliably behave in ways that justify describing them as extraordinarily dangerous.

The type of research that investigates this exact question is colloquially known as *predictive validity* studies, a term that refers to how these studies test the validity of the prognostic prediction associated with psychopathy; testing whether a PCL assessment is truly associated with future antisocial or dangerous behaviors. While there are different types of experimental designs in this area of research,[7] the most common ones use a fairly straightforward methodology where a cohort of PCL psychopathic and non-psychopathic persons are monitored for an extended period of time—sometimes years— as researchers are gathering data about their antisocial propensities. These studies may monitor individuals who have been released from prison and then focus on tracking criminal recidivism/reconviction rates as reported in police and correctional databases, or they may involve participants who are still serving a prison sentence with data collection focusing on institutional maladjustment and infraction as reported by the correctional institution. The main idea is that by surveying people's *actual* antisocial and criminal behaviors, we will be able to measure the *real* link between a PCL assessment and dangerousness (as opposed to the conceptual, assumed, or stipulated link). So, if PCL psychopathy truly is associated with extraordinary dangerousness, researchers should be able to detect a considerable difference in, for example, the future criminal reconviction rates when comparing PCL psychopathic to non-psychopathic persons.

Around a decade after the development of the first version of the PCL scales in 1980, these prediction studies began to surface in scientific journals with increasing frequency, and today, more than 40 years later, this paradigm makes up one of the largest subfields in psychopathy research with more than 200 published studies involving thousands of PCL psychopathic persons. In the following pages, I will review the main findings of this research and show how the empirical evidence from prediction studies—contrary to popular beliefs—is seriously challenging the preconception of PCL psychopathic persons as extraordinarily dangerous. These discoveries, I argue, severely undermine both the professional motivation and evidence-based justification for using the PCL scales in risk assessment contexts.

Interpreting Prediction Studies

Before we begin to survey and interpret the research from prediction studies, it will be helpful to first begin with a brief discussion about what is implied when researchers claim that psychopathic persons are extraordinarily dangerous or otherwise described as social predators. Because, after all, it is specifically this claim we want to scrutinize insofar that it is this claim that is used to justify using the PCL scales in a risk assessment context. So, we want to know if the claim is true, false, misleading, informative, and so on. However, while describing psychopathic persons as extraordinarily dangerous might be intuitively meaningful to many people, it is still my impression that these terms do entail enough ambiguities and elasticities that make it difficult to evaluate whether the empirical research supports this description of psychopathic persons.

One way to get a clearer definition of these terms is to first look at how they are used in forensic settings. When forensic psychologists describe a person as dangerous, they are usually conveying a belief about the person being a significant risk of harm to their surroundings, including *themselves* and *other* people. Within this broad definition, there are of course many individuals in the criminal justice system who would rightly be considered dangerous. For example, persons with serious substance misuse combined with homelessness and untreated debilitating mental health problems might be considered a danger to themselves due to higher risk of suicidality. However, when forensic experts invoke the term dangerousness in the context of psychopathy and prediction studies, they typically have in mind a much narrower definition of risk, namely a risk of committing aggravated violations of the criminal code or institutional rules, including violence, sexual offenses, property damage, robbery, and so forth. So, when a person is deemed *extraordinarily* dangerous, it is implied that they pose a substantial and elevated risk of these exact behaviors.

A crucial thing to notice about this claim is that it is a *relative* claim; it is not based on some clearly defined, absolute threshold of risk that a person must exceed before they are considered extraordinarily dangerous. Instead, the idea is that when we compare two criminally convicted persons—one psychopathic and one non-psychopathic—what we should find is that these two individuals pose different risk levels. The former is *relatively* much more dangerous than the latter.

However, this is where things become a bit murky and unclear because the statement about extraordinariness is not only saying there *is* a relative difference; it is implying that there is a *large* or *substantial* difference. But how much more dangerous must a person be before researchers and practitioners are warranted in using adjectives such as *extraordinarily dangerous*, *predator*, or *always deadly* to describe a person in a legal context? Must psychopathic persons on average be twice as dangerous? Is it more, for instance, five times more dangerous? What relative levels of dangerousness makes a person deserving of the label "predator"?

One way for us to answer these questions in an informal, yet precise way is to think of them in terms of how prediction studies are conducted and how the data are analyzed and reported. As mentioned, such studies typically involve surveying a cohort of persons that have been released from prison (usually over a period of months or years), where the researchers gather data about participants' post-release reconviction rates (e.g., number of probation violations, burglaries, violent assaults, illegal drug possession, and so on). Now, such a study might divide participants into two groups, comparing reconviction rates between one group of psychopathic persons who have all scored ≥ 30 on the PCL-R and another group of participants who have scored < 30 on the PCL-R. If there is a difference in reconviction rates, researchers will then conclude that there is a difference in the level of dangerousness between groups. But more specifically, the *magnitude* of this difference in dangerousness will usually be specified by comparing the difference in the average reconviction rates in each group.

In figure 2.1, I have illustrated how such a comparison is made, using a hypothetical example where I imagine the non-psychopathic group having an average of 10 new reconvictions during the study period, captured by the first "bold" bell curve (for simplicity, in this example I have assumed that reconvictions are "normally" distributed in the sample—i.e., equal dispersion around the mean). This distribution shows that those to the left in the curve have fewer than 10 reconvictions and those to the right in the curve have more than 10 (i.e., fewer people have very low/high reconviction rates). If psychopathic persons are more dangerous, what we should find is that the psychopathy group—designated by the "dotted" bell curve—is *positioned to the right*, thus illustrating that psychopathic persons on average have a higher number of reconvictions. But more tellingly, the further to the right the dotted bell curve is from the non-psychopathic bold curve, the larger a difference there is between the two groups' average reconviction rates.

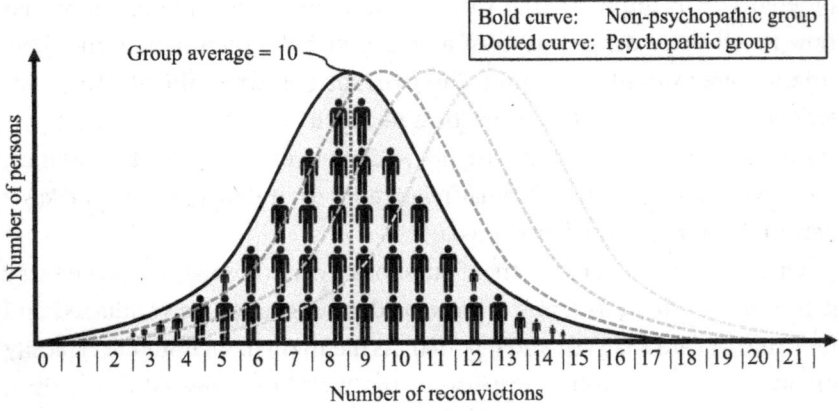

Figure 2.1
An example of a prediction study: the bell curved illustrations show distribution of the number of reconvictions in non-psychopathic (bold curve) and psychopathic groups (dotted curve).

The way this imaginary example can help us parse out our beliefs about psychopathy and dangerousness is by forcing us to answer the following question: How far to the right do you think the dotted curve must be placed before we can justify calling PCL psychopathic persons extraordinarily dangerous? By answering this question, you are directly *specifying* the relative difference between PCL psychopathic and non-psychopathic groups, and thereby determining how much more criminally active a person must be before you think they are deserving of the "social predator" label. I encourage you to take a moment to think carefully about this question, maybe even mark down which curve is the most fitting for such a label (or draw your own curve) and thus capturing your current intuitions. For instance, if you believe that social predators are those people who have around 40% more reconvictions compared to average criminals, then you should pick the last dotted curve.

People will presumably have different intuitions about how large the difference between these groups ought to be before we are warranted in using adjectives like "extraordinarily dangerous" and "social predator." But, based on my experience as a university lecturer—where I have often asked forensic psychology students this very question—there tends to be broad agreement that the difference between the curves must be substantial. It is not uncommon to have the majority of students in a lecture room willing to move the dotted curve very far to the right (as illustrated in figure 2.1). Personally, I believe that if you insist on using a strong pejorative term like *predator* to describe a group of people, you must also be prepared to position the curve so that there is a clear and very sizable gap between the two group means—or better, position the curves so there is little overlap between the groups. Whether that implies embracing the curve furthest to the right or somewhere between the third- and second-furthest ones, that I am unsure about. What I am more certain about, though, is that I think it would be an obvious abuse of our language if we use terms like *extraordinary* or *predatory* to describe a group of individuals that mostly overlaps with the comparison group (i.e., bolded curve).

Now that we have calibrated and clarified our personal intuitions about the use of these labels, let us then turn our focus onto reality and see how the groups of PCL psychopathic and non-psychopathic persons differ when we study them in real prediction experiments: how the two curves—in reality—line up against each other.

Psychopathy and Dangerousness: A Review of the Empirical Evidence

Since the creation of the PCL manual in 1980, more than 200 prediction studies have been published in peer-reviewed journals, comparing the dangerousness of PCL psychopathic to non-psychopathic persons. Instead of reviewing each of these studies, I will focus on the *review literature*—which scientists call systematic reviews or meta-analyses—which is a type of publication that methodically summarizes and synthesizes evidence across individual studies, thus providing the best overview we have of all the scientific results.

Listed in table 2.1 are all 21 review studies that have been published since 1980.[8] As displayed in the table, these reviews include studies that measure four categorically different forms of forensic risk outcomes, namely: (1) general recidivism (i.e., all forms of nonviolent reoffending), (2) violent recidivism (e.g., nonsexual violent reoffending), (3) sexual recidivism (i.e., criminal sex reoffending), and (4) institutional misconduct (e.g., maladjustments and inmate infractions). The first review was published in 1996 by James Hemphill and Robert Hare, and the latest review is by Mark Olver and colleagues from 2020. The largest review is by Anne-Marie Leistico and colleagues from 2008, which summarizes the evidence from 95 independent studies with a total of 15,826 participants.

The results or values displayed in table 2.1 are what statisticians call "effect sizes," which in this case are measures of the standardized average differences in the forensic outcome rates (e.g., general recidivism) between PCL psychopathic and non-psychopathic groups, here indicated by a statistical value called *Cohen's d*.[9] Just like our imaginary example illustrated in figure 2.1, these *d*-values tell us something about the difference between the groups in terms of their average outcome rates. As such, the *d*-value specifies the *magnitude* of the group difference; the larger the *d*-value is, the larger the magnitude. More specifically, a *d*-value ranges from 0.01 and upward, and it indicates how the first group differs from the second group relative to what statisticians call the *standard deviation* within those groups (also called "pooled" standard deviation).

So, to explain what a *d*-value signifies, I must first say something about standard deviation. A standard deviation is a measure of the amount of variation or dispersion in a sample. When we say that there is a *low* standard deviation, this means that values in a sample tend to be similar, and when there is a *high* standard deviation, the values differ greatly. Put differently,

Table 2.1

Reviews of prediction studies involving PCL samples

Study	Studies (k)	Sample size (N)	Mean recidivism differences (Cohen's d)		
			PCL total score	PCL factor 1	PCL factor 2
General recidivism					
Hemphill & Hare (1995)	7	1,275	0.56[a]	n/a	n/a
Gendreau, Little, & Goggin (1996)	9	1,040	0.58[a]	n/a	n/a
Salekin, Rogers, & Sewell (1996)	10	1,991	0.55	n/a	n/a
Hemphill, Hare, & Wong (1998)	7	1,275	0.56[a]	n/a	n/a
Gendreau, Goggin, & Smith (2002)	30	4,365	0.47[a]	0.20[a]	0.49[a]
Walters (2003a)	33	n/a	0.54	n/a	n/a
Walters (2003b)	26	n/a	n/a	0.30[a]	0.68[a]
Edens, Campbell, & Weir (2006)	20	2,787	0.49[a]	0.37[a]	0.61[a]
Leistico, Salekin, DeCoster, & Rogers (2008)	62	11,140	0.50	0.37	0.64
Olver, Stockdale, & Wormith (2009)	20	2,335	0.58[a]	n/a	n/a
Asscher et al. (2011)	39	5,853	0.43[a]	n/a	n/a
Singh et al. (2011)	20	3,854	0.40[a]	n/a	n/a
Violent recidivism					
Hemphill & Hare (1995)	6	1,374	0.56[a]	n/a	n/a
Salekin, Rogers, & Sewell (1996)	13	2,390	0.79	n/a	n/a
Hemphill, Hare, & Wong (1998)	6	1,374	0.56[a]	n/a	n/a

(continued)

Table 2.1 continued

Study	Studies (k)	Sample size (N)	Mean recidivism differences (Cohen's d)		
			PCL total score	PCL factor 1	PCL factor 2
Gendreau, Goggin, & Smith (2002)	26	4,823	0.43[a]	0.26[a]	0.39[a]
Walters (2003b)	27	n/a	n/a	0.37[a]	0.54[a]
Edens, Campbell, & Weir (2006)	14	2,067	0.52[a]	0.39[a]	0.54[a]
Leistico, Salekin, DeCoster, & Rogers (2008)	68	12,359	0.47	0.40	0.57
Campbell, French, & Gendreau (2009)	24	4,757	0.49[a]	n/a	n/a
Olver, Stockdale, & Wormith (2009)	20	2,547	0.52[a]	n/a	n/a
Kennealy, Skeem, Walters, & Camp (2010)	32	10,555	n/a	0.57[a]	0.63[a]
Yang, Wong, & Coid (2010)	16	3,854	0.55	0.22[b]	0.61
Asscher et al. (2011)	29	3,545	0.45[a]	n/a	n/a
Hawes, Boccaccini, & Murrie (2013)	10	1,701	0.63	0.06[b]	0.7
Blais, Solodukhin, & Forth (2014)	55	8,753	n/a	n/a	n/a
Instrumental violence	42	8,542	0.70[a]	0.63[a]	0.61[a]
Reactive violence	26	7,394	0.70[a]	0.63[a]	0.82[a]
Mokros, Vohs, & Habermeyer (2014)	11	2,495	n/a	n/a	n/a
Samples with PCL-R	7	1,652	0.60	n/a	n/a
Samples with PCL:SV	4	843	0.68	n/a	n/a

	k	N			
Sexual recidivism					
Salekin, Rogers, & Sewell (1996)	3		0.61	n/a	n/a
Walters (2003)	5	n/a	n/a	0.10 [a,b]	0.16 [a,b]
Edens, Campbell, & Weir (2006)	4	654	0.14 [a,b]	0.06 [a,b]	0.16 [a,b]
Olver, Stockdale, & Wormith (2009)	4	547	0.14[a]	n/a	n/a
Hawes, Boccaccini, & Murrie (2013)	20	5,239	0.40	0.17[b]	0.44
Institutional adjustment, misconduct, or violence					
Walters (2003a)	15	n/a	0.56[a]	n/a	n/a
Walters (2003b)	16	n/a	n/a	0.37[a]	0.56[a]
Physical violence	14	n/a	n/a	0.24	0.45
Guy, Edens, Anthony, & Douglas (2005)	38	5,381	0.61[a]	0.43[a]	0.56[a]
Physical violence	22	3,502	0.35	0.28	0.30
Leistico, Salekin, DeCoster, & Rogers (2008)	45	6,137	0.53	0.41	0.51
Campbell, French, & Gendreau (2009)	13	1,130	n/a	n/a	n/a
Samples with PCL-R	5	626	0.30[a]	n/a	n/a
Samples with PCL:SV	7	504	0.52[a]	n/a	n/a
Hogan & Ennis (2010)	12	1,313	0.54	n/a	n/a
Olver et al. (2020)	5	320	0.35	n/a	n/a
Physical violence	52	7,928	0.49	0.30	0.51

Note. [a] Original results were reported with a different statistics, which was converted to a *d-value*. [b] 95% confidence interval for effect size includes 0 and/or negative. Studies (k) = number of unique studies included in the review. Sample size (N) = total number of participants across studies. n/a = not applicable. Studies referenced in the table can be found in the endnotes.[10]

there is low standard deviation when each value stays close to the average within the sample; conversely, a high standard deviation means that each value is dispersed or spread out from the average. For example, if every person in a group had more or less the same reconviction rate, the group would have a low standard deviation. If we think of this in terms of a bell curve, like the example in figure 2.1, it would result in a bell shape that is tall and narrow (called leptokurtic) as opposed to short and wide (called platykurtic). Further, when we look at a bell curve, we find that the group average is represented in the center/peak of the curve, but more importantly, the curve is usually shaped in such a way that it is divided into sections or bands of standard deviation units. The most typical distribution of these units follows what scientists refer to as the *empirical rule*, where 68%, 95%, and 99.7% of all data values fall within 1, 2, and 3 standard deviation units, respectively. The shape of the bell curve is thus determined by tabulating the data into these intervals. For example, if a study reports that the average rate of criminal recidivism was 10 reconvictions with a standard deviation of two reconvictions, this means that the curve is shaped such that it shows 99.7% of the sample had between four and 16 criminal offenses (i.e., the recidivism rates are within three standard deviation units). This is roughly what our example in figure 2.1 illustrates. So, when you were earlier deciding where to position the dotted psychopathy curve, you can similarly imagine that the distance between the group averages (or curve peaks) can be measured in standard deviation units.

This brings us back to interpreting a d-value. When we use the d-value denominator of an effect size $d = 1$, this indicates that the difference in the average rate (i.e., the bell curve peak) between the two groups is one standard deviation unit. In such a case, the two groups would be positioned to the left/right of each other, so there would be a gap in the magnitude of one standard deviation unit between the two group averages (i.e., curve peaks). For instance, if we again return to our example from above, the difference in average reconviction rate between these two groups would be two criminal offenses, exactly what is illustrated by the second dotted curve in figure 2.1. We can therefore think of group differences along the lines of a difference in group averages, which correlatively also indicates how similar the two groups are (i.e., how much the curves overlap).

Let us now look at the results from the review studies listed in table 2.1. The first thing to notice is that many review studies report a d-value for

PCL total score, as well as for PCL factor 1 and PCL factor 2 scores. These results are essentially an expression of three different group comparisons. The column with PCL total score is the ordinary and perhaps most relevant one for forensic decision-makers, comparing PCL psychopathic to non-psychopathic persons. The effect sizes displayed in the factor 1 and factor 2 columns are based on an analysis where we ask what the between-group differences are if we divide participants into groups based *only* on their scores on the items included in each of the two factors of the PCL scales (i.e., interpersonal/affective vs lifestyle/antisocial items, respectively; see figures 1.2 and 1.3 in chapter 1). The main reason for providing a factor-based group analysis is that it may reveal something about which PCL items are "causing" the overall group differences (more on this later).

Across the four types of forensic outcomes, the d-values reported in review studies for PCL total score have historically ranged from: $d = 0.40$–0.58 (general recidivism); $d = 0.43$–0.79 (violent recidivism); $d = 0.14$–0.61 (sexual recidivism); and $d = 0.30$–0.61 (institutional misconduct). According to statistical convention, these effect sizes can be described as *weak* to *moderate* group differences (i.e., normative descriptions of d-values adhere to the following semantics: $d = 0.01$–0.19 is described as *very weak*; $d = 0.20$–0.49 is *weak*; $d = 0.50$–0.79 is *moderate*; $d = 0.80$–1.19 is *strong*; $d = 1.20$–1.99 is *very strong*; and $d => 2.0$ is *huge*).[11]

To illustrate what these d-values mean, I have plotted the results onto figure 2.2 such that it displays the average group differences by positioning the two group-based bell curves next to each other (similar to what I did in the hypothetical example in figure 2.1). Recall that a d-value is an expression of how many standard deviation units differentiate the two groups. If there was $d = 1$, it would mean that the difference between the two groups is one standard deviation (as illustrated by the sigma or "σ" symbols in figure 2.2). So, for example, if we look at the largest review study by Anne-Marie Leistico and colleagues from 2008, we find reported d-values ranging from $d = 0.47$–0.53 across the three forensic outcomes. With some simple napkin mathematics—adding these three effects and dividing them by three—we get an average $d = 0.50$. For the record, we are not supposed to do napkin math like this, so I am only doing it to give us a rough idea of the average effect size. In this case, it means that we should expect a real-life difference of around 0.50 standard deviation units' difference between the group averages. I have illustrated this using the right-positioned dotted curve in figure 2.2.

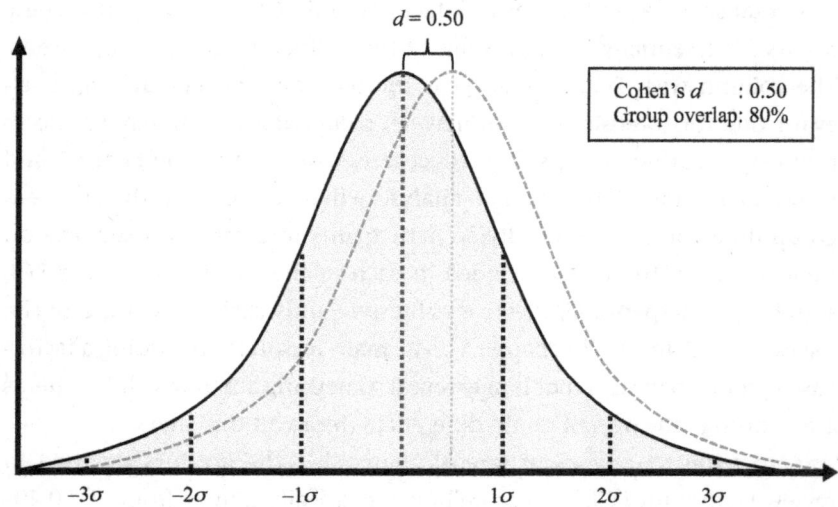

Figure 2.2
An illustration of group differences. This compares the non-psychopathic group (left)
to PCL psychopathic group (right) with an effect size of Cohen's d = 0.50. The sigma
(σ) symbol indicates one standard deviation unit.

The illustration in figure 2.2 neatly captures how a PCL psychopathy
group (really) differs from a non-psychopathy group in terms of their average levels of dangerousness when measured in empirical prediction studies.
As the graphic shows, there is substantial group similarity or overlap. In the
example in figure 2.2 with d = 0.50, there is a total of 80% overlap between
the two groups, which means that in the vast majority of cases, PCL psychopathic and non-psychopathic persons actually have *identical* forensic outcome rates (e.g., identical number of criminal reconvictions). In other words,
80% of all the times we compare psychopathic to non-psychopathic persons,
it is *not* true that individuals with clinical levels of psychopathy are more
dangerous, let alone *extraordinarily* more dangerous. Another way of describing this group overlap is to think of it in terms of the proportion of the PCL
psychopathy group that has a worse forensic outcome rate compared to the
average person in the non-psychopathic group, which in this case would be
69%. A third way to articulate this difference is in terms of the probability
that a randomly selected person from the psychopathic group has a higher
forensic outcome rate than a random person from the non-psychopathic
group. With a value of d = 0.50, there is a probability of 64% that, if we

selected one random person from each of our two samples, it would be the person from the PCL group who had the highest outcome rate, which, of course, would be 50% if there were no group differences. In short, we can think of d-values as relativizing the number of persons with a forensic outcome in each group.[12]

These results clearly show that there is a measurable difference between PCL psychopathy and non-psychopathy groups in terms of their average level of dangerousness. However, it is also equally clear that this difference is not very large; in fact, across the four forensic outcomes, statisticians would characterize this as weak to moderate. Comparatively, such a magnitude is what the mathematician who developed the equation behind the d-value, Jacob Cohen,[13] described using an analogy of the difference in height between 15- and 16-year-old girls (i.e., weak) and 14- and 18-year-old girls (i.e., moderate). In the first case, 15-year-old girls have almost the same average height as a group of 16-year-olds, and in the second case, we can easily appreciate that if we gathered a group of 14-year-old girls there would be a much larger difference in height compared to a group of 18-year-olds. Cohen described the first difference as being too small to be noticeable by the "naked eye" (i.e., weak), where in the latter case the difference could be noticed with the "naked eye" (i.e., moderate). That is, we can think of the size of these group differences as whether a normal person would be able to easily spot the difference or not, without the help of sophisticated statistical tools.

Thinking of the group differences in such a way helps us put the subject matter of this chapter in a clearer perspective, and I believe that there is a way that we can build on Cohen's illustrative analogy: imagine that we could gather everyone who had participated in a prediction study where the results were $d = 0.50$—that is, right smack in the middle between weak and moderate effect size—and thereafter asked all participants to put on a bright red t-shirt if they had committed a criminal offence during the study period. If we further asked everyone to split into the two groups—the PCL psychopathy group to the right and non-psychopathy group to the left—it would then be difficult for any normal person to easily distinguish which of the two groups contained the most red t-shirts. This is the real magnitude of difference we see between PCL psychopathic and non-psychopathic persons: on average, it is actually so meager that in order for us to accurately detect that there even is a group difference, we literally need assistance from mathematical techniques developed by people like Jacob Cohen.

A Critical Reflection on Surprising Results

The scientific results from prediction studies may strike many people as highly surprising, perhaps for obvious and understandable reasons. The PCL scales are designed to reliably identify psychopathic persons in the criminal justice system, a group of individuals who are commonly described as an extraordinarily dangerous bunch, the worst of the worst, the social predators of our society. But when we empirically measure and compare the behavior of these individuals to non-psychopathic offenders, what we find is something else: that there are only marginal differences between the groups, scarcely supportive of pejorative terms like "social predator." However, as these results might be surprising, there is additional evidence to suggest that the group differences are even more precarious than they appear, further subtracting from the narrative of psychopathic persons as extraordinarily dangerous. Here I shall briefly review a couple of such critical observations that have been discussed in the broader literature.

First, some scholars have pointed out that when results from prediction studies are summarized in systematic reviews and meta-analyses, the final aggregated overview they provide might be misleading. For example, a 2023 publication by Steven Gillespie and colleagues[14] analyzed the quality of all published review studies and found that most of these publications had serious quality issues and risk of bias in reporting. This includes some studies using flawed or problematic statistics, inconsistent reporting, and so on. Of all the studies in table 2.1, for instance, Gillespie and colleagues found that only one study had what they deemed "high" quality, namely the study by John Edens and colleagues from 2006 (table 2.1). It is interesting to notice that the effect sizes reported in this study are consistently lower than many of the other review studies: $d = 0.49$ for general recidivism; $d = 0.52$ for violent recidivism; and $d = 0.14$ for sexual recidivism (the latter d-value was indistinguishable from zero, meaning that the association might be spurious). Again, if we perform some simple napkin math (adding these three effect sizes and dividing by three) what we get is a d-value $= 0.38$ (i.e., a weak effect size), roughly 25% lower than what is reported in the study by Anne-Marie Leistico and colleagues in 2008 (table 2.1). When the effect size in a high-quality review study is substantially smaller than those reported in low-quality review studies, it suggests that the results from low-quality studies might be overestimations (i.e., it is widely acknowledged that low-quality reviews are at risk of reporting inflated effect sizes[15]).

Second, a much-debated observation about the prediction review studies listed in table 2.1 is the discovery that the observed group differences are mostly due to the lifestyle and antisocial items included in the PCL scales, the so-called factor 2 items. The interpersonal and affective, or factor 1 items—which researchers broadly view as the truly distinctive personality features of psychopathy—are much less associated with forensic risk. If we take a look at the factor columns in table 2.1, what we find is that the d-values in the factor 1 column are consistently lower than the values in the factor 2 column (in many cases, there is ~50% difference). These columns show that if we analyze the data by comparing individuals who score high vs low on factor 2 items, group differences in forensic outcomes are relatively *larger* than when they define groups by PCL total scores and/or factor 1 scores. And, correlatively, if we compare groups in terms of their factor 1 scores, what we find is that differences are relatively much *smaller* compared to groupings based on PCL total scores and/or factor 2 scores. One perplexing aspect about this observation is that factor 2 items are not unique to PCL psychopathy but are widely assumed to be present in large portions of individuals with a criminal record (e.g., factor 2 items include common criminogenic items like *parasitic lifestyle, juvenile delinquency*, and *criminal versatility*). So, if the goal is to identify a group of extraordinarily dangerous offenders, it seems that we would be better off simply grouping persons along these common criminogenic items and ignore information about who are assessed with PCL psychopathy (e.g., score ≥ 30 on the PCL-R). In short, a PCL *total score* appears to be an unreliable way to identify who are the relatively high-risk people, and certainly a very unreliable way to identify the social predators of our society.

Third, researchers have found that the weak to moderately higher risk levels observed in PCL psychopathic groups might not generalize that well across individuals with different demographic characteristics. The results we have surveyed in table 2.1 are an expression of all the research bundled together. However, when researchers analyze recidivism rates in PCL psychopathic groups more carefully (with what is called a "moderator analysis"[16]), what they find is that these reconviction rates vary significantly in psychopathic individuals based on their gender, country, and ethnicity. For example, the 2008 study by Leistico and colleagues found that recidivism rates in psychopathic men are relatively lower compared to recidivism rates in psychopathic women. They also found that recidivism rates for PCL psychopathy were higher in studies conducted in Canada vs studies conducted in the United States. And, finally, they found that recidivism rates were significantly lower in PCL samples from ethnic minorities compared to nonminority samples. This latter discovery is rather alarming insofar that it is individuals from ethnic minorities that are most likely to undergo a PCL assessment due to their relative overrepresentation in the legal system. So, not only is the empirical evidence challenging the claim that PCL psychopathic persons are extraordinarily dangerous but it is also challenging the claim that there are differences in risk levels that generalize to everyone with a high PCL score.

A fourth undermining observation has to do with the specific levels of dangerousness that are observed in PCL psychopathic samples. As various researchers have pointed out, these levels are actually significantly lower than the risk levels that are posed by other common "groupings" in a forensic context. For example, as mentioned in the beginning of the chapter, forensic psychologists have access to various tools designed to make risk assessments. These tools typically involve assessing a person for much more common and simpler risk factors such as the person's age or whether they have a history of violence and substance misuse. When prediction studies use samples that are based on these far simpler groupings, they are typically able to detect effect sizes (e.g., d-values) that exceed the effect sizes in studies comparing PCL psychopathic to non-psychopathic samples. That is, the difference in reconviction rates between these two latter groups is smaller compared to the differences found in groups selected using common and simple risk factors. For example, this was confirmed already in 2010 in a study by Min Yang and colleagues[17] but also more recently in a large-scale study by Seena Fazel and colleagues in 2022.[18] Again, if PCL psychopathic persons are supposed to be extraordinarily dangerous social predators, this evidence clearly challenges such a notion. And more pressingly, it appears to solidly undermine the idea that there should be unique motivations for using the PCL scales for risk assessment purposes.

A final (but largely unrecognized) problematic observation about psychopathy prediction studies is that they are *not* always, strictly speaking, comparing a randomly selected psychopathic group against a random non-psychopathic group. While there are many ways that the notion of randomness can be affected, what researchers sometimes do is what we (in my research group) colloquially refer to as "excluding the middle." It is a questionable research practice where researchers are essentially manipulating the sample composition. Such a study might begin by randomly selecting their psychopathy sample consisting of persons with a high PCL-R score (usually at the diagnostic threshold of ≥ 30). But when they select the comparative or control sample, they either (a) do not include individuals with PCL-R scores < 30 but instead only include individuals with much lower scores, such as PCL-R scores < 20; or (b) the researchers may have selected a comparison sample that has an unrealistically low average PCL-R score. In both cases, the researchers are effectively excluding individuals (the "middle" group) from the experiment and thus jeopardizing the randomized aspect of the study. There is currently no clear overview of how common this practice is, but I have seen too many examples for it to be ignored.[19]

To be clear, manipulating the sample composition in this way is not just a formality issue of violating the principle of randomness in scientific research sampling, but more specifically, in a prediction study, this procedure effectively guarantees that researchers will find group differences in forensic outcome rates that are larger than what is generally the case if they had compared groups of PCL-R ≥ 30 vs PCL-R < 30. There are a couple of reasons why that is, but the primary reason has to do with the way the PCL-R is partially construed of basic behavioral items that are known to be statistically correlated with higher recidivism rates (e.g., items such as criminal versatility and juvenile delinquency). So, when researchers selectively remove the *middle-scoring* group from the experiment or analysis, they are then artificially lowering the score of these items in the control sample and therefore de facto lowering the forensic outcome rates in the sample. The end result is that it generates a between-group difference (*d*-value) that is larger than if they had compared psychopathic samples to controls defined by PCL-R score < 30. In effect, this practice obscures the contextual framework within which we are ordinarily interpreting prediction studies. As mentioned earlier, prediction studies are supposed to give us a relativized comparison between random psychopathic and non-psychopathic persons in the justice system. But studies that are using gerrymandered samples give you something different. They are relativizing results between psychopathic persons and individuals who score *very low* on the PCL scales. This way of tinkering with the PCL scales during sampling is an instance of what researchers refer to as "data hacking" strategies (e.g., *scale redefinition* and/or *discretizing variables*).[20]

Concluding Remarks: Adjusting the Rhetoric to Fit with Reality

In this chapter, I have shown that the current empirical evidence from prediction studies does not clearly support the claim that PCL psychopathic persons are extraordinarily dangerous. While there appears to be a generalizable difference in risk levels between psychopathic and non-psychopathic persons, this difference is both precarious and relatively small, arguably nowhere near the magnitude that experts generally imply when they refer to psychopathic persons as dangerous social predators. Consequently, these results also challenge the notion that PCL assessments provide unique information during forensic risk evaluations; not only are effect sizes questionable and smaller than ordinarily promoted, but, rather tellingly, the PCL scales are reliably outperformed by ordinary risk assessment tools. Thus, when the aim is to evaluate and ascertain risk levels in justice-involved persons—for instance, for sentencing or parole supervision purposes—forensic practitioners and decision-makers do not appear to have any clear evidence-based reasons for using the PCL scales over other readily available tools. In fact, there appears to be strong evidence in favor of choosing these alternative purpose-built risk assessment tools.

This discovery raises the question: Why have prominent researchers and practitioners continued to reiterate the description of psychopathic persons as extraordinary offenders? I will not pretend to have a decisive answer to this question, but instead I would like to highlight some observations that I think are important to consider when trying to answer it.

One important aspect to note is that it would be wrong to think that *all* researchers have portrayed psychopathic offenders as extraordinarily dangerous. As early as the 1990s and 2000s some researchers were indeed pointing out that claims about the link between psychopathy and high risk were exaggerated and lacking empirical support. For example, in 2002, Paul Gendreau and colleagues authored a searing criticism of the PCL scales and their alleged ability to identify dangerous offenders.[21] As the authors pointed out, these claims were rhetorically inflated—they actually *mocked* their colleagues for a lack of critical thinking, which is rarely seen in peer-reviewed journals—but they also pointed out that there was plenty of evidence to suggest that PCL psychopathic persons were less dangerous compared to other subgroups in the criminal justice system (like, as mentioned above, when forensic practitioners use other simpler tools to identify risk). This sort of criticism has consistently resurfaced in various forms since the early 2000s.[22] This suggests that the issue of adjusting the rhetorical description of psychopathy to fit the scientific evidence appears to be a problem that only some researchers and practitioners are struggling with.

Another crucial observation has to do with the historical development of the evidence from prediction studies. It is tempting to think that what is driving these exaggerated and largely misleading claims is that researchers have failed to update their beliefs according to the available evidence. Maybe in the early days there was evidence to suggest that PCL psychopathy truly was linked to extraordinary risks, and only much later did researchers discover that this was not the case. However, when screening through the review studies listed in table 2.1, it is obvious that there never was any clear empirical evidence to support the notion of psychopathic persons as extreme offenders. In other words, the effect sizes reported in the 1990s have not really changed much over the three decades where the research was conducted. While the nominal d-values have fluctuated in both directions—some becoming smaller and some becoming larger—the magnitude of these changes is minimal, hardly moving the needle in terms of the inferences that can be drawn from the evidence. So, when researchers are rhetorically misrepresenting the real dangerousness of PCL psychopathy, they are not only doing this while facing criticism from esteemed colleagues (such as Paul Gendreau and others), but they have traditionally been doing this without any clear empirical evidence to back their perspectives.

To make things even more perplexing, some of the most cited literature has consistently recommended that forensic practitioners exert caution about the link between dangerousness and PCL psychopathy. For example, at least since the publication of the large and highly cited review study by Anne-Marie Leistico and colleagues in 2008 (see table 2.1), it has been widely known that the PCL scales have important caveats and limitations when it comes to their ability to identify high-risk individuals. In their conclusion, Leistico and her colleagues write that the PCL scales must be used "cautiously," insofar that they do not clearly identify increased risk levels in "minority ethnic groups, males, and prisoners," and that the PCL scales are "less reliable for shorter follow-up periods than for longer follow-up periods."[23] In my opinion, this is nothing short of a remarkable statement, because it all but disqualifies the PCL scales as a reliable tool for risk assessment purposes. For one, legal systems typically use the PCL scales under the assumption that they can tell us something about the immediate, short-term risk of an offender (there's comparatively little focus in long-term risks, which only comes up in rare cases such as capital sentencing and dangerous offender hearings). But Leistico and colleagues are here concluding that the average effect sizes are even lower for such short-term predictions. Second, the authors are pointing out that effect sizes are similarly lower when used on the individuals who make up most of the forensic population. In the United States, for instance, there are an estimated 5.5 million people under supervision in the correctional system (probation, incarceration, and so on), of which 1.7 million (31%) are held in prisons. Of these 5.5 million people, the vast majority are males (+90%) and many are from minority groups (+60%).[24] Again, on these populations, Leistico and colleagues recommended that the PCL scales should be used with caution. However, given the size of these populations, it implies that the PCL scales should almost always be used cautiously.

To me, these observations suggest that psychopathy researchers and practitioners have yet to internalize and critically appraise the results coming from prediction studies. And more importantly, it appears that researchers have struggled with disseminating the information from these studies, which might have resulted in subpar education of stakeholders in the legal system about the true risk levels associated with PCL psychopathy. The challenge for current and future generations of forensic psychologists will be to correct this record. This is not a complicated task, nor is it an unreasonable demand, as it simply requires paying closer attention to what the empirical work is showing.

3 Is Psychopathy Untreatable? A Tradition of Ignoring Empirical Research

Most clinicians and researchers are rightly pessimistic about the treatability of psychopaths with traditional methods. Unlike most other offenders, psychopaths suffer little personal distress, see little wrong with their attitudes and behavior, and seek (and remain in) treatment only when it is in their best interest to do so, such as when applying for probation or parole. It is therefore not surprising that they derive little benefit from traditional treatment programs, particularly those aimed at the development of empathy, conscience, and interpersonal skills.
—Robert Hare in *The Hare Psychopathy Checklist–Revised* (2003)[1]

In recent history, psychopathy has been and continues to be broadly perceived as a chronic and untreatable condition, meaning that the psychological and behavioral dispositions of psychopathic persons are assumed to remain stable throughout their lives, regardless of clinical intervention. This is very different from ordinary, non-psychopathic individuals where it is typically found that criminal and antisocial propensities are responsive to rehabilitation efforts, and that these propensities tend to wane considerably with age.[2] According to mainstream perceptions, psychopathic persons therefore represent as a unique type, as persons that are incorrigible and relentless in the personal harms and social havoc they cause during their lifetime.

The judicial implications of this "untreatability view" are multifaceted and complex, but evidence suggests that the view influences a variety of legal decisions, such as: (1) denying psychopathic persons access to rehabilitation programs on the premise that these programs are *ineffective*, (2) contributing to more punitive sentences based on the view that psychopathic persons are likely to remain a *permanent* threat to society due to their chronic condition, and (3) transferring juvenile psychopathic offenders to adult courts insofar that they are believed to be *resistant* to rehabilitation (for more details, see chapter 1). When a person scores high on one of the Hare Psychopathy Checklist (PCL) scales, there is a chance that the untreatability view leads decision-makers to prioritize constraint over more progressive approaches, such as recommending incarceration vs community-based sentencing. As Robert Hare and colleagues emphasized in a 2013 article, forensic practitioners must focus on the "management" of psychopathic persons as opposed to "treatment" insofar that there is "no evidence that treatment programs result in a change in the personality structure of psychopathic individuals."[3]

The view that psychopathy is an untreatable condition has a long history, but it is especially in the modern era of the 1900s that it appears to have become something akin to a widely shared belief among mental health researchers and clinicians. In earlier days of psychopathy research, many scholars held a comparatively more optimistic attitude. For example, when Benjamin Rush wrote about *moral derangement* at the turn of the eighteenth century, his work was importantly linked to a broader conversation about the role of medicine in treating and resocializing criminals, especially those who were deemed morally incapacitated,[4] a conversation that continued to thrive well into the nineteenth century.[5] However, this optimism appears to have been gradually replaced with a more pessimistic viewpoint, ostensibly driven in part by the realization that whatever psychiatric methods were in vogue in one particular era always appeared to be futile on psychopathic persons.[6]

In the fifth edition of *The Mask of Sanity*,[7] for instance, Hervey Cleckley stresses that during his lifelong career of dealing with psychopathy in his clinical work—primarily at University Hospital in Augusta, Georgia—no traditional treatments ever succeeded in changing the personality and behavior of his psychopathic patients. Cleckley writes that they appear to have not only "immunity" when it comes to the control and deterrence of the law but also a "lack of response to psychiatric treatment of any kind."[8] While the type of treatment Cleckley is referring to includes methods that have long fallen out of favor in psychiatric practices (e.g., neurosurgery like lobotomy and topectomy, and rudimentary forms of electroconvulsive therapy, combined with opaque forms of psychoanalysis), and for that reason we should hesitate placing too much weight in his experiences, he nevertheless made a central observation that some contemporary researchers still agree with. According to Cleckley, one of the more fundamental obstacles in treating psychopathic persons is that they never appear to fully comprehend—nor are they even willing to concede—that they have a problem, let alone a disorder in need of clinical care.[9] Allegedly, when psychopathic persons find themselves in a terrible predicament—such as being locked up in prison or assigned to mandatory psychiatric detainment—everybody else is always to blame but their own personality or lifestyle decisions. So, how is a clinician supposed to get through to such an individual, helping a person with a problem that they themselves flat out refuse to acknowledge?

Although there were voices during and after Cleckley's time that expressed a more optimistic and less defeating attitude about treating psychopathy,[10] it was nevertheless the untreatability view that eventually became the mainstream attitude in the 1990s, the period where psychopathy assessments slowly gained prominence in the legal system.[11] This development also dovetailed with the conventional perspective about crime that dominated the post–World War II era, a period where scholars and practitioners increasingly viewed the social improvement of justice-involved persons as a fundamentally lost cause—that clinical, disciplinary, or other sorts of interventions simply had no positive effects.[12] That researchers also concurred with this so-called *nothing works* attitude when it came to psychopathic offenders is not surprising. When reading Robert Hare's *Without Conscience* from 1993, one is provided with a quote that pretty much sums up the status of the field in the 1990s: "Many writers on the subject have commented that the shortest chapter in any book on psychopathy should be the one on treatment. A one-sentence conclusion such as, 'No effective treatment has been found,' or, 'Nothing works,' is the common wrap-up to scholarly reviews of the literature."[13]

But what exactly do researchers mean to imply with the statement that psychopathy is untreatable? According to the PCL-R manual, the untreatability view is not simply related to the mere observation that psychopathic persons seem impervious to psychiatric intervention but the view is really an expression of at least three intertwined ideas, explained in the following paragraphs.

First, the PCL-R manual states that the very act of engaging psychopathic persons in clinical therapy may be pointless as they almost always have a non-agreeable attitude toward therapeutic methods and goals, either refusing to take part, interrupting the process, or conducting themselves in ways that impede the clinical work (this is sometimes referred to as *therapeutic attrition*).

For example, it is stated in the PCL-R manual that psychopathic persons enroll and remain in treatment only "when it is their best interest to do so," and that they "put on a good show, make 'remarkable progress', convince the therapist and parole board of their reformed character, are released, and pick up where they left off before they entered prison."[14]

Second, the PCL-R manual states that even if psychopathic persons manage to complete a conventional treatment program, the effects are presumably insignificant or minor at best, meaning that whatever clinical methods work on non-psychopathic persons, these methods are alleged to have no meaningful impact on psychopathic persons (this is sometimes referred to as *unresponsiveness*). For example, it is stated that the "problem is not necessarily that psychopaths *cannot* be treated but that the treatments and interventions that do work with most offenders are inappropriate for psychopaths" (original italics).[15]

Third, the PCL-R manual also highlights that not only are clinical interventions hard to initiate and the effects elusive but treatment of psychopathy is also believed to be correlated with adverse outcomes, meaning that when psychopathic persons enroll into ordinary treatment programs, there is a real possibility of exacerbating their disorder (this is sometimes referred to as *iatrogenic effects*). For example, the PCL-R manual states that treatments such as "group therapy and insight-oriented programs may help psychopaths to develop better ways of manipulating, deceiving, and using people, but do little to help them to understand themselves."[16]

In this chapter, I will outline and critically analyze the evidence behind the untreatability view as represented by these three ideas from the PCL-R manual: that psychopathic persons (1) are unamenable to or have unusual degrees of attrition during treatment, (2) are unresponsive to ordinary treatment, and (3) become worse when exposed to such clinical intervention methods. It is these three separate ideas that have propelled the widespread belief that psychopathic persons are untreatable. Before we dive deeper into the empirical work, however, I shall first contextualize the research by addressing what is meant by the term "treatment" in forensic contexts and how this applies to persons assessed with PCL psychopathy, as well as addressing the type of methods researchers have used to study the treatability of psychopathy.

Treating Psychopathy: Definitions, Research Designs, and Clinical Methods

The question about the treatability of psychopathy faces a hurdle similar to the question about the *dangerousness* of psychopathic persons. As highlighted in chapter 2, some people might find it both intuitive and self-explainable that psychopathy is associated with dangerousness; if a person is not dangerous, then you would not assess them with psychopathy. And so, too, may the question about treatment strike some people as self-explainable: if a person is not untreatable—that is, not chronically incapable of change—then it is also not the case that this person is clinically psychopathic. Only true psychopathic persons are untreatable. Some may even claim that the proof is really in the pudding. It is as obvious as daylight that there are people in our criminal justice system that appear to be thoroughly out of clinical reach, wholly incapable of achieving any meaningful improvement to their personality and behaviors, and therefore stay as incorrigibly dangerous as they were when they were first apprehended and incarcerated. Think of well-known serial killers like Ted Bundy, John Gacy, and Edmund Kemper. Do you think they did or could have changed for the better? Should we have considered granting them parole if a prison therapist told us they had changed? The fact that virtually all legal systems invest considerable funds in building facilities and maintaining services that ensure the indefinite confinement of some offenders—the truly dangerous and incorrigible ones—is in part a testimony to a grim reality about these individuals, that there is no hope of prosocial rehabilitation. So, perhaps what explains this form of chronicity is a personality disorder like psychopathy.

While this perspective is evidently appealing among some scientists,[17] it has little to do with the question we are pursuing in this chapter. Here we are not debating whether it is reasonable to describe untreatable offenders as psychopathic. Instead, we are interested in analyzing whether those individuals we assess with psychopathy in forensic settings—that is, those individuals who score high on the PCL scales—really are unamenable to therapy programs, whether these programs have any positive effects, and lastly if there is significant risk of adverse effects. In short, and to repeat the mantra from chapter 2, we are interested in knowing if there is *truth to the prognosis*. This is a thoroughly empirical question, which can be answered through ordinary experimental research.

The Meaning of the Terms Treatment and Untreatability

As mentioned, the untreatability view is generally an expression of three separate but connected ideas: that psychopathic persons are unamenable, unresponsive, and would potentially react adversely to treatment efforts. This viewpoint raises the question: What exactly is meant by the term *treatment*? Many people intuitively understand the term treatment in relation to its usual medical connotations, as denoting the process of either curing diseases or alleviating and mitigating their symptoms. For instance, when a nurse treats a patient, there is an illness and a range of symptoms that, with the use of skills and the appropriate medication, the nurse aims to reduce or eradicate. Correspondingly, when we say that a patient is untreatable, we usually think of this as the patient being in some sort of irreversible state, but more specifically that their disease or injuries simply cannot be alleviated, resulting in either a chronic condition or cessation of the patient's life. However, when forensic practitioners speak about treatment and untreatability in the context of psychopathy, readers should note that while they still use these terms to denote something about a specific type of *act* that alleviates symptoms (i.e., treatment) and a *state* of irreversibility (i.e., *un*-treatability), the two terms nevertheless take on importantly different meanings compared to the ordinary medical use of this terminology.

There are commonly two distinct, yet complementary, understandings of the term treatment in the forensic psychiatric literature. The first use of the term refers to the clinical act of intervening or changing how psychopathic traits are manifested in a person. For example, treatment here refers to helping the individual cope with or change some of the core symptoms associated with psychopathic personality, such as mitigating their impulsiveness or grandiose sense of self, or maybe even working on developing more empathic attitudes toward other people. The second use of the term treatment designates the clinical act of aiding the person with breaking their pattern of criminal and antisocial conduct (sometimes called *desistance*). For example, in this case, treatment refers to helping a person adjust positively to the expectations in a correctional institution, reduce their risk of reoffending after release from prison, comply with probation and parole requirements, and so forth.[18]

A helpful (but oversimplified) way of thinking about psychopathy treatment is to see it as a clinical intervention that aims to reduce a person's PCL score—that is, mitigating the presence of antisocial personality traits and behaviors as represented by the different PCL items (see figures 1.2 and

1.3 in chapter 1). In extension of this observation, however, it is important to clarify that researchers rarely speak about psychopathy treatment in the strictly medical sense of eradicating or curing the disorder. Obviously, if an individual is assessed to score ≥ 30 on the PCL-R before treatment but months later is assessed by a clinician to score substantially lower than 30, we could informally say that the person is no longer clinically psychopathic, and therefore that they no longer have the personality disorder. This might be similar to how we speak of a person who suffered from major depressive disorder at one time but was later treated and cured from the disorder. While such a comparison might be meaningful,[19] many researchers are reluctant to make such a strong analogy between psychopathy treatment and medical healing, a position that is quite normal when it comes to personality disorders in general. As stressed by some researchers, perhaps personality disorders are not something we can cure but, at best, something that can be mitigated or balanced through clinical intervention.[20]

With these observations in mind, the untreatability view can therefore be seen as a commitment to the hypothesis that if a psychopathic person completes a clinical treatment program, their condition will either remain unchanged or will be exacerbated. And a logical consequence is that should they undergo a new PCL assessment after completing treatment, the score will (all things considered) either stay unchanged or have increased compared to previous assessments.

Treatment Methods in Forensic Settings

These definitions of treatment and untreatability raise another question about what sort of initiatives clinicians usually take to qualify as proper clinical intervention: What exactly is this thing we refer to as a "treatment program"? To give a thoroughly exhaustive answer to this question would go far beyond the scope of this chapter, as forensic and mental health treatment generally includes a diverse set of methods, based on an equally diverse set of theoretical views about what drives personality and behavioral change. Instead, I shall focus on describing the method of *cognitive behavioral therapy* (CBT),[21] which is broadly acknowledged as the most common clinical initiative in today's forensic practices and also happens to be the one that is most frequently used to treat psychopathic persons.[22]

This form of clinical therapy has a firmly established empirical backbone outside of forensic contexts, having been tested on a variety of mental health conditions and symptoms.[23] In forensic contexts, CBT is acknowledged for its ability to facilitate reliable change, both in terms of personality and behavior.[24] What makes CBT the preferred approach to forensic treatment and rehabilitation is its theoretical framework, which allows for a relatively straightforward and intuitive application in individuals who deal with complex and severe conduct problems and antisocial attitudes (e.g., low sense of self-worth, anger, sexual antagonism, aggression, substance misuse, and so forth).[25]

The basic tenet of CBT is the assumption that psychological and behavioral problems are rooted in faulty, unhelpful, and defeating patterns in the way a person cognitively processes, perceives, or *thinks* about their lives and social environment. This assumption expands on the observation that humans generally represent as self-aware, cognitive beings capable of executive control. Advocates of CBT suppose that one way to change problems with behavior is to empower this form of cognitive control. For example, a therapist may propose to a client that the main reason why the client repeatedly responds with overt aggression when in a specific situation (e.g., as a response to verbal provocation) is that they have adopted a cognitive pattern that involves thinking about such a scenario as something that necessarily warrants and needs an aggressive response. In a CBT program, the role of the therapist is that of a liaison who assists the client in changing this pattern of thinking, based on the theory that, on some level, such a cognitive change will likely inspire and lead to *behavioral* change (e.g., reducing aggressive behavior).

While there are multiple strategies a therapist can choose in achieving goals that are germane to CBT methods, it necessarily involves working with the patient to achieve self-awareness, assisting them in identifying problematic cognitive pattern(s) and helping them realize that their lives would be better off without these patterns. Once the therapist has made such progress—that is, established consensual compliance as well as therapeutic rapport and allegiance—they will then work on building cognitive skills and behavioral responses that can be practiced and learned by the patient. For instance, if a person has problems with reactive aggressive behavior, a therapist may use talk therapy and role-playing to practice nonaggressive responses to scenarios that have been identified as aggression triggering.

There is an open-ended combination of strategies that can be applied within the CBT framework to develop personal skills, where individual therapists might prefer their own unique set of methods and sometimes entire rehabilitation programs follow a pre-established framework. However, successful CBT therapy almost always depends on some sort of positive and constructive allegiance between the therapist and the patient, creating a positive feedback cycle in their effort to identify problems and build skills meant to mitigate them, together empowering the patient to deal with their problems on their own provisions. It is also important to highlight that forensic therapy programs are rarely organized in a "purist" fashion—for example, only making use of CBT techniques and nothing else. Instead, forensic programs might make use of several treatment methods depending on the needs of the individual clients, for instance, combining CBT private sessions with therapeutic community sessions.[26]

Forensic Treatment and Rehabilitation Research

One final aspect to consider before reviewing psychopathy treatment studies is how these types of studies are commonly designed. Most treatment studies involve an experimental framework where justice-involved persons enroll in a clinical intervention program that is designed to accomplish one or more treatment goals but always involves achieving the goal of positive personality and behavioral change as conceived from a forensic point of view (e.g., minimizing antagonism and recidivism). The experimental design is therefore relatively simple: to test whether there are changes to the targeted dependent variable (i.e., psychopathy) after the introduction of an independent variable (i.e., treatment). Data are typically collected such that it covers a specific follow-up period, where the aim of the study is to determine the effects of the program on the participants' personality traits and behavioral propensities during and/or after treatment completion.

A common type of experiment is to compare post-release recidivism rates in psychopathic individuals who have completed a specific treatment program, where a failure to detect any changes to recidivism rates is interpreted as having *zero* impact; lowering of recidivism is interpreted as an indication of a *positive* impact; and higher recidivism rates after treatment completion is interpreted as evidence supporting the hypothesis about *adverse* impact of treatment. Some studies gather more detailed data that track therapeutic progress, such as testing whether there are indications during the treatment period that the intervention is having an effect, how many treatment sessions have been completed, and whether the person dropped out during treatment. These data then allow for a more detailed analysis, such as analyzing whether these variables are correlated with changes in recidivism rates. Obviously, due to these complexities, the exact data that are collected and compared can vary greatly between studies, and direct comparison between studies and aggregation of research data can be challenging.[27]

The most compelling type of treatment evidence comes from studies that include some form of a control sample. For example, using a control sample to test for effects of treatment vs non-treatment, thus comparing results from samples of PCL psychopathic persons who received treatment and PCL psychopathic persons who did *not* enroll in a treatment program, allowing us to assess whether there are differences in recidivism rates as a factor of treatment *participation*.[28] Another example is to enroll PCL psychopathic persons into different treatment programs, allowing us to compare whether there are differences in recidivism rates as a factor of treatment *type*.[29]

Reviewing the Treatment Evidence: Consolidating an Optimistic Perspective

Just like in chapter 2, I have created an overview of the treatment review studies that have been published between the years 1980 and 2022 (see table 3.1). Each of these review studies aims to provide a general overview of the treatment literature at the time of publication by summarizing individual peer-reviewed treatment experiments, many of them explicitly designed to investigate the overall hypothesis that PCL psychopathy is untreatable. In doing so, these reviews typically explore one, two, or all of the following three sub-hypotheses: (#1) psychopathic persons are categorically unamenable to treatment, (#2) psychopathic persons show positive effects from treatment, and (#3) psychopathic persons are at risk of adverse treatment effects. The results from the review studies testing these sub-hypotheses are displayed in the last three columns to the right in table 3.1, indicating whether the review study *dismissed* or *corroborated* the hypotheses, or alternatively favored an *inconclusive* perspective (i.e., the researchers are expressing clear doubts about the evidence). Finally, the results displayed in the left column, *untreatability hypothesis*, indicate how the review authors characterize the untreatability view—that is, whether these researchers believe it is a scientifically sound hypothesis.

As we can see in table 3.1, reviewers have historically been overly dismissive of the untreatability view. Of the 21 published studies, 18 (86%) favored a dismissive conclusion, and only one review study from 2006 by Grant Harris and Marnie Rice has historically argued that the evidence corroborated the hypothesis, making it an outlier/fringe perspective among experts (two studies are inconclusive).

In the following subsections, I shall summarize and comment on the main findings that can be derived from the review literature in terms of the three sub-hypotheses concerning treatment amenability and positive vs negative treatment effects.

Evidence of Unamiability

As listed in table 3.1, all but one study—namely the 2006 study by Grant Harris and Marnie Rice—out of a total of 21 review studies *dismiss* the unamenability hypothesis, the view that psychopathic persons are categorically disagreeable to treatment and rehabilitation efforts. In short, according to the vast majority of experts in the field, psychopathic persons that enroll in treatment programs ostensibly find them worthwhile and manage to complete these programs at rates that are simply incompatible with the claim that psychopathic persons are unamenable to treatment. Although this is the conclusion we can draw from this research, it should be noted that many reviewers generally emphasize that PCL psychopathic persons can be challenging to deal with in a therapeutic context.

One broadly acknowledged challenge with PCL psychopathic patients is that they mirror many of the same type of obstacles that are present in most forensic *high-risk* persons, which makes them non-ideal candidates for treatment completion. These include problems with aggression and irritability, proneness to feeling victimized, egocentrism, poor commitment to change, negativity toward learning, and over and above a general lack of perseverance and motivation to stay engaged in a program tailored to instill personality and behavioral change.[30] On average, individuals with these attitudes—whether they have been clinically assessed with PCL psychopathy or not—have a higher likelihood (compared to low-risk, more agreeable persons) to avail themselves of treatment programs or eventually drop out before its completion. This is often due to an inability to form a meaningful therapeutic alliance with the clinical staff, exhibiting counterproductive

Table 3.1

Review studies of PCL psychopathy treatment research

Review study	Studies (k)	Untreatability hypothesis	Hypothesis #1 unamenability	Hypothesis #2 positive impact	Hypothesis #3 adverse impact
Garrido et al. (1995)	53	Dismissed	Dismissed	Corroborated	Inconclusive
Gacono et al. (2001)	34	Inconclusive	Dismissed	Corroborated	Inconclusive
Salekin (2002)	42	Dismissed	Dismissed	Corroborated	Inconclusive
D'Silva et al. (2004)	24	Inconclusive	Dismissed	Corroborated	Dismissed
Harris & Rice (2006)	16[a]	Corroborated	Corroborated	Inconclusive	Corroborated
Doren & Yates (2008)	10	Dismissed	Dismissed	Corroborated	Inconclusive
Abracen et al., (2008)	17[a]	Dismissed	Dismissed	Corroborated	Dismissed
Salekin et al. (2010)	23	Dismissed	Dismissed	Corroborated	Dismissed
Rodrigo et al. (2010)	9[a]	Dismissed	Dismissed	Corroborated	Dismissed
Shaw & Porter (2012)	10[a]	Dismissed	Dismissed	Corroborated	NR
Reidy et al. (2013)	17	Dismissed	Dismissed	Corroborated	Inconclusive
Polaschek & Daly (2013)	11[a]	Dismissed	Dismissed	Corroborated	Dismissed
Polaschek (2014)	8[a]	Dismissed	Dismissed	Corroborated	Dismissed
Bailey et al. (2015)	6	Dismissed	Dismissed	Corroborated	NR
Rosenfeld et al. (2015)	17[a]	Dismissed	Dismissed	Corroborated	Dismissed
de Ruiter et al. (2016)	13[a]	Dismissed	Dismissed	Corroborated	Dismissed

Hecht et al. (2018)	15[a]	Dismissed	Corroborated	Dismissed
Polaschek & Skeem (2018)	17[a]	Dismissed	Corroborated	Dismissed
Olver (2018)	24[a]	Dismissed	Corroborated	Dismissed
Polaschek (2019)	48[a]	Dismissed	Corroborated	Dismissed
Olver (2022)	17[a]	Dismissed	Corroborated	Dismissed

Note. Adapted from Larsen et al. (2020) with new three studies added (Olver 2018; 2022; Polaschek 2019). Studies (k) = number of unique studies included in the review. NR = not reported (i.e., information not provided by authors). Corroborated = authors interpret the evidence to be consistent with the hypothesis. Inconclusive = authors interpret the evidence as too weak to inform the hypothesis. Dismissed = authors interpret the evidence as incongruent with the hypothesis. [a]The number is estimated as it was not openly reported. Studies referenced in the table can be found in the endnotes.[31]

and disruptive behavior, or something as basic as failing to realize that the program is designed to help them break a self-defeating cycle of antisocial behavior.[32]

However, even though PCL psychopathic persons present with these sorts of well-known challenges, there is virtually no evidence to suggest that these challenges are unique to them as a patient group, nor should a psychopathy assessment be interpreted as categorically defeating the potential for positive treatment outcomes. Such a perspective/recommendation might sound counterintuitive to some readers: If a person is fundamentally difficult to treat, why should we waste resources on such individuals? The answer is that by focusing treatment efforts on these highly difficult individuals, we are actually optimizing the combined relative effect of treatment programs. This conclusion logically follows from the observation that there is really no need to spend resources treating low-risk persons because these individuals are likely to socially reintegrate on their own provision (hence the reason why they are described as *low risk*). If the goal is to optimize the social impact of treatment and rehabilitation programs, what we need to do is to identify high-risk persons and focus on therapeutically engaging these individuals.

This basic strategy is also known in forensic practices as the *risk-need-responsivity* model of rehabilitation and treatment.[33] This model posits that treatment must be tailored to the individual person such that it is (1) importantly guided by risk levels (i.e., high-risk individuals are prioritized), (2) targets the specific needs associated with risk (i.e., reduces criminogenic factors), and (3) uses strategies that have been empirically corroborated to instill responsiveness (i.e., reducing risk in persons with the identified criminogenic needs). In short, the risk-need-responsivity model is built on the assumption that effective treatment must be targeted at high-risk individuals and further tailored to the individual's unique criminogenic needs.[34] As noted by some of the review studies in table 3.1, according to the risk-need-responsivity model, psychopathic persons might therefore be described as ideal candidates for treatment and rehabilitation programs. So, instead of disqualifying them from treatment efforts, we should design programs that are equipped to respond to the needs of individuals that have high PCL scores.[35]

As an example of a study that has tested the treatment amenability of psychopathy, consider a 2020 study by Danielle DeSorcy and colleagues,[36] who analyzed the performance of 111 high-risk sex offenders enrolled in a high-intensity treatment program in Canada (i.e., 15 to 20 contact hours per week combining community and CBT therapy). Participants were divided into a psychopathy group (PCL-R score ≥ 25) and non-psychopathy group (PCL-R score < 25) and were routinely assessed on their level of working alliance (i.e., a self-report tool that keeps track of the strength and quality of the therapeutic relationship from the point of view of both the client and clinician). The study also tested for between-group differences in treatment program completion rates. The results indicated that there was no difference in overall working alliance between psychopathic and non-psychopathic groups, indicating that those who remained in treatment had similar levels of treatment amenability regardless of PCL-R score. The treatment completion rate for the entire group was 89.2%; however, there was a small difference between the psychopathic (84.3%) and the non-psychopathic group (93.3%). For obvious reasons, the observation that psychopathic persons have an average completion rate of 84% and normal levels of therapeutic alliance is not compatible with the view that they are categorically unamenable to treatment.[37]

To conclude, the vast majority of review studies emphasize that we can dismiss the hypothesis that psychopathic persons represent unique challenges to treatment compared to other groups in forensic contexts (e.g., ordinary, non-psychopathic high-risk offenders). However, this is not the same as saying that PCL psychopathy is easy to deal with. As stressed by Devon Polaschek: "Psychopathy is associated with a number of treatment factors that differ from some therapists' ideals or perhaps their training. For high-psychopathy people, their high-risk status on its own ensures that the treatment process will be bumpy, that therapists will need to work hard to engage clients, that it may be hard to judge progress, and that a certain amount of antisocial behavior and attrition is unavoidable."[38] While this difficulty is undoubtedly expressed in the data from a variety of studies that found elevated rates of non-completion and treatment attrition,[39] it nevertheless does not follow that psychopathic persons are categorically unamenable to treatment, nor does it remove from the fact that PCL psychopathic persons should be considered legitimate or even ideal candidates for treatment targets within the widely accepted risk-need-responsivity framework (i.e., as they are generally associated with moderately increased risks of reoffending compared to offenders who score low on the PCL scales; see chapter 2).

Evidence of Positive Impact

In table 3.1, the positive impact column summarizes the conclusion from each review study on whether there is evidence to support the hypothesis that PCL psychopathic persons can achieve positive change from treatment. In all but one review study (again, the 2006 review by Grant Harris and Marnie Rice), the authors report that there is evidence to corroborate the hypothesis. So, it turns out that according to contemporary experts—and contrary to the view expressed in the PCL-R manual—psychopathic persons *can* be successfully treated.

The positive evidence that is highlighted in these review studies primarily comes from 19 independent treatment studies involving PCL samples listed in table 3.2 (note that some review studies survey many more experiments, but this is because they also review studies with no effects and studies with subclinical and/or non-PCL samples). In the following, I shall review three studies that are highlighted by most reviewers to support the more optimistic conclusion about the positive impact of treatment, which also happens to be three studies that are *not* mentioned in the PCL-R manual (two of the studies were published after the release of the PCL-R manual in 2003). Lastly, I shall outline and comment on why the study by Grant Harris and Marnie Rice from 2006—contrary to all other reviews—has advocated for a more cautious conclusion.

One study that is broadly characterized as a trail-blazing project in the review literature is a study from 2002 by Jennifer Skeem and colleagues[40] that analyzes the effect of treatment on violent behavior in a civil sample of 871 individuals released from psychiatric hospitalization and assigned to a variety of *standard* treatment programs based on the individual's needs (e.g., substance misuse and mental health problems). A total of 195 individuals (out of 871 participants) were assessed with either potentially or high-scoring psychopathy based on the PCL:SV (screening version). During the follow-up period of a maximum of 50 weeks, participants were contacted in 10-week intervals to ascertain treatment participation and violence rates. The results indicated that psychopathic persons who remained in therapy responded positively to treatment by exhibiting less violent behavior compared to psychopathic persons who received no or less therapy. For example, of the high-scoring psychopathic persons who received seven or more treatment sessions during the first 10 weeks of hospital discharge, only 8% were involved in violent episodes during the second 10-week follow-up interval. This number

Table 3.2
Treatment studies with evidence of positive impact on PCL samples

Study	Population	Sample	Treatment type	Use of control
Skeem et al. (2002)	Psychiatric	871	Mixed	Yes
Hildebrand et al. (2004)	Sex offender	95	CBT	No
Langton et al. (2006)	Sex offender	418	CBT	No
Caldwell & Van Rybroek (2005)	Juvenile	248	Mixed	Yes
Looman et al. (2005)	Sex offender	154	CBT	No
Caldwell et al. (2006b)	Juvenile	202	Mixed	Yes
Caldwell et al. (2006a)	Juvenile	141	Mixed	Yes
Caldwell et al. (2007)	Juvenile	86	Mixed	No
Olver & Wong (2009)	Sex offender	156	CBT	No
Caldwell (2011)	Juvenile	248	Mixed	Yes
Abracen et al. (2011)	Sex offender	119	CBT	Yes
Wong at al. (2012)	Adult	64	CBT	Yes
Caldwell et al. (2012)	Juvenile	127	Mixed	No
Olver et al. (2013)	Adult	152	CBT	No
Lewis et al. (2013)	Adult	150	CBT	No
Burt et al. (2016)	Adult	103	CBT	No
Sewall & Olver (2019)	Sex offender	302	CBT	No
Rojas & Olver (2021)	Juvenile sex	100	CBT	No
Olver & Riemer (2021)	Sex offender	461	CBT	No

Note. CBT = cognitive behavioral therapy. Mixed = treatment involved a program with multiple treatment methods. Studies referenced in the table can be found in the endnotes.[41]

was 24% in psychopathic samples who received fewer than six treatment sessions during the first 10 weeks. To test whether these results were merely an effect of the individuals attending more sessions also on average having a lower likelihood of violent behavior (i.e., regardless of the impact of treatment), the data were controlled for various moderating factors, such as age, substance abuse, alcoholism, and education. After this analysis, the data still indicated a positive relationship between the number of treatment sessions and lower violence rates in psychopathic persons (this study was part of the MacArthur Violence Risk Assessment Study[42]).

Another broadly discussed study is the one by Mark Olver and Stephen Wong from 2009,[43] which analyzes the effect of a high-intensity CBT program in a sample of 156 sex offenders, testing its effect on long-term sexual and violent recidivism rates. A total of 45 persons were assessed with psychopathy using the PCL-R, of which 73% completed the treatment program. During an average follow-up period of 9.9 years, a total of 61% of psychopathic offenders who completed the program violently reoffended compared to 92% of psychopathic persons who did not complete the program, suggesting positive effects of treatment (although treatment completion was not correlated with lower rates of *sexual* reoffending). The study also examined the relationship between therapeutic change in psychopathic persons and reconviction rates—that is, whether observed improvement during therapeutic sessions (determined by a therapist using a separate standardized measure) was a better predictor of recidivism rates compared to the initial PCL-R score. Results indicated that psychopathic persons who made therapeutic progress were less likely to both sexually and violently recidivate.

A third experiment that has received broad attention is a study by Michael Caldwell and colleagues from 2006,[44] which investigated the effect of two different treatment programs involving 141 juvenile psychopathic persons on general and violent recidivism. The first program was a mixed, high-intensity, and resourceful CBT initiative, and the second program was characterized as a standard "treatment as usual" initiative. Psychopathy assessments were based on the PCL:YV (youth version), and the maximum follow-up period was 730 days after release from custody. The recidivism rates for participants in the high-intensity program were 57% for general recidivism and 21% for violent recidivism during the follow-up period, compared to the higher 78% and 49%, respectively, for the participants in the standard program, suggesting that the high-intensity program can cut risk levels in psychopathic youths in half. The study also found that participants in the high-intensity program had slower rates of reoffending (i.e., took longer before their first violation) as well as less serious forms of reoffending (e.g., caused less damage to victims). For instance, after analyzing data from an extended follow-up period, 0% of the high-intensity participants were accused of homicide, whereas this number was 10.6% in the standard-treatment group.

While the conclusions from the review studies listed in table 3.1 express ubiquitous optimism, it should be noted that most reviewers also emphasize that their conclusion should be interpreted with caution. This is because the treatment literature is still relatively underdeveloped. For example, where the treatment literature on PCL psychopathy only includes around two-three dozen experiments, the research from prediction studies we reviewed in chapter 2 makes up more than 200 independent studies. As noted by Devon Polaschek and Jennifer Skeem in their 2018 review, the field still struggles with a "serious dearth" of quality studies on central questions about treatability.[45] Despite this small evidence base, the results demonstrate that the hypothesis about psychopathic persons being incapable of making positive gains following treatment has little empirical basis, and therefore has been almost uniformly rejected by leading scientists. In short, it is contrary to the available scientific evidence to make the sweeping claim that PCL psychopathic persons cannot benefit from treatment.

Although the review studies from table 3.1 also mention studies that have found that psychopathic persons (on average) benefited little from treatment initiatives,[46] these publications are relatively few and must therefore be balanced by the fact that many more studies have found evidence of positive results from treatment. As is stated in one of the more recent reviews by Mark Olver, "Psychopathic offenders have routinely been treated, released, and managed in the community by correctional systems, and treatment advocates assert that there is little evidence to suggest that psychopathic offenders are any less treatable than other high risk-high need offenders."[47] The viewpoint Olver is expressing in this quote is similar to the conclusion drawn in the previous section: that individuals who score high on the PCL scales do not represent with (generalizable) treatment difficulties that are categorically different from that of other high-risk (non-psychopathic) offenders. But Olver is also alluding to the common-sensical conclusion that since treatment and rehabilitation programs across the world for decades have been targeting high-risk offenders with measurable and documented success, it is not unreasonable to assume that these success stories also include individuals with high PCL scores.

Finally, it should be noted that the review study by Grant Harris and Marnie Rice from 2006 goes against the conclusion of the remaining 20 review studies, arguing instead that the evidence is *inconclusive*. The authors' view is that we still cannot know whether PCL psychopathic persons can positively improve from treatment programs. In their reading of the scientific evidence, the authors state "that there is absolutely no basis for optimism regarding treatment to reduce the risk of criminal or violent recidivism" in psychopathic individuals.[48] One aspect that they highlight as counting in favor of this more pessimistic attitude is a lack of methodologically rigorous treatment studies (an observation iterated in many review studies). However, this conclusion is no longer a fair viewpoint, as more than a dozen rigorous studies have been published since Harris and Rice wrote their review back in 2006, and we might therefore wonder whether the authors would still advocate for similar caution today (we will never know, as both authors have since passed away).

However, Harris and Rice do make a fair and important observation that no psychopathy treatment experiment has ever been conducted within a *blinded randomized control trial* framework, which is normally seen as the gold standard of medical trials.[49] In such a trial, all participants are randomly distributed into two or more groups that, again, are randomly assigned different types of treatments. Further, the researchers involved in the study are "blinded," meaning that they do not know who is receiving what type of treatment. There are various advantages of such studies, one obvious advantage being that it can minimize *confirmation bias*, where a medical practitioner has a preconception that unknowingly, yet significantly, impacts the results of the study. For example, if a practitioner has a preconception that psychopathic persons cannot benefit from the treatment program, then it is possible that this preconception will affect their actual treatment of the client in a negative way—that is, if they know that their patient has scored high on the PCL scales. One way to conduct a randomized control trial is to enroll high-risk persons in two different treatment programs and ensure that the clinicians who run the program are unaware of whether someone is assessed with PCL psychopathy, and only after treatment completion will researchers retrospectively unveil the PCL assessment scores and analyze data for any differences between psychopathic and non-psychopathic persons across treatment methods and contexts. But, as Harris and Rice point out in their 2006 article, such a study has never been conducted (which is still true as of this writing in 2024).

However, although a lack of blinded randomized trials would normally be interpreted as a serious problem in any medical field, it must be noted that these trials are not necessarily superior to other types of controlled studies when it comes to measuring the effects of *psychotherapy*. In a medical trial, a randomized control framework is explicitly designed to minimize external factors that can distort results (e.g., bias), which might be easily achieved if the treatment that is tested can be ordinated in a discrete and controlled way, such as distributing two different pharmaceutical drugs between groups (e.g., drug vs placebo). In this scenario, we might safely assume that it is possible to isolate and mask the independent variable we are testing for (i.e., pharmaceutical treatment). But in a psychotherapeutic experiment, the situation is drastically different as the treatment itself (e.g., CBT) involves an element of uncontrolled interaction between the therapist and the client, which is further complicated by the fact that different therapists might not have equal abilities, may interact differently with certain types of patients, may have good and bad days, and so on. Due to these uncontrollable variables, it is perhaps naïve to think that an attempt to instill a blinded randomized framework will bring enormous improvements to the research setting.[50]

With this in mind, it is interesting to observe that while Harris and Rice propose in their 2006 paper that we should be cautious with the conclusions we draw about positive treatment effects, the same authors draw the surprising conclusion that the hypothesis about adverse effects has been *corroborated* (i.e., more on this in the next subsection). With this particular issue on adverse effects, they appear to be less bothered by the methodological shortcomings like the lack of randomized control trials. For obvious reasons, if methodological issues are a problem for interpreting positive treatment results, then it should also be a problem for interpreting the evidence of all types of effects.[51]

Evidence of Adverse Effects

As displayed in table 3.1 under the *adverse impact* column, the majority of review studies *dismiss* the third sub-hypothesis that psychopathy is associated with adverse treatment effects. A total of five studies—and notably only one study since 2010—has expressed caution, rendering the evidence *inconclusive*, and of all published reviews since 1995, only one study has argued that the hypothesis is *corroborated* (i.e., the Harris-Rice review from 2006).

The main reason review studies give for dismissing the hypothesis about adverse effects is—unsurprisingly—that there is a complete lack of empirical evidence to support it and, correlatively, plenty of evidence that adverse effects essentially never occur in any reliable way during therapeutic services. In fact, only one single experiment has been published with results that can be reasonably interpreted as showing adverse effects, namely a 1992 study by Marnie Rice, Grant Harris, and their colleague Catherine Cormier.[52]

Although there is another study by Michael Seto and Howard Barbaree from 1999[53] with results that were initially interpreted to suggest adverse effects, this conclusion was later recanted by Howard Barbaree, the senior author, when he published a follow-up study in 2005 with more detailed and complete data.[54] Another study by Gareth Hughes and colleagues from 1997[55] has occasionally been described as indicating adverse effects, though this study does not analyze treatment effects (it is a report about participants' therapeutic progress) nor does it show negative effects (analysis showed that PCL-R scores predicted lower progress). This leaves the field with only *one* study that has suggested adverse effects—namely the 1992 study by Marnie Rice and colleagues—but, as I shall demonstrate here, this study should be emphatically dismissed.

The 1992 study by Marnie Rice and colleagues is a retrospective analysis of data coded from file records that were obtained from the authors' workplace at the Oak Ridge Maximum Security Division at the Mental Health Centre in Penetanguishene, Ontario, Canada, a forensic psychiatric institution that ran a nonconsensual and peer-led therapeutic community program from 1965 to 1983 named the *Social Therapy Unit*. Based on additional data acquired from the Canadian authorities, Rice and colleagues analyzed post-release behavior during an average follow-up period of 10.5 years, comparing the performance of 169 patients who had participated in the therapeutic community program to a "matched" control group of 136 persons with similar demographics and criminal backgrounds who had not participated in the treatment program. The authors calculated PCL-R scores from reviewing file records, juxtaposing a group of 53 *treated* psychopathic persons with a group of 29 *untreated* psychopathic persons. Surprisingly, the rate of violent recidivism was higher for the psychopathy group who *had* received treatment (77%) than the psychopathy group who *had not* been assigned to the therapeutic community program (55%). This led the authors to conclude that the results were "especially surprising," because, as they interpreted it, the treatment program "was explicitly designed to effect positive changes in the psychopathic personality based on a solid theoretical background provided by the existing literature, and it provided extensive opportunities for patients to gain insight into their own behavior and to learn to be caring and empathic."[56]

However, as is evident from reading the review literature, there are today few scholars in the scientific community who take Rice and colleagues' study seriously. This is mainly because the data on which they base their analysis were obtained from the Social Therapy Unit at the Oak Ridge facility, which we today know was a notorious and scandalous forensic program, broadly acknowledged for adding an incredibly dark chapter to the history of Canadian psychiatry.[57] Although Rice and colleagues described the Social Therapy Unit in commendable terms—stating that the program was "especially suitable for psychopaths," that it had received "worldwide attention for its novel approach to treatment," and that it had a "solid theoretical backbone"[58]— the truth is that these are euphemisms obscuring what the program really was: an indiscriminate and systematic misuse of clinical-medical authority.

The clients who were involuntarily referred to the Social Therapy Unit were subjected to a broad variety of practices that have absolutely nothing to do with treatment and are certainly counterproductive to common therapeutic aims. This included forcing patients to participate in systematic experimentation with stimulant and hallucinogenic drugs such as LSD and amphetamine[59] as well as a practice of chaining patients together naked and locking them up in what the program referred to as a *Total Encounter Capsule*.[60] This capsule was a bright eight-by-ten-foot room, purposely designed as a sense-deprecating chamber. It had no windows and no natural light, the temperature was at times set chillingly low or sweltering high, and the capsule had 24-hour continuous artificial illumination. Through a one-way mirror and closed-circuit video recording, practitioners and other inmates would watch as patients went to the bathroom, ate liquid food through tubes in the wall, suffered emergency mental health incidents, and acted out as a result of the mind-altering drugs they were forcibly administered.[61]

We know today that the various therapeutic programs that were conducted during the two decades the Social Therapy Unit operated were essentially a harrowing effort to methodically denigrate patients, yet it was preposterously disguised as a treatment program designed to help the patients. Absurdly enough, the program was not operating in the dark—as a covert operation hidden from the public and deprived of bureaucratic oversight—but the practices and theories were vulgarly discussed in the open, leading to a number of grotesque publications in prominent Canadian academic journals by the program creators.[62]

Thus, it is simply a distortion of the facts when Marnie Rice and colleagues wrote in 1992 that the treatment program was "designed" for psychopathic persons, that it had a "solid" theoretical backbone, and that the program gave participants opportunities to "learn to be caring and empathic." Indeed, one would only know if the program was suitable for psychopathy treatment if it had been empirically tested (which it was not). To claim that the program had a solid theoretical backbone is, frankly, to make a mockery of what counts as a "theoretical backbone" in the behavioral and therapeutic sciences. The program was motivated by obscure and flip theories about the human mind—adopted from the existentialist philosopher Martin Buber— paired with early theories about community therapy. The end-product was a sort of Frankensteinian program that was certainly novel but perhaps better understood as a reckless experimentation based on nonsensical ideas taken from the fringes of 1970s clinical academic discourse. Lastly, the patients did not have "opportunities" to adopt caring and empathic attitudes—as stated by Rice and colleagues—but were instead daily denigrated and physically violated with impunity. As the forensic psychiatrist John Gunn recalls from when he visited the facility back in 1970:

> What I saw left a very deep impression. I talked to two patients who were hand-cuffed together and had been so for several days and nights; they had to do every-thing together, eating, sleeping, toileting and so on. This was to teach them how to look after somebody else. Another patient was on his own on the floor but delirious. When I asked why he was delirious, I was told that he had been given increasing doses of scopolamine in order to regress him to helplessness, including double incontinence, as if he were a baby. The drug would then be stopped, and he would gradually return to adult life in the company of the other patients, who would care for him and his every need.

The evidence that finally exposed the murkier details of the Social Therapy Unit was recently brought to full recognition in public records through a large class-action lawsuit raised by former residents against the province of Ontario and two of the program's lead psychiatrists Elliott T. Barker and Gary Maier. This lawsuit was initiated in the year 2000—roughly 15 years after the program was axed from the public budget, and well before the publication of Grant Harris and Marnie Rice's 2006 review article—and the class lawsuit concluded with rulings in 2017 and 2020 with the Ontario Supreme Court siding with the plaintiffs (in *Barker v. Barker*[63]). In short, the court agreed with the accusations of assault and battery, and that the Social Therapy Unit had caused short- and long-term harm to the patients' physical and mental health. The judge who ruled in the case stated in the written summary that the program should not be characterized as treatment but as "torture and a degradation of human dignity."[64]

Due to these appalling circumstances, it is only natural to be fundamen-tally suspicious of any data taken from the archives of the Social Therapy

Unit, as they suffer from an obvious lack of integrity, veracity, and credibility. Such unreliable data are not only disqualified as a proper source of information but also unsuitable for a scientific enterprise. Any conclusions that are drawn from such a source must therefore be rendered scientifically invalid, and, for similar reasons, they ought to be recanted and/or ignored. (Note that there are many more problems with Rice and colleagues' study, which I shall not go into detail here, but to readers interested in this issue my research team has written extensively about it elsewhere.[65])

Despite these obvious problems, the 1992 study by Marnie Rice and colleagues has nevertheless had an enormous impact on psychopathy research. For example, the study is discussed in detail in the PCL-R manual, where it is referenced as the main source of evidence to support the hypothesis about adverse treatment effects. Notice also that, according to Google Scholar, the study has been cited 1,019 times (~32 per year). Compare this to the study by Jennifer Skeem and colleagues from 2002—perhaps the second most influential study—that has only mustered around half of the same impact of 437 citations (~20 per year). As noted by many reviewers, the study by Rice and colleagues has been instrumental in shaping the overall pessimistic view of psychopathy treatment,[66] where a review by Devon Polaschek and Tadhg Daly from 2013 stated that it virtually "slammed the lid shut for many on the advisability of even attempting treatment" of psychopathic persons.[67]

Finally, we should consider the review study by Grant Harris and Marnie Rice from 2006, the only review study from table 3.1 that claims that the adverse effect hypothesis has been corroborated. In this piece, the authors offer some speculations to explain the data from their 1992 study:

> Why did the therapeutic community program have such different effects on the two offender groups? We speculate that both the psychopaths and nonpsychopaths who participated in the program learned more about the feelings of others, taking others' perspective, using emotional language, behaving in socially skilled ways, and delaying gratification. For the nonpsychopaths, these new skills helped them behave in prosocial and noncriminal ways. For the psychopaths, however, the new skills emboldened them to manipulate and exploit others.[68]

In short, the authors' speculative conclusion is not that there must have been something wrong with the therapeutic program itself but rather that psychopathic persons learned new skills that emboldened their criminal tendencies. What might be bothersome to many readers is that their review refrains from discussing the controversial nature of the Social Therapy

Unity, let alone the details concerning denigration and torturous behavior from the clinicians who ran the program. This is particularly troublesome as it was widely known by the early 1990s that the Social Therapy Unit had engaged in harmful practices—for example, the Canadian scholar Richard Weisman had published a tell-all critical article in 1995.[69] And the province of Ontario had also completed a very critical review of the program in 1984: the so-called *Hucker Report*.[70] Although this report was never made public, most people who worked at Oak Ridge or had been involved with its practices and research knew about it,[71] and there is certainly no reason to think that Grant Harris and Marnie Rice were unaware of it, as they worked at Oak Ridge and disclose detailed familiarity with the program in other publications. For example, the authors wrote a 2007 paper that provides an adamant response to mounting criticism of their 1992 study, indicating that they knew of the horrors inside the Social Therapy Unit.[72] In addition, there was also plenty of academic discussion of what had happened and what had gone wrong at Oak Ridge, which already by the 1990s had undergone drastic revisions and restructuring to its program and facilities (now called Waypoint Centre for Mental Health Care). It is difficult to say *why* these worrisome facts about the Social Therapy Unit were left out of the conversation when Harris and Rice published their 2006 review, as it would certainly have helped contextualize and assess their argument.

But how should the results from the 1992 study by Marnie Rice and colleagues have ideally been interpreted? In my opinion, and as mentioned earlier, there is no reason to think that we can draw any sound conclusions from data of this sort, as the data undoubtedly suffer from reliability and integrity concerns, thus disqualifying the study as proper science. The institution was functioning as a far cry from what can be reasonably seen as a clinical environment, and interpreting the data from such a program is antithetical to science. Dismissing the 1992 study in this way also appears to be a very uncontroversial view, as most of the review studies listed in table 3.1 express some version of this perspective.

Concluding Remarks: A Case for Optimism on Treatment

A review of the psychopathy treatment literature shows that the untreatability claim expressed in the PCL-R manual and elsewhere in the historic and current literature has little scientific merit. The conclusion that can be drawn from the past decades of peer-reviewed studies is comparatively much more optimistic. Indeed, all the three common claims that are subsumed in the untreatability view have been dismissed by leading scientists, where there now appears to be growing consensus that PCL psychopathic persons (a) are amenable to treatment, (b) appear to gain from treatment programs in similar ways as non-psychopathic persons, and (c) do not have any adverse effects associated with treatment initiatives. These empirical discoveries fundamentally undermine the practice of interpreting a high PCL assessment as an indicator of untreatability.

While it appears that the untreatability view can be straightforwardly rejected on empirical grounds, it could still be argued that it is too early to be *optimistic* about the treatment of PCL psychopathic persons. As I pointed out earlier, various authors (and especially those of the more recent publications) emphasize that the research paradigm suffers from notable limitations, which in turn weakens the conclusions we draw from it. For instance, it has been observed that there are relatively few published studies on psychopathy treatment and much of this research has methodological shortcomings. Although the field has published around 50–100 treatment studies during the past five decades, review authors agree that only a couple dozen of these studies meet the sort of methodological standards that would qualify them as decent science (i.e., many of which are included in table 3.2). This is admittedly not a very impressive evidence base. As Liza Hecht and colleagues pointed out in their 2018 review, the field has only yielded "minimal" evidence to support *any* view about the treatment prospects of PCL psychopathic persons.[73] Similarly, a review study by Devon Polaschek and Jennifer Skeem from 2018 states that there is a "serious dearth" of high-quality studies.[74]

However, I am not convinced that these observations about limitations are compelling reasons to be overall skeptical or agnostic about the prospects of treating PCL psychopathic persons. Instead, I think there are plenty of reasons to maintain a healthy optimism. Perhaps one of the strongest reasons for optimism is the fact that researchers have already produced a consistent track record of studies that all demonstrate positive treatment effects (e.g., studies listed in table 3.2), and this track record has been steadily growing in recent years.[75] While these studies use differing methods with different strengths and weaknesses, it is undeniable that they provide reasons to believe that PCL psychopathic persons benefit from psychological intervention. Moreover, and as emphasized by many researchers, there is no evidence that PCL psychopathic persons represent any unique challenges compared to other high-risk individuals. This suggests that if we are optimistic about the role of forensic psychological interventions in general, then we should also be optimistic about their effects in PCL psychopathic persons. Finally, I think it is important to underscore that these promising results have been obtained in a research environment that we might speculate has been rife with confirmation biases. That is, even in a field where many researchers must have held biased anticipations of finding evidence of untreatability, the research output has overwhelmingly falsified these expectations. This is not a small feat as confirmation bias in the sciences is a well-recognized problem that has been shown to skew scientific results.[76] As these attitudes change, we might expect that future research will further cement current conclusions.

In any case, it might not matter that much to forensic practitioners and decision-makers whether scientists decide to interpret the psychopathy treatment research in an optimistic or agnostic way. What seems to matter in a forensic context is whether there is evidence that psychopathic persons are untreatable, if there is evidence that a high PCL score can be interpreted as a unique inference about chronicity and incorrigibility. If such a pessimistic inference is unwarranted, then it becomes less relevant if a PCL assessment suggests optimistic or agnostic expectations about treatment outcomes. Why? Because optimism and agnosticism about treatment outcome are the basic attitudes that forensic practitioners take toward justice-involved persons in general. For example, within the risk-need-responsivity model—perhaps the most common approach to treatment and rehabilitation in North America—practitioners assume that positive change is possible, but they are generally advised to lower their expectations in terms of the magnitude of the impact of psychological intervention. And decision-makers are aware that the average person is not going to make huge gains from completing such a program. For all we know, the only evidence-based inference that can be made from a high PCL assessment is that treatment and rehabilitation is going to be challenging, the same as with other justice-involved persons who enroll in these programs.

In addition to these conclusions, the research I reviewed in this chapter also raises a curious question about *why* the untreatability view emerged as a prevalent perspective among researchers and practitioners in the first place. If we glance over the conclusions drawn from review studies in the past decades (table 3.1), it seems that there was never any reasonably convincing evidence to suggest that psychopathic persons were untreatable. For instance, this was the unmistakable conclusion drawn from one of the first large review studies by Randall Salekin published in 2002. And ever since, leading scientists have released one review study after the other stating—unambiguously—that there is plenty of evidence to suggest that psychopathic persons can benefit from treatment. Furthermore, we might ask that if there never was any clear evidence to support the untreatability view, why does the PCL-R manual—which was published in 2003—elaborate so one-sidedly on this view, ostensibly going against the established research at the time of publication?

I shall not pretend to have any definitive answers to this question. But I do believe that it is interesting to observe that the PCL-R manual only references one treatment review study—namely Randall Salekin's 2002 study—which the manual dismisses without any serious argumentation, stating that: "Although some reviewers (e.g., Salekin, 2002) have suggested that clinical pessimism might be replaced with clinical optimism, most clinicians and researchers are rightly pessimistic about the treatability of psychopaths with traditional methods."[77] In the year 2003, it might have been true that most clinicians and researchers were pessimistic about treatment. But that does not make it the *right* view from a scientific standpoint, as the PCL-R manual is here suggesting. Already in 2003 it was becoming increasingly clear that the untreatability view lacked empirical support, a conclusion that is now resoundingly obvious.

A final comment I would like to make pertains to how damaging the untreatability view might have been, and how difficult it seemingly is to mitigate and reverse these negative effects in the legal system. As noted by Howard Barbaree already in 2005, the impact of the untreatability view is "profound" as it puts justice-involved persons who scored high on the PCL scales in a "serious double bind." If a psychopathic person shows signs of improvement during treatment, this might be interpreted by decision-makers as "increased risk of serious re-offence," as if they have acquired new methods for antisocial behavior. And, on the other hand, the untreatability view might eventually prevent justice-involved persons from "taking psychological treatments," which is going to reflect negatively in their forensic evaluation.[78] This is why Barbaree's 2005 piece went so far as to reanalyze and subsequently recant the results from his group's earlier 1999 study that alleged to have found *negative* treatment outcomes (i.e., that psychopaths who did well during treatment had higher reconviction rates),[79] stating that their research project provided two important lessons:

> First, the experience highlights the importance of replicating single studies before making major changes in policy and practice. Decision makers and practitioners who, after reading Seto and Barbaree (1999), inferred higher risk based on good

treatment behavior in sex offenders scoring higher on psychopathy will have to revise their practice again. Second, the experience underlines the importance of data quality. Recidivism studies in sex offenders have often used data as they are available, and these data sets are often of unknown quality. As our experience shows, analyses of incomplete data sets can lead to erroneous conclusions.

It is arguably admirable when researchers like Barbaree go out of their way to set the record straight and recant their own conclusions from earlier studies. But I do wonder how efficient this strategy is, whether it actually mitigates or reverses the "intellectual damage" already caused by the misinformation in their 1999 study. For example, this earlier original study was never retracted by the journal, nor has the journal flagged the study to indicate that the conclusion about negative treatment effects is misleading. So, when you access the article through the journal's website, readers have few opportunities to learn that the results are invalid. What is even more worrisome is that when we look at the citation data of the two studies, the 1999 study with misleading results has been cited 680 times, compared to 176 citations of the 2005 study (according to Google Scholar). That is, researchers have cited the misleading research almost four times as much. Recall that there is a similar citation pattern with the discredited study by Marnie Rice and colleagues from 1992, which has been cited 1,019 times, many more than any other treatment study. To put things in perspective, consider that the 1999 study by Seto and Barbaree and the 1992 study by Rice and colleagues have received a total of 1,699 citations, which is more than half the total number of citations received by the 19 "positive" studies listed in table 3.2 (cited 3,166 times).

It is difficult to say why there is a preference among researchers to emphasize and cite studies with negative effects, especially because these studies are so clearly and widely known to be distortions (consider that Rice and colleagues' study has been cited more than 200 times since 2017, the year the class-action lawsuit documented the horrors in the Social Therapy Unit at Oak Ridge). Perhaps this tells us something about how difficult it is to reel in scientific misinformation once it has been spread throughout the academic ecosystem. And similarly, it tells us something about the challenges that current and future researchers face in their effort to (re)educate stakeholders about the treatment prospects of psychopathic persons.

4 Are Psychopathic Persons Morally Colorblind?
Insights from Empathy and Moral Reasoning Research

> He is unfamiliar with the primary facts or data of what might be called personal values and is altogether incapable of understanding such matters. It is impossible for him to take even a slight interest in the tragedy or joy or the striving of humanity as presented in serious literature or art. He is also indifferent to all these matters in life itself. Beauty and ugliness, except in a very superficial sense, goodness, evil, love, horror, and humor have no actual meaning, no power to move him. He is, furthermore, lacking in the ability to see that others are moved. It is as though he were colorblind, despite his sharp intelligence, to this aspect of human existence. It cannot be explained to him because there is nothing in his orbit of awareness that can bridge the gap with comparison. He can repeat the words and say glibly that he understands, and there is no way for him to realize that he does not understand.
>
> —Hervey Cleckley describing one of his psychopathic patients in
> *The Mask of Sanity* (1976)[1]

Throughout its research history, psychopathy has been described as a condition associated with profound impairments of moral psychological capacities, an impairment of the human disposition to comprehend *why* it is wrong to harm other people. When an ordinary, non-psychopathic person does something morally wrong (for example, commits a violent crime), that person will not only know that society disapproves—such as being aware that there are laws against violence—but the person will also have an intuitive comprehension about why their behavior can be characterized as immoral. With psychopathy, this sort of *moral intuition* is usually described as either totally absent or at least noticeably muted. When psychopathic persons commit violent crimes, they might know that there is a law against it, but they allegedly do not understand why violence is such a big (moral) deal.

We see various versions of this idea surface throughout history. As early as 1786, Benjamin Rush speculated that psychopathic persons have an inability to perceive, appreciate, and be guided by what is morally right and wrong, characterizing psychopathy as a disease of the "moral faculty."[2] Virtually the same explanation was repeated by scholars in the nineteenth century such as Philippe Pinel, Jean-Étienne Dominique Esquirol, James Prichard, Henry Maudsley, and many others.[3] Similarly, in the modern period, Hervey Cleckley promotes the view that psychopathy is a form of *moral colorblindness*—that these individuals are incapable of perceiving moral values—famously drawing analogies between a colorblind person who knows the words people use to describe colors but is nevertheless oblivious to their qualitative meaning.[4] Does a colorblind person truly understand what is meant by terms like *salmon pink* and *vermilion* if they have never seen these different shades of red? A psychopathic person can observe people speaking about morality, but, according to Cleckley, they have an incomplete understanding of what ordinary people mean when they make moral statements, just like the term vermilion is incomprehensible to a colorblind person. In Robert Hare's influential 1993 book *Without Conscience*, the colorblind metaphor is retold, adding that a psychopathic person "sees the world in shades of gray," and while they may have "learned all sorts of ways to compensate for this problem," it remains obvious to a trained clinician that their moral intuition is fundamentally different.[5]

While many of today's leading researchers agree with this basic idea—that psychopathy is associated with profound moral psychological impairments—there remains considerable variance in how researchers characterize and explain this deficiency. Some have argued that psychopathic persons have an inability to *empathize* with others, that they are incapable of mirroring, recognizing, and sensing what other people are feeling and thinking, and therefore that they see other people as mere objects.[6] Similarly, some researchers have described psychopathic persons' moral incapacities as a fundamental lack of conscience, that they are incapable of feeling guilt and remorse, which presumably play a crucial role in shaping moral perspectives.[7] Finally, some researchers have described the impairment on comparatively much broader terms, proposing that it is linked to a fundamental incapacity to process emotions, which they believe is an important driver in moral learning during childhood, thus resulting in a vastly different (or lack of) moral outlook when adolescence kicks in.[8] What these varying explanations have in common, though, is that they link psychopathy to a profound deficiency in comprehending moral values, which inevitably leads to the kind of morally unrestrained behavior that they associate with psychopathy.

The idea that psychopathy is linked to these kinds of moral incapacities is central to the Hare Psychopathy Checklist (PCL). For example, the PCL-R manual uses many of the different theories mentioned above as part of its basic explanation of what psychopathy *is*, but more specifically, the 20-item PCL-R checklist describes several traits that are direct expressions of this view, such as item 6, *lack of remorse or guilt*, item 7, *shallow affect*, and item 8, *callous/lack of empathy*. In addition, the checklist also includes traits that are indirectly associated with morality, such as item 5, *conning/manipulative*, and item 15, *irresponsibility* (see figures 1.2 and 1.3).[9] In short, when a person is clinically assessed with psychopathy—such as receiving a PCL-R score ≥ 30—the assessment is believed to indicate that we are dealing with a person who is largely incapable of appreciating morality, including a lack of empathy, guilt, and remorse. As explained in the PCL-R manual, this is a morally unrestrained person who "readily violates social norms and fails to fulfill social obligations and responsibilities, both explicit and implied."[10]

In this chapter, we shall take a closer look at what the scientific evidence tells us about the link between PCL psychopathy and moral psychological deficiencies. This research has primarily focused on answering two separate questions: whether PCL psychopathic persons have impaired (a) *empathy* and/or (b) *moral reasoning* capacities. It was not before the mid-1990s that researchers began to empirically test these long-held assumptions in a systematic way, and since then several dozens of studies have been published in academic journals. However, once again we will be confronted with a pattern that is becoming all too familiar in this book: initially, in the 1990s, there was a lot of enthusiasm in the research community about a few early studies seemingly supporting the idea that PCL psychopathic persons are morally deficient, only later to be replaced by a mountain of evidence directly refuting these early conclusions. As I will aim to show in this chapter, the idea that PCL psychopathic persons have impaired empathy and moral reasoning capacities might be one of the most unsubstantiated yet hard-lived tropes to ever emerge from the field. Unfortunately, and similar to the "untreatability view" we looked at in chapter 3, it is also a trope that is artificially kept alive in the field by a combination of selectively emphasizing a few experiments that appear to support it, while uncritically ignoring obvious quality issues pertaining to these few experiments.

How to Interpret (Moral) Behavioral Scientific Experiments

In the previous two chapters, the general discussion was mostly based on information that I derived from the review literature: a type of academic publication where scientists methodologically summarize the relevant experimental results in a given field. In this section, I will take a slightly different approach and provide a more detailed discussion of the individual experiments. However, anyone who has read a scientific publication knows that researchers use all sorts of complicated and nonintuitive terminology to describe their results, making it difficult for non-scientists to interpret and understand what is being said in these studies. So, to avoid unnecessary confusion, I will first explain the meaning of a handful of central terms used in psychological research. (Note that in doing so, I have aimed for brevity and clarity, knowing very well that some of my explanations are crude simplifications.)

Across the studies we will be reviewing in this chapter, there are two central concepts that researchers use to describe their results—namely when they are referring to them as either *null* or *statistically significant*. When researchers use the term null—or *null finding* or *null result*—to describe the outcome of a study, they are usually referring to the fact that they did not find a meaningful signal in the data: that there was statistically no difference in test performance across participants. For example, in an empathy study, a null result would indicate that the researchers did not find anything unusual about how PCL psychopathic persons performed, thus resembling what we assume to be *normal* empathic capacities (i.e., the word null is derived from what researchers call a "null hypothesis"). Conversely, when researchers describe their results as statistically significant—or sometimes in short as *significant results*—they are commonly referring to the fact that they did indeed find a meaningful signal in the data, that test performance across participants had robust differences. For example, in an empathy test, it would indicate that the researchers found evidence that PCL psychopathic persons' performance was unusual and that there was a pattern in their performance that did not resemble what is assumed to be normal empathy. In an oversimplified way, we can therefore view psychological studies as first and foremost testing for whether there *is* or *is not* a meaningful difference in performance across test participants.[11]

But what sort of differences are researchers generally looking for? The studies we will be surveying in this chapter generally have participants complete some form of assessment or a performance-based test—for example, tests that measure either empathy or moral reasoning capacities—whereafter the researchers will analyze the data for two different types of patterns or signals (i.e., in their attempt to discover any meaningful "differences"):

1. The first type is a *group-comparison analysis*, which is when researchers analyze the data for differences in the average test-performance between a group of PCL psychopathic persons (e.g., individuals with PCL-R scores ≥ 30) and a group of non-psychopathic persons (e.g., individuals with PCL-R score < 30). This is perhaps the most intuitive type of analysis because it directly pitches PCL psychopathic persons (as a cohesive group) against non-psychopathic individuals to check if there are any obvious patterns or differences in their average performance.

2. The second type is a *correlational analysis*, which is when researchers analyze the test data in terms of its linear (or monotonic) relationship with a PCL score—that is, whether there is a relationship such that a higher/lower PCL score is associated or correlated with increased/reduced test performance. Here researchers take a group of individuals who have all received a PCL assessment score and then analyze if there is an obvious pattern in test performance relative to each individual's PCL score.

Whenever a study makes a group-comparison or correlational analysis, researchers refer to its analytic outcome as an *effect*, indicating that the researchers have looked at one or more performance measure(s) to determine if there is (or is not) a meaningful signal. It is important to note that researchers often carry out several individual analyses in one single study, often generating dozens of effects. We might therefore say that a study has an overall result or conclusion, which is derived from its many effects. For example, as we will see, a common empathy study might involve test participants answering a self-report questionnaire where this assessment yields a final total score but also a score on multiple subscales. So, a group-comparison analysis between PCL psychopathic and non-psychopathic persons may include one analysis of overall group difference but also multiple post hoc analyses of total score and subscale differences. Consider that studies may use various tests or compare various groupings (e.g., low- vs middle- vs high-PCL psychopathy). So, the number of effects can swiftly multiply into the dozens. The more elaborate the study is, the more total effects and the more information for us to interpret. An important aspect to notice is that when we begin to draw inferences from studies like these—asking ourselves what the overall conclusion or result of the study is—we will have to balance and interpret all the information from these many effects, and not just a selective subset of effects. For example, as I will show later, one central piece of information is the proportion of null vs statistically significant effects. If there truly is a meaningful difference between psychopathic and non-psychopathic persons, null effects should only make up a relatively small proportion of the total number of effects (i.e., indicating that there are clear group differences).

Regardless of whether the researchers are conducting a group-comparison or correlational analysis, they eventually move on to analyze if a given effect is statistically significant or a null—that is, whether the effect is signaling a robust and meaningful difference or not. In making this distinction, researchers are faced with the conundrum that their experiment will always show *some* differences across test participants (for many simple reasons, one being that human behavior is rarely identical). In other words, there will almost always be some difference in average group performance, and there will often be some linear correlation between a PCL score and test performance. So, how do researchers determine whether an effect is signaling a robust and meaningful difference? Traditionally, researchers have been treating this question as a question about probability, where an effect is seen as statistically significant if it is believed to be probabilistically unusual. The rationale behind this approach is relatively straightforward. By making some assumptions about what is *normal* in the general population, we can also define some parameters for when something looks *unusual*. And if an effect looks unusual, researchers will conclude that it is signaling something meaningful.

But how improbable must test performance be before they are deemed statistically significant? The short answer is that it must be so unusual that researchers would only expect to find it less than 5% of the time if the test was conducted on a sample where there are no real, population-wide differences (i.e., where everyone in the experiment is "normal"). That is, if the scientist repeated the same test 100 times in randomly selected people from the general population, they would only expect to find the same test performance less than five times. Another way to look at this is that scientists will only declare a test result unusual—that is, statistically significant—if they have good reasons to believe that there is less than 5% chance or risk that their conclusion is a false positive: the case where they interpret the results as signaling robust differences but where it is just an instance of normal variation.

The calculation scientists traditionally rely on to make this determination is called a *probability value*—or a *p-value*—which is a simple estimation of how probable an effect is relative to what is assumed to be normal.[12] So, for example, when an effect has obtained a *p*-value of less than 5% ($p < 0.05$), what this value is an expression of is the degree to which the effect is probabilistically unusual or the probability that it is a false-positive observation. In short, when scientists conclude that an effect is statistically significant, this is almost always the same as saying that they obtained a *p*-value (on that single effect/measure) that is less than 5%.

Notice that this 5% threshold level—called the *alpha* level—is set by convention, and researchers mostly use it to make this basic distinction between which effects can be (conventionally) characterized as a null vs statistically significant. But when using these two labels, it is important to have in mind that they say nothing about the actual effect itself, such as how big or important a specific difference is between two groups. One common misconception of the concept of statistical significance is that it indicates that something big and important has been detected in an experiment (because this is ordinarily how people use the English word "significant," hence the common misunderstanding). However, whether experimental effects are important depends much more on the context and on what is being measured. For example, a statistically significant finding can easily turn out to be small and maybe even utterly unimportant (i.e., without any real-life implications). Therefore, a common way to further analyze if an effect is (practically) important is to calculate an effect size, which quantifies how large the difference really is.

There are many different ways effect sizes can be calculated. Though, it is usually decided on the basis of what type of analysis a researcher is conducting. For example, in a group-comparison analysis, it is common to calculate a *Cohen's d* or *d*-value, just like we saw in chapter 2.[13] As such, a *d*-value specifies the magnitude of an observed difference between the average group performance, indicating how much one group differs from another group relative to the "standard deviation" within those groups. In chapter 2, I illustrated this by positioning two bell curves (or normal distributions) relative to each other, displaying both their overlap and differences.

In correlational analyses, effect sizes are typically reported in *Pearson's* or *Spearman's r*—or an *r-value*—which is an analysis of the correlational *strength* and *direction* between two values: whether the increase of one variable leads to an increase or decrease of the other (sometimes referred to as a *zero-order* or *bivariate* correlation).[14] An *r*-value can range from –1 to 1, where the value *r* = 1 indicates a perfect (positive) linear relationship and the value of *r* = –1 indicates a perfect (negative) linear relationship. For example, a perfect linear relationship would indicate that a lower/higher PCL score is always correlated with a corresponding lower/higher test performance (an *r*-value of zero indicates that there is no linear correlation between data variables). Although it can be tricky to translate statistical values into ordinary language, *r*-values are often described within the following normative framework: *r* = < 0.10 is described as *near-zero* or *negligible*; *r* = 0.10–0.39 is *weak*; *r* = 0.40–0.69 is *moderate*; *r* = 0.70–0.89 is *strong*; and *r* = > 0.90 is *very strong*. The important aspect to understand is that the closer an *r* value is to 1 or –1, the stronger the positive/negative correlation; and the stronger the correlation, the more support there is for hypothesis that PCL psychopathic persons have practically meaningful differences in their empathy and moral reasoning capacities (i.e., strongly positive means better performance and strongly negative means worse performance).

Do Psychopathic Persons Lack Empathy?

The question that has attracted the most attention among researchers studying morality in PCL psychopathy is the question about empathy. This area of research generally aims to test two separate hypotheses: that psychopathic persons are differently disposed in terms of their (1) *cognitive empathy*,[15] an empathy subtype defined as the capacity to intuitively understand or perceive (without verbal cues) what another person is feeling and thinking, and/ or (2) *affective empathy*,[16] a subtype defined as the capacity to emotionally respond to or mirror another person's feelings.[17] These two types of empathic capacities have been investigated using two different test paradigms, namely the use of (a) *empathy assessment* tools, which are psychological tools specifically designed to quantify and measure both cognitive and affective empathic capacities, and (b) *emotion recognition* tests, a performance-based psychological test that measures the ability to accurately detect nonverbalized emotional expressions (i.e., a main component of cognitive empathy).

In what follows, I will briefly review the research results published in these two test paradigms between the years 1980 and 2023, whereafter I discuss some important shortcomings of this research. This summary of the evidence is based on a systematic review conducted by my research group and published in 2024.[18] In this report, we concluded—contrary to popular beliefs—that there is *no* scientific basis for inferring that PCL psychopathy is linked to an impairment or lack of empathy.

Overview of Paradigm (a): Empathy Assessment Studies

Empathy assessment tools are typically designed as self-report tests, consisting of an elaborate combination of narrative, multiple-choice, and task-based questionnaires that survey the person on a long list of empathy-related issues. For example, a popular tool is the *interpersonal reactivity index* (IRI),[19] a self-report assessment consisting of 28 statements such as "I really get involved with the feelings of the characters in a novel" or "Being in a tense emotional situation scares me," which participants rate on a five-point Likert scale ranging from "Does not describe me well" to "Describes me very well." After the test, the answers are accumulated into a final total score consisting of four different seven-item subscales: *perspective taking* and *fantasy*, which are indicative of cognitive empathy, and *empathic concern* and *personal distress*, which are indicative of affective empathy. Although tools like the IRI are far from perfect measures of empathy, what makes them popular among researchers is that they still come with considerable empirical support, having been tested in diverse populations, and are therefore acknowledged as the best ways of measuring of empathic capacities. So, if PCL psychopathic persons truly have a profound lack of empathy, it is only reasonable to assume that it would be easily detectable using tools like the IRI.[20]

In our 2024 systematic review, we found that 34 empathy assessment studies were published between the years 1980 and 2023. These studies involve a total of 2,712 test participants and have yielded 559 effects—that is, individual results where an empathy assessment score (total or subscale) was tested for statistical significance in group-comparison or correlational analyses. However, across these 34 studies there is vanishingly little evidence to suggest that PCL psychopathic persons lack empathy. Of the 559 effects, a total of 474 effects (84.79%) are nulls, meaning researchers found no evidence of empathy-related differences in PCL samples. In fact, the most common overall result from these studies was 100% null-findings, reported in 15 (44.12%) studies. Consider that only 85 effects (15.21%) are statistically significant, meaning that these results are representing probabilistically unusual empathy assessment in PCL psychopathic persons. However, of these effects, 66 signaled negative correlation or worse empathic capacities, but there were also 19 effects that signaled positive correlation or *better* empathic capacities in PCL samples.

Overview of Paradigm (b): Emotion Recognition Studies

Emotion recognition studies involve a performance-based test designed to measure a person's ability to identify nonverbally expressed emotions. These tests usually have participants view dozens of photographic pictures of human faces—but also morphing pictures, cartoons, or videos—each expressing between five to ten different prototypical emotional states (e.g., fear, anger, surprise, or pain). After viewing each picture, participants are asked to identify the expressed emotion using a multiple-choice format. The scoring system generally tracks accuracy and error rates, where scores might be averaged out across emotion types to indicate an overall performance, or a score might be calculated for each individual emotion to see if the participants were better at identifying some emotions than others. Emotion recognition tests have been used in many different clinical and nonclinical research settings and are conventionally believed to measure cognitive empathic capacities insofar that the task is to infer what a person is feeling or thinking based on nonverbal clues. So, if PCL psychopathic persons really have impaired cognitive empathy, this should be clearly reflected in their test performance.

In our 2024 systematic review, we found that 37 emotion recognition studies have been published between the years 1980 and 2023. These studies involve a total of 3,398 test participants and have yielded 1,113 effects—that is, individual results where an emotion recognition performance metric was tested for statistical significance in group-comparison or correlational analyses. However, just like the studies using empathy assessment tools, these emotion recognition tests have also *not* found compelling evidence to suggest that PCL psychopathy is associated with impaired cognitive empathy. Of the 1,113 effects, a total of 1,016 effects (91.28%) are nulls, meaning researchers found no evidence of performance differences in PCL samples. Again, similar to empathy assessment studies the most common overall result was 100% null-findings, reported in 16 (43.24%) studies. Only 97 effects (8.72%) are statistically significant, where 76 effects signaled negative correlation or worse emotion recognition performance, and 21 effects signaled positive correlation or better performance by PCL samples.

Interpreting Empathy Research in PCL Samples

When we conducted our 2024 systematic review, we made a couple of crucial observations about the research, which should be carefully considered before anyone begins to interpret the results from these two research paradigms. One important observation is that many of the empathy studies were designed as *exploratory* studies, which is a type of approach researchers take when they aim to make a preliminary probing of a theory or hypothesis (as opposed to what researchers call a *confirmatory* or *critical* experiment[21]). When researchers do exploratory work, it is conventionally accepted to use less rigorous methodologies and apply a more lenient statistical threshold during significance testing of experimental data—essentially accepting higher risks of false-positive effects. It is therefore curious to observe that even when researchers increase the risk threshold for false-positive discoveries, the proportion of null findings remains high across both paradigms (i.e., 84.79% and 91.38%).

To demonstrate how this exploratory approach has impacted the research, consider that of all the 1,672 effects, only 516 effects (30.86%) were controlled for what is called the *family-wise error rate* (FWER), or sometimes referred to as controlling for *multiple comparisons*.[22] As mentioned in the previous section, an experimental effect is determined to be statistically significant if researchers have evidence to conclude that the effect is probabilistically unusual. The calculation behind this determination involves estimating the probability of the effect being a false-positive discovery, usually by calculating a *p*-value. The accepted probability ratio for false positives in the sciences is set at 5% by convention. However, when researchers perform multiple separate analyses of the same dataset, they will ordinarily have to control for the fact that every time they make additional (post hoc) analyses of the data, the probability of making a false-positive discovery accumulates. As a metaphor, imagine that you put five bullets in a 100-chamber revolver. The first time you pull the trigger, there is a 5% chance that the hammer strikes a chamber with a bullet in it. But if you keep pulling the trigger, the likelihood of striking the chamber with a bullet increases. A similar logic applies when conducting separate (or post hoc) analyses of experimental data. If your dataset contains five false positives (i.e., metaphorical bullets), then every time you go looking—that is, analyze the data for an effect—the chances of you discovering one of the false-positive effects increases. For instance, if you perform 15 analyses on a given dataset, the FWER jumps from the initial 5% of a single analysis to a whopping 54% chance that your analyses will make at least one false-positive discovery.

In ordinary scientific experimentation, researchers can use various statistical tools to *correct* or keep the FWER at a 5% threshold. Perhaps the most common approach is to perform a *Bonferroni* correction, which is done by dividing your initial alpha threshold (of 5%) by the number of (post hoc) analyses you are running on the dataset. For instance, with 15 analyses, the new alpha value for each individual test will be 0.003 (0.05/15). This means that researchers will only reject the null hypothesis if the *p*-value is less than 0.3% (as opposed to less than 5%).[23] So, all things being equal, it is only fair to assume that if researchers rerun the uncorrected analyses in empathy research using the Bonferroni adjustment, the overall proportion of null effects would increase.

In our 2024 systematic review, we tested this exact assumption by comparing effects that were FWER correct vs uncorrected. Unsurprisingly, in uncorrected studies, the proportion of null effects across both paradigms was 86.59%, but in the FWER corrected studies the null proportion jumps to 94.77%. What is particularly interesting about this discovery—that the null proportion increases with FWER correction—is that it suggests that significant effects might actually be false positives. Indeed, recall that when researchers analyze experimental results for statistical significance, the conventionally defined alpha threshold is set at 5% in a *p*-value calculation. This means that researchers are expecting or tolerating a 5% risk that significant effects might be false positives. So, when we find that FWER-corrected studies have a null proportion of ~95%, this strongly indicates that the remaining 5% significant effects are a by-product of their methods tolerating a 5% false-positive ratio. In other words, if the ratio of significant effects is equal to or lower than 5%, this means that, in reality, all effects are likely nulls.

In addition to lacking FWER corrections, we also found that many studies were of questionable quality, often deploying data "hacking" strategies that effectively increase the likelihood of false-positive discoveries. For example, 22.89% of all studies had "excluded the middle" during sample selection—a hacking strategy that I explained in chapter 2—thus only comparing "low" scoring to "high" scoring PCL samples. Furthermore, almost half of the studies (44.58%) did not conduct a moderator analysis where they explore the possibility that significant effects are caused by confounding factors (e.g., psychotropic medicine). Another problematic discovery we made was that many studies had incomplete reporting, often omitting some of the results of their analysis in the published article (e.g., only reporting facial recognition effects from a few emotions when the study tested many more according to their methodology). Such selective reporting is not very helpful, as it essentially distorts the true proportion of nulls vs significant effects (i.e., it is probably safe to assume that non-reported effects are nulls).

Finally, it is important to stress that even if we assume that the few statistically significant results in empathy assessment and emotion recognition studies are not false positives, this still does not indicate any form of *deficiency* or *abnormality*. We can infer this because there is a large body of research that investigates empathy in normal (non-psychopathic) samples using the same methods as reviewed above. And this research has reliably demonstrated that empathy assessment scores and emotion recognition performance can vary considerably within the general population and across contexts.[24] So, if you are not convinced that the significant effects are a byproduct of using exploratory research designs, there might be another obvious explanation for why a few of the studies from paradigm (a) and (b) found indications of empathy differences: these are simply instances of how there is context-dependent and normal variation in the general population. Moreover, this research therefore strongly suggests that PCL psychopathic persons have ordinary empathic capacities.

Do Psychopathic Persons Have Impaired Moral Reasoning?

The second question that has been pursued by researchers (though to a much lesser extent than the empathy question) is whether PCL psychopathy is associated with clear and reliable differences in how moral issues are reasoned about. If PCL psychopathic persons truly are morally colorblind—to use Hervey Cleckley and Robert Hare's famous metaphor—this might show up in how they think about the moral landscape. To my knowledge, a total of 21 studies have been published between the years 1980 and 2023, which investigates moral reasoning in PCL psychopathic samples. This research involves a great diversity of methods, the most popular ones being experiments testing participants' ability to make a so-called *moral/conventional* distinction, as well as experiments testing their responses to ordinary moral dilemma questions. However, since this research tends to have great methodological diversity, it makes little sense to directly compare results across studies (as I did with the empathy research). Therefore, I will review the moral reasoning research by first explaining the basic features of each experimental paradigm followed by a concise discussion of the individual studies.

Research on the Moral/Conventional Test

Many developmental and social psychologists argue that humans from a very early age begin to intuitively distinguish between two different types of social rules, namely what they call *moral* and *conventional* rules.[25] The basic idea is that humans readily recognize when something is (socially) right or wrong, but at some stage in our lives, we acquire a deeper intuitive understanding that something can be right/wrong for different reasons: either because of morally salient reasons or because of social convention. For example, consider two different instances of wrongful behavior, such as a child speaking in class without permission and a group of children bullying an innocent girl in the schoolyard. While people might agree that both types of behaviors are wrong, psychologists claim to have discovered that we also tend to agree on *why* these behaviors are wrong and, importantly, that we agree that both behaviors are wrong for *different* reasons. One important aspect is that people appear to agree that the former issue is wrong primarily because there is a legitimate authority (e.g., the schoolteacher) that has imposed or established the rule, where the latter case is wrong for largely authority-independent reasons (e.g., personal harm). For instance, even if a schoolteacher decided that it is now "okay to bully," people might not change their opinion about it being wrong. Though, if a schoolteacher decides that it is now "okay to speak without permission in class," perhaps we are much more inclined to accept this rule change and no longer consider it wrong to speak without permission.

Psychologists have developed all sorts of experiments that can test whether individuals are making the moral/conventional distinction.[26] Typically, these experiments have participants read narratives about moral and conventional transgressions (e.g., bullying or speaking in class without permission) whereafter they are asked to rate to what degree the transgression was *permissible*, how *serious* it was, and lastly whether they will change their permissibility rating when a relevant *authority* permits the transgression (e.g., schoolteacher allows bullying or speaking without raising a hand). If individuals make the moral/conventional distinction, what we should see in the data is that they clearly rate moral issues as comparatively less permissible, more serious, and unaffected by authority modification.

Since the moral/conventional research has been recognized for identifying reliable patterns in what might be characterized as normal moral reasoning—namely drawing a distinction between moral and conventional issues—it then provides an excellent opportunity to test whether PCL psychopathic persons deviate from non-psychopathic persons along this parameter. If psychopathy truly is defined by moral colorblindness, it is at least not unreasonable to expect that they would fail to draw clear distinctions between moral and conventional scenarios. For example, it would seem that they would view all social rules as essentially conventional, insofar that they are hypothesized to have no real moral insights and, thus, no basis for distinguishing the nuances that separate moral issues from conventional issues.

A total of five studies have tested this hypothesis using moral/conventional tests. However, while there was some early enthusiasm about an initial (exploratory, low-quality) study from 1995 seemingly showing PCL psychopathic persons failing the test, subsequent methodologically improved experiments have all demonstrated equal capacities in PCL samples.

The first study is by James Blair published in 1995,[27] which issued a moral/conventional test to a small cohort of 10 PCL psychopathic (mean PCL-R score 31.6) and 10 low-scoring non-psychopathic persons (mean PCL-R score 16.1) sampled from a forensic psychiatric institution in the UK. The results indicated that the PCL group did not properly distinguish between moral and conventional transgressions, as this group was *less* inclined to change their initial permissibility rating on conventional stories if an authority allowed the transgression. That is, PCL psychopathic persons appeared to respond to conventional transgressions as if they were moral, rating both types of transgressions as equally serious and authority independent. It should be noted that there are notable quality issues with this study. For one, the sample is extremely small, which introduces high risk for biases in the data. Second, there is little information about sample selection, and it appears that participants were not randomly selected (e.g., evidenced by the large difference in mean PCL-R scores; it looks like the author "excluded the middle," see chapter 2). Third, the study does not control for FWER. Fourth, and perhaps the most disqualifying reason, the author mentions (in the manuscript) that he suspects that some participants in the PCL group deliberately responded in a way that would make themselves look good (i.e., they were in a treatment program and "motivated to be released"[28]). In other words, the entire study may be compromised due to unreliable data.

As a side note, it is interesting to observe that despite the obvious problems and limitations with James Blair's 1995 study, it still went on to becoming enormously popular, today counting more than 2,300 citations (according to Google Scholar). Not only is it one of Blair's most cited publications but it is also one of the most cited studies in the history of psychopathy research (according to the Web of Science citation index). And while some researchers in recent years have been critical about how popular this study has become,[29] and given the fact that psychologists generally do not believe that anything important can be inferred from studies with extremely small samples and reliability issues, it is still odd that so many scientists keep citing the study. For instance, more than half of all citations of Blair's study have occurred in the last 10 years, after the year 2014.

Another moral/conventional study was published by James Blair and his colleagues also in 1995,[30] this time involving 20 PCL psychopathic (PCL-R ≥ 30) and 20 low-scoring non-psychopathic persons (PCL-R ≤ 20). Again, the PCL group appeared to have some differences in how they distinguished between the moral/conventional stories, though not as pronounced as in the previous study and therefore not a replication of the earlier results. For example, this time the PCL group did rate moral transgressions as less permissible, more serious, and authority independent compared to conventional transgressions. Nevertheless, their permissibility and authority ratings were still significantly different compared to non-psychopathic persons. However, notice that this study suffers from exactly the same limitations as the previous 1995 study: it did not correct for FWER, the researchers excluded the middle, and they did not compensate for the risk that some participants might be responding in a way that would impress the prison management (i.e., the methodology and experimental procedures were essentially identical to James Blair's initial 1995 study).

A third moral/conventional study was published by Mairead Dolan and Rachel Fullam in 2010,[31] which included 115 juvenile offenders divided into three groups of 45 high-scoring (PCL:YV > 24), 31 medium-scoring (18–24), and 39 low-scoring individuals (< 18). Participants were administered the same transgression stories and questions from James Blair's initial 1995 study. However, the results did not reveal any statistically significant differences between the groups in their distinction between moral and conventional issues, thus amounting to a second failure to replicate Blair's initial 1995 results.

Finally, two moral/conventional studies were published by Eyal Aharoni and colleagues in 2012[32] and in 2014.[33] These two studies made important changes to Blair's former methodology, such as changing the instructions given to participants to avoid the risk of them answering in a way that would make them *look good* in front of prison management. Instead of simply answering the questions, they were instructed to identify which stories were about moral issues and which were about conventional issues. Their 2012 study involved a sample of 109 adult offenders, on which the authors made a group-comparison and correlational analysis. There was no evidence of PCL psychopathic persons having problems with moral/conventional distinctions, nor were there any correlations with PCL scores and distinction performance. In their 2014 study, Aharoni and colleagues issued the test to a sample of 139 mixed adult and juvenile offenders (mean PCL score 23.2), and a correlational analysis found no link between PCL scores and test performance accuracy. These two studies by Eyal Aharoni and colleagues are of higher quality than the previous studies by James Blair and colleagues. Not only did they improve the method by removing the opportunity to impress prison management with phony answers but they also have much larger samples, removed the risk of bias during sampling, and deployed a more rigorous data analysis. With this in mind, it is worth emphasizing that under these more stringent conditions, there is no evidence of problems with moral/conventional distinctions in PCL samples.

It is curious to notice that even as the two Aharoni studies have been widely recognized by experts in the field as "debunking" the hypothesis about PCL psychopathic persons being incapable of making the moral/conventional distinction,[34] these two studies are still cited far less than the two original studies by James Blair. According to Google Scholar, their 2012 experiment has received 254 citations (only 18 in 2022 and 17 in 2023) and their 2014 study has received 62 citations (6 in 2022 and 10 in 2023). Since the publication of the 2012 Aharoni study, James Blair's 1995 experiment has received 1,470 citations, more than four times as many citations as the 2012 and 2014 studies have received *altogether*.

Research on Moral Dilemma Tests

Another popular type of moral reasoning experiments are tests that query participants about how they would act if they were faced with a variety of moral dilemmas (i.e., moral issues where it is broadly agreed that there is no obviously true or best answer). One of the most famous and utilized dilemmas in these studies is the so-called trolley problem (invented by philosopher Philippa Foot), which is commonly issued in two different versions of the same dilemma, a detached *impersonal* and a *personal* (here, I have adopted a description used in a 2001 study by Joshua Greene and colleagues[35]):

Impersonal version: Suppose a runaway trolley is about to run over and kill five people. Suppose further that you can hit a switch that will divert the trolley onto a different set of tracks where it will kill only one person instead of five. Is it okay to hit the switch?

Personal version: What if the only way to save the five people were to push a large person (larger than yourself) in front of the trolley, killing him but saving the others? Would that be okay?

In these moral dilemma studies, participants are asked to respond to both dilemmas by answering *yes* or *no* to whether they would act or intervene (e.g., hitting the switch or pushing the person). To be sure, researchers generally agree that there are no satisfying solutions to the scenario; hence, why it is characterized as a dilemma. Is it justified to kill one person to save many? Or must we constrain our options and watch as the majority die, knowing well that we could have acted? Or maybe you think it is ok to pull the switch and save the many but not ok to push a person to their certain death to save the people down the tracks (even though the same number of people die in both cases). Thought experiments like these are often designed to force people to make a swift decision by probing their own moral intuitions, presumably an important component in moral reasoning.

Where these tests become interesting to a psychopathy researcher is that they allow us to check whether PCL psychopathic persons respond differently to these scenarios compared to non-psychopathic individuals. One hypothesis is that psychopathic persons—due to their alleged incapacity to sense and comprehend morality—will be more inclined to accept higher levels of personal harm. For example, as noted by various researchers, when we ask ordinary people about the impersonal vs personal versions of the trolley problem, people often realize that even though the calculation of lost lives is identical in both versions (if you decide to intervene), the personal version nevertheless appears to be much more sinister as it involves you directly harming another individual. And some experiments have indeed found that ordinary people may be more inclined to intervene in the impersonal version but less inclined in the personal version.[36] So, if we assume that PCL psychopathic persons are morally colorblind, that they have no real comprehension of why it is wrong to hurt another person, we might then expect that they would have no real basis for preferring the impersonal over the personal: to them it might just all come down to a cold calculation of the number of lives saved. Pulling a switch or pushing a person off a bridge, perhaps it is all the same to them.

However, a total of five studies have tested PCL psychopathic persons in moral dilemma experiments, and all studies have failed to show any statistically significant differences in their response patterns. The largest and most cited experiment is a 2010 study by Maaike Cima and colleagues,[37] which compared answers to seven impersonal and 14 personal moral dilemmas in a sample of 62 participants divided into groups of 35 community controls, 23 non-psychopathic offenders (PCL-R < 26), and 14 PCL psychopathic offenders (PCL-R ≥ 26). Answers were also compared to a prior independent survey of 672 community participants on the same moral dilemmas. All three groups rated the moral dilemmas similarly as the large civil control group: namely being more willing to intervene in impersonal dilemmas as opposed to personal dilemmas. There was no statistically significant difference between the PCL psychopathic group and the two other groups.

A 2012 study by Michael Koenigs and colleagues[38] issued the 24 personal and impersonal dilemmas to a group of 24 PCL psychopathic (PCL-R ≥ 30) and a group of low-scoring non-psychopathic offenders (PCL-R ≤ 20). Both groups rated impersonal dilemmas as more permissible than personal dilemmas, the only difference being that the PCL psychopathic persons were slightly more inclined to interact in impersonal dilemmas compared to those in the non-psychopathic group (but no difference in personal dilemmas). Notice that the study excluded the middle and still found no important differences.

Similarly, a study by Andrea Glenn and colleagues from 2009[39] issued a moral dilemma test to a group of 17 individuals assessed with PCL-R psychopathy (average score not reported) and found no statistically significant correlation between a PCL score and answers to impersonal and personal dilemmas. A study from 2012 by Jesus Pujol and colleagues[40] issued a moral dilemma test to 22 PCL psychopathic persons (PCL-R > 20) and 22 community participants and found no statistically significant differences in response patterns. And finally, a slightly different study by Nicole Hauser and colleagues from 2021[41] tested the correlation between PCL scores and responses to 22 personal dilemmas in a sample of 60 male offenders (mean PCL-R 23.03), probing whether a PCL total score is associated with increased willingness to intervene in personal dilemmas (i.e., the thesis about psychopathy and cold utilitarian calculation). The researchers found no correlation between a PCL score and such decisions. If anything, the researchers found evidence *against* the hypothesis, as the only significant effect was a negative correlation between PCL-R facet 2 and decisions to intervene in two different dilemma types.

Other Research of Interest

In addition to these moral/conventional and moral dilemma experiments, a total of 11 mostly exploratory studies have tested the moral reasoning capacities of PCL psychopathic persons using a variety of experimental paradigms. But like the research reviewed above, these studies have also failed to demonstrate any fundamental differences in PCL samples. Next, I will briefly survey the results from these studies and offer a few interpretive comments.

An early exploratory study by Aisling O'Kane and colleagues from 1996[42] (not corrected for FWER) issued the *defining issue test* (DIT)[43] to a group of 40 adult forensic psychiatric patients in the UK assessed for PCL-R psychopathy (average PCL-R score not reported). The DIT consists of having participants read five different morally loaded stories (e.g., about a husband who steals an expensive drug to save his dying wife), whereafter participants must rate 12 issues related to the scenario in terms of how important they believe these issues to be (e.g., if it matters that there is a law against stealing, whether the husband could go to prison, and so on). The idea behind the DIT is that people with a fully developed moral psychology (e.g., adults) will rate different issues higher compared to persons with less developed moral reasoning capacities (e.g., children). So, if PCL psychopathic persons have underdeveloped moral reasoning, they should deviate in how they perceive these issues compared to non-psychopathic individuals. The authors initially found a significant weak to moderate correlation between PCL-R scores and answer types; however, this effect turned insignificant after controlling for intelligence scores (i.e., the effect was caused by differences in intelligence, not psychopathy).

An exploratory study (which was not corrected for FWER and excluded the middle) from 2010 by Carla Harenski and colleagues[44] tested the "moral sensitivity" between a group of 16 PCL psychopathic offenders (PCL-R mean

31.8) and 16 low-scoring non-psychopathic offenders (PCL-R mean 13.3). Participants were shown a series of 75 different pictures, 25 representing unpleasant moral scenes (e.g., a picture of a person attacking another), 25 with unpleasant non-moral scenes (e.g., two people arguing), and 25 with emotionally neutral content (e.g., a hand being fingerprinted). Participants were instructed to rate if the picture portrayed a moral violation and, if so, rate the violation on a five-point scale in terms of severity (e.g., score 5 if very serious). The results indicated no significant differences between the two groups, and the researchers thus concluded that psychopathic and non-psychopathic persons are "similarly able to identify moral violations and rate their severity."[45] Notice that these results hold even as the researchers excluded the middle and did not correct for the FWER.

In 2014, Carla Harenski and colleagues[46] published another exploratory study (not corrected for FWER), administering a slightly different moral sensitivity test. This time participants were asked to rate the three categories of moral, non-moral, and neutral scenes on a scale from one to five in terms of the severity (if any) of the violation in the pictures. The experiment included 119 juvenile offenders assessed with the PCL:YV (mean score 23.70). Across the entire sample, participants clearly distinguished between moral and non-moral violations, rating the former significantly higher. However, the results indicated a weak, negative correlation ($r = -0.16$ to -0.23) between PCL total/factor scores and violation ratings—that is, higher PCL scores were correlated with lower rating of violation severity. However, the p-values reported in this study are relatively high (lowest p-value was 0.02), suggesting that they would not *survive* a correction for FWER.

A 2011 exploratory study by Eyal Aharoni and colleagues[47] (not corrected for FWER) administered a so-called *moral foundations questionnaire*[48] to 222 adult male offenders assessed with the PCL-R (mean score 21.54). The moral foundations questionnaire is designed to measure the ways people differ in their moral perspectives, based on the assumption that humans have a small number of consistent, fundamental values—or moral foundations—that underpin their moral views. The test consists of 30 questions designed to measure how individuals prioritize six of the following foundations: (1) harm, (2) fairness, (3) loyalty, (4) authority, (5) purity, and (6) liberty. For example, when making moral judgments, some individuals might prioritize individual liberty (#6) over the importance of religious dogma (#5), which may explain why/how people disagree on moral issues. The results in Eyal

Aharoni and colleagues' study were considerably mixed, but there was evidence of weak inverse correlations between PCL-R scores and the foundations of *harm* and *fairness*. That is, the higher a PCL-R score the less likely participants were to invoke these foundations in their moral reasoning. Another moral foundation study was published by Maya Irvin-Vitela and colleagues in 2021,[49] which administered the test to a group of 316 female offenders (mean PCL-R score 18.47). In this study, the researchers found evidence that PCL-R scores had a weak, negative correlation with the foundations of harm, fairness, and authority.

Initially, these findings seem to indicate important differences and perhaps even signs of abnormal moral reasoning. However, there are several reasons why these results show no such thing. It must be underlined that the moral foundations questionnaire is not designed to measure abnormalities but only meant to demonstrate how *all people* have different priorities when they reason about moral issues. So, when we read through the moral foundation literature, it is not uncommon to find results similar to the two studies above when the test is issued to ordinary non-psychopathic community samples.[50] An iconic illustration of this is a study that found political conservatives to be less inclined to endorse harm and fairness principles compared to liberals.[51] In short, the findings from Aharoni and colleagues, and Maya Irvin-Vitela and colleagues, is nothing out of the ordinary. Just because certain groups prioritize some principles over others it does not automatically imply abnormality, nor does it demonstrate deficiencies in moral values. At best, it demonstrates how there is normal variation in a population.

A 2016 exploratory study (not corrected for FWER) by Samantha Fede and colleagues[52] administered a simple test to a group of 245 incarcerated adult offenders, who had all been assessed with the PCL-R (mean score 20.73). The test consisted of showing a word on a screen that described a type of behavior that people generally understand as either *morally wrong* (stealing, murder, lying, and so on), or *morally not wrong* (e.g., charity or kindness). After viewing the word, participants would push a button indicating whether they believed the word depicted something that was morally wrong or not wrong. The results indicated a weak, positive correlation ($r = 0.20$) between PCL-R scores and items identified as not wrong, and a similarly weak, negative correlation ($r = -0.19$) between items identified as wrong. That is, the higher a PCL-R score the more likely a person was to rate an item as not wrong, and/or less likely to rate an item as wrong. However, these correlations were almost entirely driven by scores on the PCL-R factor 2 items (which non-psychopathic offenders often score high on). Notice also that across the entire sample, 93% of all questions were answered correctly (i.e., participants accurately rated items as wrong/not wrong), and the standard deviation is incredibly small (0.11), indicating a near identical performance across the sample (i.e., the difference in correctness in 99% of all participants is less than 1%). So, when the researchers found a statistically significant weak correlation with PCL-R scores, this correlation has no real practical importance as it conclusively pertains to *percentage-decimal* differences in responses.

An exploratory study (not corrected for FWER and excluded the middle) by Liane Young and colleagues from 2012 administered a test to 20 PCL psychopathic (PCL-R $\geq$ 30) and 25 low-scoring non-psychopathic persons (PCL-R $\leq$ 20), asking participants to rate the degree of "moral permissibility" in 42 imaginary moral scenarios where the behaviors in the stories had different intentions and consequences. Each scenario was designed to portray one of four features: (1) accidental harms (i.e., an act with neutral intentions but harmful outcomes); (2) attempted harms (i.e., an act with harmful intentions but neutral outcomes); (3) intentional harms (i.e., harmful intentions with harmful outcomes); and (4) neutral acts (i.e., neutral intentions with neutral outcomes). For example, one story would depict a person pouring a cup of coffee for a colleague but accidentally and unknowingly adding a white-powdered poison instead of sugar in the cup, whereafter the colleague dies from intoxication (i.e., the story has neutral intention but a harmful outcome). The results showed that psychopathic and non-psychopathic persons rated actions similarly, except when comparing ratings of (1) accidental harms (i.e., harmful acts with neutral intentions). Here, PCL psychopathic persons rated these actions to be comparatively more permissible. On a scale from 1 to 7 (1 = morally forbidden, 7 = morally acceptable), psychopathic persons rated these scenarios 4.63, where non-psychopathic persons rated them 3.84. Notice that this difference has no real practical implication due to how a seven-point Likert scale is ordinarily interpreted, where scores 4.63 and 3.84 are customarily grouped together, as if they are reflecting the same general type of answer (i.e., both answers fall within a less than one-point difference from the neutral middle of score 4).[53] What is more, it is rather normal to see that average permissibility ratings of accidental harm deviates across groups in these kind of studies.[54]

An exploratory study (not corrected for FWER) by Jana Borg and colleagues from 2013 issued a so-called *deserved punishment test*[55] to a group of 100 adult offenders assessed with the PCL-R (mean score 22.60). The test is designed to investigate whether individuals have similar intuitions about punishment, and it presents participants with 24 imaginary scenarios where some form of moral/legal transgression has occurred, whereafter the participants must rank each scenario in terms of which ones deserve relatively more punishment (e.g., a man kills another man in self-defense might be ranked less than a premeditated homicide). The researchers found that PCL-R scores were not correlated with any particular response patterns. And when comparing the responses from the entire offender sample to data from a noncriminal community sample, responses were near identical.

An exploratory study (not corrected for FWER and excluded the middle) by Keith Yoder and colleagues from 2015[56] issued a simple moral content test to 28 PCL psychopathic (PCL-R score ≥ 30) and 32 low-scoring nonpsychopathic persons (PCL-R < 20). The test consisted of viewing pictures displaying moral violations (e.g., interpersonal harm) and morally good behaviors (e.g., a person helping another), and participants would rate whether they believed the content in the picture was morally wrong. There was no statistically significant difference between group performances.

An exploratory study (not corrected for FWER) by Ghitta Weizmann-Henelius and colleagues from 2002[57] tested the association between PCL-R scores and self-reported feelings of guilt and blame in a sample of 58 violent female offenders (mean PCL-R score 16). Although feelings of guilt and blame might not be clear indicators of moral reasoning capacities, it is at least possible that if a person is incapable of moral reasoning, they would also not feel any guilt for their convicted crimes, and perhaps they would blame others (and not themselves) for their crimes. Participants completed a questionnaire where they rated true/false statements such as "I feel very ashamed of the crime(s) I committed" or "I am entirely to blame for my crime(s)." The researchers found a significant moderate negative correlation between PCL-R total scores and guilt-related statements ($r = -0.40$), meaning that the higher a PCL-R score the less likely persons were to agree with guilt-statements. However, PCL-R scores were not associated with perception of blame.

Finally, an exploratory 2015 study (not corrected for FWER) by Andrew Spice and colleagues[58] analyzed the link between PCL:YV scores and self-reported feelings of shame and guilt in a sample of 97 juvenile offenders (average PCL:YV score 20). Participants completed two different self-report assessments of guilt/shame, which involve reading scenarios where such feelings would be appropriate (e.g., accidentally throwing a ball that hits your friend in the face), whereafter participants rate the degree to which they would feel shame/guilt if they engaged in such behavior. The researchers found a statistically significant negative correlation between PCL:YV total scores and guilt. However, this correlation was weak for both assessment tools ($r = -0.37$ and $r = -0.39$), meaning that there is little indication that PCL samples have substantially different experiences of guilt, let alone that they are incapable of feeling guilt. The researchers also found a small statistically significant *positive* correlation ($r = 0.23$) between PCL:YV facet 3 scores (but not total scores), indicating minimally higher feelings of shame in persons scoring high in these items. Though, again, it is not clear what such a result tells us (if anything) about moral reasoning in PCL psychopathic samples.

Concluding Remarks: The Slow Process of Killing a Trope

In this chapter, I reviewed all the research that has tested the hypothesis that PCL psychopathic persons are morally colorblind—the mainstream claim that they have profoundly impaired empathy and moral reasoning capacities. However, after conducting more than 80 studies—involving more than 6,000 participants—researchers have yet to find any compelling evidence that PCL psychopathy is associated with these sort of grave moral psychological impairments. In short, there is a sizable evidence base of dozens of studies consistently demonstrating intact or "normal" empathic and moral reasoning capacities in PCL psychopathic persons, thus rendering the moral colorblindness idea little more than a trope; a cliché deprived of empirical support.

Historically, a couple of early studies from the 1990s did appear to corroborate the moral colorblindness thesis, such as James Blair's popular and widely cited 1995 moral/conventional study, which might have given the idea enough empirical support to keep it "alive" in the scientific community (remember, before the 1990s, psychiatrists and psychologists were largely skeptical about the idea of psychopathy). But these early studies, like the many that followed, have deep quality issues (such as selective reporting, biased sampling, and non-rigorous statistics), which seriously subtracts from their inferential value. Perhaps the strongest evidence why these early studies are not credible is that it did not take long before researchers tried and eventually failed to replicate their results. For example, in empathy-research, where there is considerable consistency and similarity in research methods, the most consistently replicated finding are studies with 100% null effects. And the early Blair studies about the moral/conventional distinction were eventually dismissed by the larger and more rigorous studies by Eyal Aharoni and colleagues in the 2010s.

It is interesting to notice that the accumulation of evidence from moral psychological experiments only really began to take speed in the years after the publication of the PCL-R manual in 2003 (which is different from the research on dangerousness and treatment we explored in chapters 2 and 3). For example, out of the many empathy studies we surveyed in this chapter, only six studies (~10%) are published before 2003. And of all the 21 moral reasoning studies, only four experiments are published prior to 2003 (i.e., James Blair's two moral/conventional studies from 1995, Aisling O'Kane and colleagues' study from 1996, and Ghitta Weizmann-Henelius and colleagues' study from 2002). This might explain why there is little critical discussion of the moral colorblindness metaphor in the PCL-R manual, where psychopathic persons are described as lacking empathy and moral reasoning capacities.

However, even as the scientific evidence began to accumulate in the 2000s, there appears to have been (and still is) an odd tendency in the field to avoid a critical discussion of the moral colorblindness thesis, combined with a similarly odd tendency to uncritically accept it. Even as it was becoming increasingly clear that PCL psychopathic persons did not have any problems with empathy or moral reasoning, somehow the myth was kept alive. For example, in a book chapter from 2007, James Blair writes that "there can be no doubt that psychopathy is associated with empathic dysfunction."[59] Sweeping claims like these were common in the 2000s, frequently iterated by prominent experts in the field.[60] But the problem is that if one takes a closer look at the timeline for when empathy studies are published—for instance, as we do in our 2024 systematic review[61]—it is clear that already in the mid-2000s the evidence was strongly suggesting that researchers had no basis for expressing confidence in the thesis. By the mid-2000s, there were more than two dozen published empathy studies, and the results from these studies were unmistakably suggesting a strong pattern of null effects.

If we fast-forward a decade, when it is now abundantly clear that there is no basis for claiming that PCL psychopathic persons are morally incapacitated, we still see prominent researchers pushing the moral colorblindness trope. For instance, in a 2019 book published by Oxford University Press,

Psychopathy: A Very Short Introduction, psychopathy is described as being defined by a "profound absence of empathy."[62] Similarly, in a 2019 book titled *No Remorse*, the author writes that psychopathic persons "epitomize the antihuman who is incapable of caring about anyone."[63] And in a 2021 review article published in the prestigious medical journal *Nature Reviews Disease Primers*, the authors write that psychopathy is characterized by a "lack of empathy" and "compromised" moral judgments.[64] Find any random contemporary article about psychopathy, even those written by high-profile leading researchers, and the odds are that you will find an iteration of the moral colorblindness trope.

It will be up to future researchers to change the narrative—to kill the trope, so to speak—a process that undoubtedly begins with clearly communicating what is discovered across the entirety of the research record instead of focusing narrowly on a few methodologically weak studies, regardless of how intriguing their results might be.

5 The Psychopathic Brain: Neuroimaging and Scientific Spin

Psychopathy is characterized by structural and functional brain abnormalities in cortical (such as the prefrontal and insular cortices) and subcortical (for example, the amygdala and striatum) regions leading to neurocognitive disruption in emotional responsiveness, reinforcement-based decision-making and attention.
—Stephane De Brito and colleagues in *Psychopathy* (2021)[1]

For as long as we can track its research history, top scientists have continuously portrayed psychopathy as a *biological* condition. They have described and studied psychopathy under the assumption that it is a disorder with discrete physiological causes and mechanisms, just like any other medical condition.[2] For example, the earliest account by Benjamin Rush from 1786 describes psychopathy as a *disease* of the moral faculty, and he speculates that this disease might be caused by external factors such as warm climates, poor diet, and foul odors.[3] The nineteenth-century Italian criminologist Cesare Lombroso claimed that psychopathy was the result of hereditary *degeneration*, framing the disorder as an instance of a genetic mutation gone awry.[4] This focus on genetics subsisted well into the twentieth century. For instance, when *karyotyping* technology—a biological test for visualizing human chromosomes—became mainstream in the 1950s, some researchers began entertaining a theory that psychopathy might be linked to an extra Y chromosome in the male genome.[5] And when DNA sequencing became possible toward the end of the twentieth century, researchers investigated whether psychopathy was a hereditary condition.[6] This is a theory that some contemporary researchers are still testing but nevertheless—like the other speculations—has found no clear empirical support of.[7]

However, among these many biological theories, one of the most accepted and long-lived assumptions is that psychopathy is a brain-based, neurological condition. In essence, this theory entails two interrelated postulations: (1) that there are discrete regions or systems in the brain that are centrally involved in moral reasoning and prosocial behavior; and (2) that psychopathy is causally linked to abnormalities or dysfunctions in these very regions/systems.[8]

This theory of psychopathy as a brain-based disorder is already expressed in the earliest work from the eighteenth and nineteenth centuries, but it is perhaps not before the contemporary period that we find its most consistent and meaningful iteration, notably in the publications by Hervey Cleckley,[9] David Henderson, Ben Karpman,[10] David Lykken,[11] Robert Hare,[12] and others.[13] During this period, researchers took inspiration from decades of preceding innovation in neurological research, where experiments were suggesting that the brain was compartmentalized in such a way that, like a Swiss army knife, discrete areas of the brain controlled similarly discrete cognitive and behavioral functions (though many of these assumptions have later been challenged and disproven[14]). For example, the work by neurologists like Pierre Paul Broca and Carl Wernicke paved the way around the 1860s by showing how damage or lesions to certain brain regions could severely impair language capabilities.[15] So, just like there was a region in the brain "responsible" for our language capabilities, researchers in the twentieth century increasingly began to theorize that psychopathy is linked to developmental abnormalities in specific brain areas that control moral reasoning and prosocial behavior.

This assumption figures centrally in Hervey Cleckley's influential book *The Mask of Sanity*, where he discusses the work by the contemporary (and aptly named) neurologist Henry Head, who had documented how severe brain trauma (e.g., following a stroke) could cause all sorts of strange psychological symptoms. One line of research that was particularly inspiring to Cleckley was evidence regarding a type of *aphasia* where people lose the ability to process inner thought and language yet remain normal in other respects, such as being able to understand simple communication. These patients may not be able to say the word *pen*, but they still respond to the word—for instance, being able to pick out a pen if they are asked to identify it among other objects such as pencils. Drawing parallels to such peculiar clinical cases, Cleckley speculated that just like aphasia patients have lost a discrete component of their cognitive function—but remain normal in all other aspects—so too might psychopathy be a brain-based defect that deprives the person of the subjective experience of emotions.[16]

Even though many researchers from the early contemporary period—like Cleckley and his colleagues—were convinced that psychopathy is linked to brain abnormalities, their work generally falls short of any robust experimental evidence, rendering their theories little more than rudimentary conjectures. But it would be unfair to criticize them for this crucial shortcoming, as there was little practical opportunity for testing brain-based theories. This limitation was partially due to an incomplete understanding of neurobiology but also due to the absence of technology powerful enough to systematically measure and analyze the brain at the cellular level. Even when twentieth-century scientists successfully linked certain brain regions to specific mental capabilities—such as Pierre Paul Broca, Carl Wernicke, and Henry Head's pioneering work on language processing—their experimental methods were not readily transferrable to psychopathy research. It is one thing to make inferences about how damage to a *known* brain area impacts cognitive capabilities; it is a completely different task to link a personality disorder (defined by personality traits and behavior) to *unknown* brain systems. In the former, you search for abnormalities in *behavior*, and in the latter, you must search for abnormalities in the *brain*. It is exactly the opposite type of discovery process.[17]

However, there was some newfound enthusiasm in the 1940s as innovative technologies began to emerge in the early days of the third industrial revolution. For example, the emergence of *electroencephalography* (EEG)—which measures electrical potentials in the brain by placing electrodes on the scalp—made scientists hopeful that they would soon be able to detect brain wave abnormalities in psychopathic patients.[18] But, as Hervey Cleckley stresses in the final 1976 edition of *The Mask of Sanity*, the evidence from EEG studies was inconsistent and difficult to interpret,[19] and he eventually cautions the reader in the closing parts of the book when he voices a sobering perspective: "I do not believe that the cause of the psychopath's disorder has yet been discovered and demonstrated. Until we have more and better evidence than is at present available, let us admit the incompleteness of our knowledge and modestly pursue our inquiry."[20]

Today, the attitude among researchers has changed considerably. Theories are no longer described as speculative, and the deep uncertainties expressed in Cleckley's iconic work are long gone. What has taken its place is a widely shared confidence that psychopathy has been decisively linked to specific brain abnormalities. For example, in their acclaimed 2014 book, *Psychopathy: An Introduction to Biological Findings and Its Implications*, Andrea Glenn and Adrian Raine write that neurological studies have shown that certain brain regions "function differently" in psychopathic persons.[21] Similarly, in a comprehensive review from 2011, Kent Kiehl and Morris Hoffman write that "neuroscience is beginning to open the hood on psychopathy," and the accumulation of evidence "strongly suggest that all psychopaths share common neurological traits that are becoming relatively easy to diagnose."[22] More recently, a 2021 review article by Stephane De Brito and colleagues, published in the prestigious journal *Nature Reviews Disease Primers*, describes psychopathy as a complex "neurodevelopmental" disorder "characterized by structural and functional brain abnormalities" in the cortical (e.g., prefrontal) and subcortical (e.g., amygdala) regions of the brain.[23]

The scientific and clinical implications of linking psychopathy to discrete brain abnormalities are multifaceted, but one central consequence is that the evidence elevates the epistemic integrity of the diagnosis. That is, the evidence fundamentally changes and therefore enhances our knowledge about psychopathy, from first being a speculative syndrome to now, allegedly, being an empirically validated diagnosis with known neurological causal mechanisms. Recall that around the 1990s, prominent scientists were still deeply skeptical about psychopathy, such as Ronald Blackburn stating in 1988 that psychopathy is a *myth*.[24] So, when some scientists today speak of psychopathy as a validated diagnosis, they do so largely because they believe that there is empirical evidence robustly linking the overt clinical signs and symptoms (sometimes referred to as the *exophenotype*) with dysfunctions in underlying neurological systems (sometimes referred to as the *endophenotype*).[25]

Further, there are also potential legal implications of linking psychopathy to brain abnormalities, though these are less clear and perhaps not yet fully understood.[26] To be sure, we do have a fairly good understanding of how a psychopathy assessment is used to inform and currently do impact legal decisions (as outlined in chapter 1), but there is comparatively little knowledge on how the neuroscientific evidence, on its own, does or should complement these decisions. Of course, when psychopathy is presented as a scientifically validated diagnosis—such as being accompanied with *hard* neuroscientific evidence—it follows that it is indirectly enshrined in a layer of epistemic veracity: it is no longer speculative, it is *scientific*. In a legal context, this means that it has cleared a major step in earning a rite of passage to inform legal decisions, insofar that valid science is almost uniformly seen as an appropriate type of evidence across legal institutions (such as courts, parole boards, and correctional institutions).

The aim of this chapter is to provide a critical review of the neuroscientific research on PCL psychopathy. During the past three decades, the field has seen a dramatic increase in neurological experiments primarily driven by the advent of radiological technologies that have greatly innovated the way we study the cellular structure and function of the living human brain. However, while these studies ignited a vocal optimism among researchers in the 2000s about discovering the biological roots of psychopathy—as exemplified by the quotes above—they also generated a tidal wave of exaggerated claims that, in hindsight, appears to have distorted the fact that neurological experiments have mostly yielded null effects, as well as inconsistent and non-replicated findings. The narrative about *psychopathy-as-a-brain-disorder* is not only unsupported by empirical evidence—contrary to what is regularly claimed in prestigious scientific journals—but this narrative appears to be primarily driven by scientific spin: the act of selectively constructing a narrative around a few spurious scientific results while ignoring mountains of evidence that contradict these results.

The Psychopathic Brain in the Age of Neuroimaging

When studying the brain of PCL psychopathic persons, the broadest question that researchers aim to answer is this: Do psychopathic persons have measurable and generalizable neurological differences compared to non-psychopathic individuals? That is, if we study the brain organ—that lump of soft but extremely complex tissue located inside the human skull—of individuals who score high on the PCL scales, will we then find that their brains on average reveal an unusual pattern of biological differences that are distinguishable in predictable ways from ordinary non-psychopathic controls?[27] As we have seen, many of today's leading scientists have responded to this question with a decisive affirmative answer. For example, if we follow Kent Kiehl and Morris Hoffman's conclusion in their 2011 review, it is now already "relatively easy to diagnose" psychopathy by looking at a brain.[28]

The scientists pursuing the question about brain abnormalities have traditionally done so by breaking it down into two separate analyses, namely: (1) probing whether there are any *structural* brain differences—that is, universal differences in the anatomical state of the brain—and (2) probing whether there are any *functional* brain differences—that is, universal differences in the cellular network activations during mental processing. In short, and to put it in lay terms, researchers want to know whether PCL psychopathic persons' brains are differently *wired*, and whether this wired circuitry is *firing* or *operating* differently.

There are many different experimental methods and technologies that researchers may choose from when investigating brain structure and functioning. For example, we could design a study where we collect the brain tissue from deceased psychopathic persons and explore these samples under a microscope to see if we can detect any abnormalities, searching for developmental and cellular irregularities (e.g., neurological damage or underdevelopment). This is how early research on many brain diseases was conducted, such as Alzheimer's disease and chronic traumatic encephalopathy. However, this approach would not only be laborious and practically difficult to realize—for one, you would need informed consent from psychopathic persons to donate their brain organ before they die—but these experiments would also be incapable of informing us about the crucial functional properties of the brain. And it is especially the functioning of the brain that researchers generally assume is implicated in personality disorders like psychopathy. Therefore, the preferred approach among researchers is to utilize noninvasive technologies that can yield information about the *living* human brain (sometimes referred to as *in vivo* experiments).[29]

One popular and relatively inexpensive approach to study functional properties involves the aforementioned EEG technology, which can test for abnormalities in cellular activation by measuring electrical brain signals through electrodes placed on the cranial scalp. The human brain consists of roughly 86 billion neurons,[30] the type of nerve cells that are believed to be central to mental processes. When the brain is operating—such as performing a cognitive task or processing sensory stimulus—the neurons give off electrical wave impulses, which the EEG technology can intercept and represent in different wave-motion graphs, thus allowing for a comparison across experimental tests. However, while EEG is an effective tool in diagnosing, treating, and studying various neurological conditions (e.g., epilepsy and sleep disorders),[31] its efficacy in studying psychopathy is both limited and questionable. For example, a 2019 review by Abby Clark and colleagues found no evidence of generalizable, replicated results across a total of 68 studies.[32] Furthermore, many researchers have challenged the usefulness of EEG for studying personality disorders, in which brain abnormalities are often hypothesized to be much more subtle and complex compared to other neurological disorders and diseases—complexities that the EEG technology might not be able to reliably detect.[33]

To be sure, for us to study the structure and functioning of brains in an effective and reliable way, it is crucial to deploy the right kind of technology: a technology powerful enough to detect brain differences, even if these are subtle and intricate. To achieve that aim, it is widely agreed that the best approach is to use neuroimaging techniques, a cohort of different technologies and methods designed to create an estimated digital visualization of the anatomical composition (structure) and activation (functioning) of a living human brain. What makes neuroimaging techniques particularly useful is their ability to yield information about the *spatiotemporal* state of the brain; in other words, it can generate data that tell us something about the physiological properties at specific locations in the brain (spatial) but also the state of these physiological properties over the course of time (temporal). In comparison, EEG cannot provide reliable spatial information about the brain, but mostly temporal information: it can tell us *when* electrical signals occur, but it is less useful in telling us *where* (EEG has some spatial information and is often described as a neuroimaging technology, but this is a bit misleading as it cannot create a visualization of the brain per se).

The quality of imaging technologies can be weighted in terms of their spatiotemporal *resolution*—that is, the level of three-dimensional detail they can provide about the physical properties throughout the brain and the degree to which this information is directly corresponding to its real-time states. As an analogy to spatiotemporal resolution, consider doing a science experiment where you film the trajectory of a bullet fired from a rifle. Since a film is just a sequence of still images or "frames," the *spatial* resolution thus corresponds to how much detail of the moving bullet the camera is able to capture in each image, and the *temporal* resolution refers to how many images the camera is able to take during the recorded time frame. So, to generate a high-quality film of the bullet, we will need a camera with high spatial resolution (i.e., capturing details about the bullet in its spatial location), but we would also need a camera with high temporal resolution (i.e., capturing many images per second). The lower the spatiotemporal resolution, the blurrier the bullet will appear in each frame, resulting in a blurry film. Today, we have high-speed cameras that make it possible to produce enormous spatial detail during each frame while also recording hundreds of frames per second, tracking the spatiotemporal state of the bullet with breathtaking detail.

In an ideal world, we would have a technology that could capture similar information about a brain all the way down to the cellular level, capturing where different types of cells (neurons, microglia, and so on) are located in the brain, paired with information about their physiological processes at any given time. However, such technology currently does not exist. Instead, the technology with one of the highest spatiotemporal resolutions—and also the most frequently deployed brain-imaging technology in psychopathy research—is *magnetic resonance imaging* (MRI). An MRI device (or "brain scanner," as it is sometimes dubbed) is a large tube-shaped machine capable of detecting and measuring nuclear *resonance*, the vibration frequency signal released from subatomic particles inside the brain. Similar to how a microphone can detect the sound coming from a piano—the acoustic resonance produced when a key strikes the strings on the soundboard causing them to vibrate—MRI technology can measure a resonance signal from the particles that are spinning and bouncing in your brain.[34]

But how exactly can this resonance signal be used to create an image of the brain's structural anatomy, let alone a visualization of its functional activity? In essence, an MRI image is created by measuring the relative strength of the resonance signal coming from tiny different parts of the brain. Think of how a grayscale photograph is a two-dimensional surface made up of small pixels, each representing the relative intensity of the light waves captured by the camera. An MRI device does something similar in a three-dimensional framework, by parsing out and measuring the relative strength of the resonance signal coming from thousands of three-dimensional coordinates in the brain (x, y, z axis). These cubical areas are called "voxels"—a blended term from pixel and volume—and in an MRI scanner they are typically 1–3 mm on each side depending on the spatial resolution of the scanner. We can then think of these voxels as individual pixels or building blocks in creating the three-dimensional digital image of the brain, its color-scale corresponding to the relative intensity of the resonance signal.

MRI technology can be utilized to generate a variety of different types of images—depending on machine capabilities and software packages—each designed to visualize different properties of the brain. However, the two most common types are images that give information about the anatomical *structure* of the human brain and images that tell us something about when and where in the brain there is increased *functional* activity. Fittingly, these types of visualizations are colloquially known as structural vs functional imaging (or *morphometric* vs *task-based* imaging).

A structural MRI image (sMRI) is a digital representation of the *static anatomical* state of the brain, usually containing information about the estimated volume of the brain, including variation in cellular density and material. The most common method behind an sMRI image is *voxel-based morphometry* (VBM), which estimates the spatial locations and volumes of *gray* and *white* matter, the main organic material that makes up the human brain (these tissue types really have a gray and white tone; the terms are not just a reference to how they show up as gray scale in the MRI image). We know from autopsies that gray matter contains more cellular bodies such as neurons (which gives it its gray tone), where white matter is primarily made up of elongated nerve fibers called axons covered in a protective (fatty) layer of myelin (which gives it its white tone). In an sMRI image, the contrast between gray and white matter is made possible because the two tissue types have different magnetic properties—due to their distinctive organic composition and density—which therefore gives off resonance signals with different intensities. And because of the difference in resonance signal, it is then possible to use it to make inferences or approximations about the distribution of gray vs white matter in the imaged brain.[35]

A functional MRI image (fMRI) is a digital representation of the estimated average *dynamic physiological* processes in the brain, typically measured over 30–60 minutes as the test-subject performs and repeats a host of simple cognitive tasks or is being exposed to various visual stimuli inside the MRI scanner. Where an sMRI image is an estimation of the organic composition of the brain, fMRI imaging aims to inform us about the brain's peak activity, namely when and where its cellular activity fluctuates during mental processing (sometimes referred to as "peak coordinates"). The most common approach to visualize cellular activity is a method called *blood oxygenation level dependent* (BOLD) imaging, which uses an MRI signal to estimate the flow and changes in the local oxygenation of blood in the brain. When a neuron becomes actively involved in a mental process, we are assuming that its energy needs will increase, leading to an uptake of oxygen, which the neuron receives from the red blood cells flowing in the brain. As the neuron consumes oxygen from a blood cell, there is a corresponding change in the blood cell's basic magnetic properties (since it releases oxygen), which is then detected in the MRI signal. So, when a study participant is repeating an experimental task inside the MRI scanner—which is collecting five to ten images per second—this imaging data will tell us (approximately) where and when cell activity is averagely concentrated during the task, and whether this activity was relatively low or high (e.g., large relative change in the MRI signal).[36]

Although it should be clear from this summary, it is important to emphasize that the images we create using sMRI and fMRI methods are far from perfect visualizations of brain structures and functioning. A central aspect to note is that an MRI image is not a *photographic* image—the process of capturing light waves, like the filming of a moving bullet—but instead crude *cartographic* approximations of the soft tissues and liquids inside the human skull. We can photograph brain tissue when placed under a microscope, and we can even create photograph-like images of cellular tissues smaller than light waves using an electron microscope. But this is scarcely comparable to what we are doing when creating an image with an MRI machine.[37]

Consider, for example, that a whole-brain sMRI image with a spatial resolution of 1 mm is made up of around 16.7 million voxels (e.g., in a $256 \times 256 \times 256$ voxel matrix), each of which shows up as a grayscale cube in the image, its color corresponding to the intensity of the resonance signal in those many individual coordinates. However, the information contained in such an image has virtually nothing to do with the actual biochemical and biophysiological systems contained *inside* each of the voxel coordinates in the brain. If you placed a typical 1 mm cubic gray matter under a powerful microscope, what you would discover is a gaspingly complex organic structure, containing thousands of neurons and other types of cellular and extracellular material. So far, there are no existing noninvasive neuroimaging techniques that can visualize cellular structures in this way.[38]

MRI Experiments on PCL Psychopathy: A Summary of the Evidence

The first MRI experiment on PCL psychopathy was published in 2000 by Adrian Raine and colleagues,[39] which analyzed and compared gray matter morphometry (i.e., sMRI) in groups of offenders assessed with the PCL-R. Since then, dozens of MRI studies have been published in scientific journals, making up one of the most comprehensive subfields in psychopathy research.[40] However, one challenge with summarizing the results from these studies is that the evidence—contrary to popular belief—is highly heterogenous, showing no obvious patterns or consistency. This observation was first made by Michael Koenigs and colleagues in a 2011 review,[41] it was demonstrated in greater detail in a 2017 systematic review by Stephanie Griffiths and Jarkko Jalava,[42] and, in 2023, our research group published a study demonstrating that the field has yet to find any clearly replicated evidence of functional differences in PCL samples.[43]

When research results are heterogenous, it means that there are many cases where experiments find different and sometimes conflicting results. For example, one sMRI experiment might find *reduced* gray matter in a specific brain region, where a subsequent sMRI study finds the opposite effect of *increased* gray matter in the same region. Furthermore, many studies have even reported effects from brain regions that researchers generally believe are completely unrelated to the symptoms associated with psychopathy (e.g., effects from the cerebellum and occipital regions, which are normally thought to be involved in motor functions and in forming sense impressions such as vision). All in all, the past 20 years of MRI research has shown that there are no consistent patterns in the experimental results.

Faced with this challenge, researchers have instead tried to make sense of the experimental results by weighing them up against various theoretical understandings of psychopathy, where they first ask themselves what sort of brain regions they would expect—or hypothesize—to be causally implemented in psychopathy, and thereafter analyze if there are any consistent results to support such a hypothesis. There are two main brain regions that researchers have traditionally hypothesized to be associated with PCL psychopathy:

The amygdala region: A very small (~2 cubic centimeters) cellular cluster in the limbic system. Despite its small size, the amygdala is believed to play a central role in emotion processing and regulation, where some researchers hypothesize that the emotional abnormalities that are conceptually associated with psychopathy could be linked to abnormalities in the amygdala. Thus, if psychopathic persons truly are deprived of emotions, researchers should be able to detect this in the MRI signals, either as a signal of structural aberration in the amygdala or as deviating patterns in the functional peak coordinates during emotion-related tasks.[44]

The prefrontal cortex (PFC): A large (~300 cubic centimeters) forward-facing part of the frontal lobe, usually divided into three larger medial subregions: ventromedial prefrontal cortex (vmPFC), dorsomedial prefrontal cortex (dmPFC), and anterior cingulate cortex (ACC). The PFC as a whole is believed to play a central role in personality, executive function, and impulse control, where some researchers hypothesize that the aggressive and irresponsible behaviors conceptually associated with psychopathy could be linked to abnormalities across the PFC subregions. Thus, if psychopathic persons truly are linked to such disabilities, this should be detectable in the structural and functional MRI signals.[45]

In what follows, I will summarize the results from neuroimaging studies that have investigated hypotheses that involve the amygdala and PFC subregions. In doing so, I will rely mostly on two recently published systematic reviews. In the subsection on amygdala effects, I will draw on a study by Philip Deming and colleagues from 2022,[46] which makes up the most comprehensive review of sMRI and fMRI experiments involving the amygdala in PCL samples. In summarizing the results from the PFC subsection, I draw on a 2024 study also by Philip Deming and colleagues (which I was personally involved in) that similarly makes up the most comprehensive summary of evidence involving PCL samples.[47] As in chapter 4, I will be cataloging all the evidence in terms of null and statistically significant effects, where the latter is described as either positive or negative effects. These two labels refer to evidence linking PCL psychopathy with increased/decreased gray matter volumes in sMRI VBM experiments or peak coordinate activity in fMRI task-based or BOLD experiments. I have limited my summary to experiments on adult samples, as the evidence involving juvenile samples is limited (to my knowledge, only four studies have investigated the amygdala and PFC regions in juvenile PCL:YV samples, and the results from these four studies are inconsistent[48]).

Summary of Evidence from the Amygdala Region

In their systematic review from 2022, Philip Deming and colleagues reviewed the evidence from all 60 studies published between the years 2000 and 2022 that have investigated the amygdala region in adult PCL samples (note that their review also examined results from experiments based on non-PCL samples, which I have excluded from the following summary). In total, there are

20 sMRI experiments with 54 effects involving 1,177 participants, and 40 fMRI experiments with 197 effects involving 2,652 participants. Of all the 60 studies, 35 (58.33%) yielded 100% null effects (15 sMRI and 20 fMRI), making this the most common discovery. Here is an overview of how effects are divided between sMRI and fMRI studies (notice that the numbers I report here are more detailed than the original 2022 publication thanks to Philip Deming sharing the raw data).

In sMRI experiments, 38 effects were nulls (70.37%), meaning that researchers found no unusual relationship between gray matter volumes in the amygdala in PCL samples. A total of 16 effects were statistically significant (29.63%), of which 11 effects were negative (68.75%), indicating relatively lower gray matter volumes, and five effects were positive (31.25%), indicating a relationship with increased gray matter volumes in PCL samples.

In fMRI experiments, 152 effects were nulls (77.16%), meaning that there was no unusual relationship in BOLD activation patterns in the amygdala region during experimental tasks in PCL samples. A total of 45 effects were statistically significant (22.96%), of which 29 effects were negative (64.44%), indicating relatively lower BOLD activity, and 16 effects were positive (35.56%), indicating a relationship with increased activity in PCL samples.

By this simple categorization of experimental results, it seems that one would not be justified in making sweeping conclusions about PCL psychopathy and a potential relationship with structural and functional abnormalities in the amygdala region. Indeed, the most replicated result is 100% null findings, and statistically significant results were broadly contradictory (both negative and positive direction). However, there are more reasons to think that the amygdala thesis is much less substantiated than it appears.

In a somewhat unorthodox analysis, Philip Deming and colleagues double-checked the data from each fMRI experiment to verify if the MRI signals were truly coming from the amygdala region. As explained earlier, fMRI imaging data are reported as peak signals, which are usually conveyed in a simple three-dimensional (x,y,z) coordinate system, meant to indicate "where" in the brain the peak signal is coming from. The way researchers conventionally decide whether their peak signal coordinates fall within or outside the amygdala—or any other brain region for that matter—is to manually overlay these coordinates on a standardized anatomical map of a human brain. These maps have demarcated by consensus where all subregions (like the amygdala) are located, defined by specific coordinates that clearly indicate their three-dimensional areas. So, only when researchers have found peak signals in coordinates that correspond with the conventional anatomical areas of the amygdala (as defined by the anatomical map) should the researchers report their results as coming from the amygdala. Metaphorically, it is similar to how a seismologist reports where an earthquake has occurred. They first estimate the coordinates of the epicenter and thereafter they look at a map to determine if the earthquake was in, say, Alaska or the Yukon Territory. Now, the lead author, Philip Deming—a good colleague of mine—has told me that as a neuroscience graduate student he had grown suspicious about the reliability of these fMRI amygdala experiments as he had discovered by coincidence that some researchers had mislabeled their results (i.e., reporting an effect in the amygdala, where the fMRI coordinate is not actually corresponding to the brain region on the map)—hence the motivation of their unorthodox analysis.

When reanalyzing the coordinates of fMRI signals, Deming and his colleagues found that the coordinates from 33% of the statistically significant negative effects were mislabeled by researchers, having *zero* percent overlap with the amygdala. Furthermore, 40% of the negative effects were partially mislabeled, such that only between zero and half of the signal was from the amygdala region (i.e., fMRI peak signals are typically bundles of several voxels, so when they speak of overlap, it is referring to how many of the voxels from the cluster are detected within the amygdala region). Just 27% of the negative fMRI effects reported in amygdala studies are detected in the amygdala. The positive fMRI effects were slightly more reliable, where 50% of the effects had some overlap and the other 50% was mostly overlapping with the amygdala. But overall, to use the metaphor from above, it would be similar to if a semiologist detected an earthquake in Alaska and then reported that it had occurred in the Yukon Territory.

Another interesting discovery by Deming and his colleagues is that statistically significant effects in fMRI experiments are disproportionately reported in studies with smaller sample sizes—that is, across these experiments, the smaller a sample size the higher the proportion of significant effects. For example, if we divide all the fMRI studies into three groups based on sample size, what we find is that the group with the largest sample sizes has a null proportion of 86.67%, which is higher than the average of 77.16% across all fMRI studies. To scientists, such a trend where effects are correlated with sample size might say something about the reliability of the overall results, as experiments on larger sample sizes are usually seen as better evidence than experiments on small sample sizes. In other words, when there is a trend of statistically significant effects disproportionately coming from smaller experiments, this subtracts from the overall inferential value of the evidence.

But why do experiments with larger sample sizes find more null results? Is it just an odd coincidence, or is there something driving this pattern? One obvious (and I think broadly acknowledged[49]) explanation promoted by Deming and colleagues is that these larger sampled studies are also the experiments with the highest quality in terms of what statisticians call *statistical power*.[50] This concept refers to how likely an experiment is to accurately detect a true effect. For instance, if PCL psychopathy is truly linked with specific brain abnormalities, some studies will—by the quality of their research method or design—have a higher likelihood of detecting/finding these true abnormalities, namely the studies with high statistical power. Estimating the statistical power of a particular experiment can be tricky, as it depends on many variables and assumptions. However, one main variable is *sample size*, such that the larger your sample the higher the statistical power. On the contrary, an interesting thing about low-powered studies is that they are not only unlikely to detect true effects but they also have—by their design—an increased probability of making false-positive discoveries (i.e., analyses that yield a statistically significant effect but where the effect is not true or "real"). Conversely, since high-powered studies will be more likely to correctly detect true differences (again, if they exist), the fact that null effects are overrepresented in large sample studies tells us that the few significant results from amygdala studies may be nothing but spurious findings with questionable inferential value.

In sum, and to echo the conclusion from Deming and colleagues' 2022 study, the evidence from these 60 MRI experiments does not support the claim that PCL psychopathy is related to generalizable abnormalities in the amygdala.

Summary of Evidence from the Prefrontal Cortex (PFC) Region

In our meta-analysis from 2024, a project spearheaded by Philip Deming and Stephanie Griffiths,[51] we summarize the research published between the years 2000 and 2023 that investigated the three different subregions of the PFC—vmPFC, dmPFC, and ACC—in adult PCL psychopathic samples (we also examined results from experiments based on non-PCL samples, but I will not address these here). These studies included 69 experiments, totaling 20 sMRI experiments with 202 effects involving 1,755 participants and 49 fMRI experiments with 951 effects involving 3,246 participants. Similar to the amygdala research, the most common and replicated finding was 100% nulls. Across the three subregions, almost two-thirds of all experiments (i.e., 58.82% of sMRI and 61.31% of fMRI) reported zero statistically significant effects.

Table 5.1 provides an overview of these effects divided between the three different subregions and whether the effects indicated a null, negative, or positive relationship in PCL samples. In sMRI studies, 164 effects were nulls (81.19%), meaning that researchers found no unusual relationship between gray matter volumes in the PFC regions in psychopathic samples. There were 38 statistically significant effects (18.81%), of which 27 effects were negative (71.05%), indicating relatively lower gray matter volumes, and 11 effects were positive (28.95%), indicating a relationship with increased gray

Table 5.1
Overview of results from studies involving PFC regions

	Total effects	Null	Negative	Positive	Null proportion
Effects from sMRI experiments (n = 20)					
vmPFC	79	57	15	7	72.15%
dmPFC	52	45	7	0	86.54%
ACC	71	62	5	4	87.32%
Total effects PFC	202	164	27	11	
Percent		*81.19%*	*13.37%*	*5.45%*	
Effects from fMRI experiments (n = 49)					
vmPFC	317	281	28	8	88.64%
dmPFC	313	261	30	22	83.39%
ACC	321	263	39	19	81.93%
Total effects PFC	951	805	97	49	
Percent		*84.65%*	*10.20%*	*5.15%*	

Notes. Abbreviations: PFC = prefrontal cortex; vmPFC = ventromedial prefrontal cortex; dmPFC = dorsomedial prefrontal cortex; ACC = anterior cingulate cortex. The table is based on results reported in the study by Philip Deming and colleagues from 2024.[52]

matter volumes in PCL samples. In fMRI experiments, 805 effects were nulls (84.65%), meaning that there was no unusual relationship in BOLD activation patterns in the PFC regions during experimental tasks in PCL samples. A total of 146 effects were statistically significant (15.35%), of which 97 effects were negative (66.44%), indicating relatively lower BOLD activity, and 49 effects were positive (33.56%), indicating a relationship with increased activity in PCL samples. Across the different PFC regions, the null proportion was consistently high, ranging from 72% to 87% in sMRI, and 82% to 89% in fMRI experiments.

An important observation to emphasize here is that the few statistically significant effects are diverse, meaning that they are giving us conflicting information of both increased vs decreased gray matter volume and task-based activity levels. Roughly two-thirds of significant effects indicate increased volume/activity, where the last third effects indicate decreased volume/activity. Since there is no obvious pattern in significant effects, it suggests that they are spurious and perhaps caused by confounds unrelated to psychopathy (e.g., substance misuse and head trauma). Indeed, the issue of confounds in MRI research is a well-recognized problem. For example, it has been documented that drugs or psychiatric medication can alter the magnetic properties in the brain, which then shows up in an MRI scan.[53] The way researchers get around this issue is to control for such known confounds, for instance, by conducting a moderator analysis.[54] Unfortunately, we found in our 2024 review that experiments generally did not perform such an analysis.

Similar to the results reported in Deming and colleagues' 2022 review of amygdala experiments, we found that null proportions and significant effects were associated with sample sizes, although this relationship was less pronounced and inconsistent. For example, in sMRI studies, the upper third studies with the largest samples had an average null proportion of 88.46%, which is higher than the average of 81.19% across all studies (+7.27 percentage points difference). However, the relationship was in the opposite direction in fMRI studies, where the upper third studies with largest samples had an average null proportion of 78.78%, which is *lower* than the average of 84.65% (–5.87 percentage points difference).

There are many potential explanations for the inconsistent relationship between sample size and null proportions. In my perspective, one compelling (though speculative) explanation is that *all* of the 69 PFC experiments are severely underpowered, and for that reason they have essentially an equally high risk of making false-positive discoveries, which is why they generate a similar proportion of null and significant effects regardless of their specific sample size. This would explain why null proportions only deviate with few percentage points.

The reasons why I think the PFC experiments on average might have lower statistical power compared to experiments targeting the amygdala are multifaceted, but one major issue is that the PFC is a much larger brain region than the amygdala—around 150 times larger—and therefore these experiments are analyzing a much larger dataset (i.e., more datapoints). So, for a PFC study to be sufficiently powered, it generally needs a sample size that is proportionately larger than studies investigating a smaller region like the amygdala (with much less data).[55] There are some important exceptions to this rule of thumb, for example, if a PFC study is limited to only investigating a predefined small subregion of the PFC (called a *region of interest* study[56]). But overall, most PFC experiments are conducted in a way where they analyze a much larger portion of the brain compared to amygdala studies (often because they actually analyze the whole brain, called a *brain-wide association* study[57]).

There are two pieces of evidence that suggest that all or most of the 69 PFC experiments might be severely underpowered and that statistically significant effects might be nothing but spurious findings. First, even when we look at the five (top) outlier fMRI studies in terms of sample size—that is, studies with over 150 participants, twice as high as the average sample, and therefore in theory higher statistical power—the null proportion is still very high (86.36%), but it is only slightly higher than the average null proportion (84.65%) across all fMRI studies. Second, in our 2024 review study, we mapped the dispersion of significant effects in the PFC (i.e., we tracked *where* in the PFC regions the peak coordinates were reported), and we found that there was very little overlap between effects. That is, statistically significant effects appeared to be mostly one-off discoveries, suggesting that they might have been false-positive effects (i.e., effects that do not generalize onto PCL psychopathic persons).

In sum, just like the amygdala studies, the evidence from these 69 MRI experiments does not support the claim that PCL psychopathy is related to structural or functional aberrations in any of the three PFC subregions.

Concluding Remarks: The Psychopathic Brain as Scientific Spin?

In 2009, MRI research on psychopathy was presented for the first time in a US courtroom in the homicide case *State v. Brian Dugan*.[58] Since then, neuroimaging research on psychopathy appears to have made its way into the court system with increasing frequency (although little is known of its true extent).[59] How exactly neuroimaging evidence impacts decisions about psychopathic persons in the broader criminal justice system—and how it impacts practitioners' general understanding of psychopathy—is difficult to say for sure. But MRI experiments have historically been presented by scientists as a validating form of information, as a line of evidence that legitimizes the use of psychopathy assessments as an evidence-based practice. And we may speculate that once a disorder is described as having been validated by "hard" technology-driven science like neuroimaging research, this might bolster the claims that are ordinarily associated with the condition.[60]

However, in this chapter, I reviewed the MRI research that has investigated the claim about psychopathy as a disorder with neurological bases, and what emerged is a perspective that is markedly different from the one that is typically encountered in the mainstream psychopathy literature. While the idea of psychopathy as a "brain disorder" goes far back in history and has been studied using a variety of different technologies, it was not before the year 2000 that scientists began to rigorously test it using structural and functional MRI methods. Since then, dozens of MRI studies have been published, but the most reasonable conclusion to draw from this research is that no reliable evidence has emerged to corroborate the idea that PCL psychopathy is correlated with brain abnormalities of any kind. Overall, experimental results are predominately nulls with few inconsistent statistically significant effects (e.g., effects in contradicting directions), which might be better explained as a byproduct of confounding variables unrelated to psychopathy such the presence of substance misuse, medication, and head trauma.

This conclusion raises an important question: If there has never been any clear evidence of brain abnormalities in PCL psychopathic persons, why do so many scientists keep portraying psychopathy as a neurodevelopmental disorder? As mentioned in the introduction to this chapter, if you read recent review studies from scientific journals, what you will soon discover is that authors appear to draw conclusions that are markedly different from the ones we have arrived at in this chapter, repeatedly, yet wrongly, claiming that psychopathy *is* linked to brain abnormalities. For example, in a study I conducted with Jarkko Jalava, Stephanie Griffths, and Emma Alcott, we found that between the years 2000 and 2022 at least 45 review studies have been published, of which the vast majority describe the MRI neuroimaging evidence as supportive of the brain-disorder view of psychopathy.[61] Even as some reviewers began to voice hesitancy about the inferential value of the MRI research during the 2010s—for instance, a review by Michael Koenigs and colleagues in 2011 and later by Stephanie Griffiths and Jarkko Jalava in 2017—review authors appear to have kept repeating statements that express little doubt about linking psychopathy to brain abnormalities. At least 30 review studies have been published since 2010, but only a few of these state that the evidence is (at best) inconclusive.

What I find particularly curious about this observation is that it appears to deviate from the research I surveyed in chapters 2–4 on the topics of dangerousness, treatment, and moral capacities. In these three paradigms, the review literature has generally been very accurate, historically reporting a narrative that goes counter to the mainstream depiction of psychopathic persons as untreatable, morally colorblind, and dangerous social predators. The problem seems to be that some researchers and practitioners are promoting views that conflict with what has been documented in scientific review studies. For instance, in 2002, Paul Gendreau and colleagues pointed out that there was no empirical evidence to suggest that PCL psychopathic persons were associated with extraordinary risk levels.[62] And in the same year, 2002, Randall Salekin demonstrated that there was little evidence to support the claim that psychopathic persons were untreatable.[63] Large-scale reviews of the empathy and moral reasoning literature came a bit later (because the research did not develop before the mid-2000s), with the first quality review study published in 2013 by Jana Borg and Walter Sinnott-Armstrong, which concluded that there was no basis for describing psychopathic persons as severely morally incapacitated.[64] In short, with the three questions covered in chapters 2–4, the science has always been accurately summarized in the review literature, where the misleading depictions of psychopathy have originated elsewhere. But with the brain-disorder question, the review literature appears to have gotten it systematically wrong. This is particularly problematic because it is usually this type of literature that forensic practitioners and legal decision-makers rely on when they seek information about psychopathy. So, these professional groups might be wondering why the empirical evidence is systematically misrepresented in the review literature.

In trying to find answers to this question, it seems to me that we cannot rule out the possibility that these tendencies are driven (at least in part) by the phenomenon of *scientific spin*, the practice where scientists make their experimental work look "better" than it actually is.[65] One reason why we cannot rule out scientific spin is that it has already been documented to have broadly impacted the behavioral and social sciences; so, it would be odd if it had not also impacted psychopathy research. For example, a 2012 study by Leslie John and colleagues surveyed 2,155 anonymous psychology researchers and found that *more than half* of respondents admitted to having used several spin tactics (e.g., selectively excluding experimental data that contradicts the researchers' theory).[66] There might be many reasons why scientists resort to spin tactics, but a central motivation appears to be simple "careerism," where scientists are disproportionately rewarded if they find/report statistically significant and exciting results as opposed to using their work to promote cautious messages and (boring) null findings.[67] Thus, maybe the neuroscience of psychopathy has a problem with scientific spin?

To probe this theory, Jarkko Jalava, Stephanie Griffths, Emma Alcott, and myself designed a study where we wanted to test if authors of review studies were selectively reporting results that made the neuroscientific research, as a whole, look more robust than it actually is; that is, we wanted to test whether there was only a problem with *interpreting* the neuroscientific evidence as opposed to *selectively making* the evidence look better than it is. To do this, we first identified all the sMRI research on PCL psychopathy published since the year 2000, which amounted to 38 studies, and we then catalogued all the effects from each study to calculate the proportion of null findings across these studies. Out of a total of 791 effects, we found that 64.10% were nulls. Thereafter, we took all the review studies published in the same period and calculated the proportion of null findings reported in these publications. If the review literature did a decent job at communicating the basic insights from the field, review studies should then report numbers that mirror the true proportion of nulls. However, what we found was that the nulls had virtually disappeared from the review literature, as if the review literature acted as a sift, like a miner panning for gold in a river. Across a total of 45 review studies, authors were on average only reporting 8.99% null effects. The picture was even more skewed in focused reviews, such as those investigating the effects from the amygdala region, where the average null reporting dropped to 2.59% (where we know from Deming and colleagues' 2022 study that the true proportion of sMRI nulls is 70.37%). In reviews tailored to making recommendations to forensic practitioners and legal decision-makers, the null proportion was 5.39%. An equally interesting finding was that 19 review studies (42%) did not report a single null finding, which should have been a major red flag for experts in charge of the peer-review process (i.e., it is well known that neuroimaging studies almost always find some null effects). This latter observation suggests more systemic problems with spin, where peer-reviewers appear to tolerate the practice.

Of course, it is inherently difficult to determine whether the systematic omission of null effects from the review literature is an act of scientific spin, or simply an honest mistake of overlooking null effects (or a mixture of both). And perhaps it does not matter what the reason is because, whether the problem boils down to an issue with scientific spin or honest mistakes, the reality is the same, namely that for the past two decades forensic practitioners and legal decision-makers would have been misled if they had followed due diligence and relied on the review literature when seeking information about neuroimaging research about psychopathy. The good news is that in the past years, we have seen the publication of high-quality review studies, where authors are now paying more attention to the extent of nulls (a practice that was initiated in the field in 2017 by Stephanie Griffiths and Jarkko Jalava[68]).

However, even as the review literature is slowly correcting, readers should be aware that spin about the brain-disorder view of psychopathy is not a problem limited to the scientific peer-reviewed literature, but it is arguably much more rampant in public media like op-ed articles or journalistic interviews,[69] as well as in popular books about psychopathy (sometimes) written by leading scientists.[70] A search on media platforms like YouTube and TikTok will readily yield hundreds of videos (amassing millions of views) where experts are explaining how psychopathy is caused by a brain abnormality, and it is not uncommon for TV documentaries about psychopathy to include a segment with MRI research. A prototypical example of a documentary that I often show to my students (as an example of scientific spin) is a 2014 production by the *Canadian Broadcasting Corporation*—the national television station in Canada—titled "The Psychopath Next Door,"[71] which figures various leading scientists in the field. In this documentary, one expert is on record saying that psychopathic persons have "degraded connections in the brain," which causes their brains to be "hypersensitive to reward" and "undersensitive to the suffering of others." What is interesting about this comment is that a couple of years before this documentary was released, the same expert coauthored a review article that concluded that neuroimaging research is "remarkably heterogenous."[72]

Perhaps the strongest evidence of spin comes from the book publishing industry, where some researchers have signed *popular science* book contracts on psychopathy, a format where they are encouraged to write more excitingly so the books can move in the market. This may have created a tension as excitement very easily morphs into scientific spin (it is arguably a very fine line to balance). I do not mean to say that there is something problematic about simplifying science for the sake of making it more entertaining, but the problem with this popular genre is that the way psychopathy is described in these books often migrates into the scientific disciplines and legal practices as students and professionals may use these sources as a steppingstone in their quest to better understand the subject matter. There are many examples of what appears to be spin from this domain, but it is interesting to observe that the most cited and (I suppose) most widely distributed titles all build on the narrative about psychopathic persons having brain abnormalities.

For instance, perhaps the most popular book about psychopathy, Robert Hare's *Without Conscience* (from 1993), begins with an alluring anecdote about how EEG neuroimaging technology has documented extreme brain abnormalities in psychopathic persons, so much so that Hare claims to have had one of his scientific papers rejected because the EEG data from psychopathic persons were so abnormal that the journal editor would not believe that it was retrieved from human beings. Similarly, the book *The Psychopath Inside* by James Fallon (from 2013) conveys an admittedly absorbing story about how Fallon himself, a neuroscientist, accidentally discovered that he was a psychopath by studying his own MRIs. Lastly, in *The Psychopath Whisperer* by Kent Kiehl (from 2014), the narrative centers around neuroimaging research on psychopathic persons, of which the author, a leading expert, writes that the "consistency of their [psychopathic persons'] brain abnormalities never ceased to amaze me."[73]

The ideas conveyed in these popular books are, of course, scientifically untenable, and they arguably border on sheer make-believe. While they are undoubtedly entertaining, they come across as a form of spin that ends up doing a disservice to forensic practitioners and legal decision-makers as they perpetuate scientifically misleading views.

6   Psychopathy Theories as Falsified Explanations

Clinicians and researchers often are stunned by the apparent ease with which psychopaths engage in cold-blooded, instrumental behavior. They also are puzzled by the apparently candid—yet superficial and mechanistical—manner in which many of these individuals describe their actions, their feelings about what they have done, and the impact their behavior might have had on others. Their expressions of remorse are unconvincing to astute observers, and their use of emotional words and phrases seem like mere mimicry.
—Robert Hare in *Psychopathy, Affect, and Behavior* (1998)[1]

In the previous four chapters, we have surveyed the empirical evidence behind four mainstream claims about psychopathy—more specifically, three claims related to its *signs* and *symptoms*, and one about the alleged *causes* of the disorder. In chapter 2, we scrutinized the idea that psychopathic persons are extraordinarily dangerous; in chapter 3, we looked at the claim that they are chronically untreatable; in chapter 4, we examined the claim that they are morally colorblind, lacking empathy and moral reasoning capacities; and finally, in chapter 5, we analyzed the widespread view that psychopathy is caused by distinct brain abnormalities. By examining all the research published since 1980, I demonstrated that not only is there remarkably little evidence to support these four popular claims but they have also been robustly challenged if not refuted by the experimental data.

In this chapter, I shall survey a comparatively more fundamental question about psychopathy that goes beyond the conversation about prognosis and causes, namely whether scientific *theories* of psychopathy are meaningful and informative. More specifically, I will examine whether these theories provide satisfactory, evidence-based *explanations* of what makes psychopathic individuals psychologically different from non-psychopathic persons.

When legal and forensic practitioners use PCL psychopathy assessments to inform their decision-making, they are (as we saw in chapter 1) using these scales under the assumption that they provide empirically verified prognostic information (e.g., they are predictive of risk levels and treatment outcomes). As I have shown throughout this book, these assumptions are at best dubious if not just plain wrong. But one of the more fundamental professional justifications for even considering the use of the PCL scales hinges on the belief that psychopathy—the personality disorder that the scales are designed to measure—is vetted by scientific evidence (in colloquial terms, when a diagnosis is vetted by science, we refer to it as a "valid" diagnosis). While there are many components to this scientific vetting procedure,[2] a necessary and central part of it is scientific theorizing and testing: that researchers have evidence-supported explanations of what makes psychopathic persons *psychologically* different from ordinary, non-psychopathic people (aside from the fact that they have been selected or labeled by a PCL assessment).

Although the term *theory* is used in many different ways in the sciences, one common way is to think of theories as formal explanations of an observable/measurable aspect of reality.[3] That is, a theory is a formal articulation of how scientists account for and make sense of the portion of reality they study. A scientific theory may have descriptive elements—for instance, it may incorporate models that map causal relationships between entities it aims to account for—but the main aspiration of a theory is to explain why things are the way they are (so to speak). To paraphrase the philosopher of science Carl Hempel, to have a theory is to have an answer to the *why* of the *what*.[4] To use the term theory in this way is also to connect it to one of the most basic aims of working scientists, namely to better understand the portion of reality that they study. When we achieve a good explanation of whatever it is we study, we can thereby claim to have a good understanding of it.

A trademark example of a scientific theory is the *theory of evolution* (initially proposed by Charles Darwin in 1859), which is the preferred explanation among scientists of the genotypic and phenotypic diversity observed across organic lifeforms. This includes the theory's ability to give an overarching explanation of why birds have beaks, why *Homo sapiens* have a tailbone, or why coronaviruses have spike proteins. If we read a biology textbook, we will learn that the current version of evolutionary theory posits that such organic variation emerges because of a complex interaction between environmental and genetic factors, particularly four processes referred to as genetic mutation, gene flow, genetic drift, and natural selection. It is with reference to these processes that contemporary scientists explain the immense diversity observed in and across all organisms (i.e., *why* there is organic diversity), and, importantly, they prefer this explanation because it has been extensively corroborated in empirical studies. Wherever scientists have scrutinized organic lifeforms, they have found unmistakable evidence of the four types of gene-environment interactions.[5]

Similarly, when we speak of a scientific *theory of psychopathy*, what we are really addressing is that there is some observable aspect of reality or a *what*— namely individuals that clinicians identify as stereotypical psychopathic persons—that scientists claim to have a sound, evidence-based explanation of *why* they are the way they are (i.e., why they are psychologically different from ordinary people). Notice that there are two core aspects to this orientation:

1. First, researchers are implying that there is a stereotypical person commonly encountered in clinical settings—colloquially knows as "psychopaths"— who presents with unusual psychological traits that clearly separate them from the rest of us (this would be analogous to how we can *observe* phenotypic and genetic variation in organisms).

2. Secondly, researchers uphold that they have a compelling explanation of the overarching reasons why these individuals are different from ordinary people (analogous to how we *explain* organismal diversity with reference to gene-environment interactions).

So, for a personality disorder like psychopathy to be scientifically vetted, it is not enough to simply describe the *what*—for example, what signs and symptoms typically characterize psychopathic persons—the crucial element is that scientists have a compelling explanation for *why* these traits occur or cluster in these individuals.

Throughout history, there has been long-lasting disagreement in psychopathy research on how they account for both of these theory-related pillars (i.e., the *what* and the *why*). For example, as covered in chapter 1, there was until recently widespread confusion and disagreement about what was the most accurate description of a stereotypical psychopathic person. In 1974, the British psychiatrist Aubrey Lewis declared that psychopathy was the "most elusive category" in the entire profession of psychiatry,[6] and in 1978, Robert Hare and Daisy Schalling edited an anthology with contributions from leading experts, which above all demonstrated profound incongruities in clinical descriptions.[7] As Hare reminisced in 2007, it was almost as if the experts were talking about different disorders.[8] A central implication of these confusions is that researchers in the 1970s disagreed on what aspect of reality they needed to explain; they disagreed about the *what* in need of a *why*. And it was partially this type of confusion that led prominent scientists to declare that the idea of psychopathy should be discarded by the scientific community.

However, many of these disagreements have now been largely resolved, where a broad consensus has formed around the clinical stereotype initially proposed by Hervey Cleckley and later refined by Robert Hare's work on the PCL scales (although, some long-lived disagreements still exist[9]). According to this consensus, a psychopathic person is a person who, from a clinician's point of view, appears to be incapable of comprehending and appreciating moral and normative values, as well as showing an uncanny pattern of self-defeating and careless antisocial tendencies. A stereotypical psychopathic person, we are told, comes across as charming, cunning, and with no obvious mental defect. But, on closer scrutiny, clinicians report that something appears—that is, in their professional opinion—to be fundamentally abnormal about these individuals' basic attitudes and behavior toward other people. As Hervey Cleckley famously put it, a psychopathic person behaves like a "downright madman" yet maintains a deceptive outer mask of near perfect sanity.[10] The proper aim of scientific theoretical work, then, is to explore and provide explanations for why these individuals appear the way they do—or why they are the way they are—which is another way of saying that the goal is to explain, at the psychological level, what makes PCL psychopathic persons different from non-psychopathic persons.

Over the years, researchers have proposed and investigated a variety of such basic theories of psychopathy,[11] including many that never really caught on in mainstream scientific circles (such as theories from psychoanalytic[12] and evolutionary psychology traditions[13]). However, in the modern period, there are two main theories that have been both broadly endorsed and investigated by leading scientists: one theory that explains psychopathy as a disorder related to *emotional* dysfunctions and another theory that explains psychopathy as a disorder related to *cognitive* dysfunctions.[14] Proponents of the *emotion theory* believe that psychopathy can be explained as a complex impairment of the biological systems that generate emotions, leading to an inability to feel some or all types of emotions. According to this theory, psychopathic persons are abnormally cold, fearless, and largely unemotional.[15] Proponents of the *cognitive theory* believe that psychopathy is best explained by its association with deep impairments of executive functions, such as the disposition to control and modulate behavioral impulses. According to this theory, psychopathic persons are extraordinarily impulsive and disinhibited, and behave mostly without forethought and executive control.[16]

It should be noted that while these two theories have historically been pitched by scientists as separate and competing explanations of psychopathy,[17] some researchers are today endorsing both theories (or some aspects of them), arguing that psychopathy is best explained as a *multifaceted* or *nonunitary* condition related to emotional *and* cognitive impairments.[18] This view is promoted by Robert Hare in the 2003 version of the PCL-R manual, and it is often highlighted in popular review articles.[19] Perhaps one of the main reasons that researchers have shifted toward a more multifaceted explanation of psychopathy is the discovery that it is somewhat problematic to think of emotion and cognitive processes as separate entities at the psychological level.[20] That is, emotions evidently have an enormous impact on how we think and behave, but the way we orient our thinking about certain affairs may also impact our emotional reaction to them. But despite this integrated explanation of psychopathy as related to a *dual deficit* of emotion and cognitive functions, researchers can still empirically test the emotion and cognitive theory *as if* they are separate explanations insofar that the theories are hypothesizing different types of impairments (i.e., lack of emotion vs poor impulse control).

In what follows, I will review the empirical research that has tested both the emotion and the cognitive theory of psychopathy. While we have already reviewed some research that indirectly addresses these theories (e.g., the neuroimaging experiments in chapter 5), there is a large body of experimental work that has tested PCL psychopathic persons' biological functions involved in emotion processes (e.g., autonomic nervous system responses) as well as studies that probe their ability to control behavioral impulses (e.g., attention tests). However, as we shall see, it has become increasingly clear in recent years that none of these two popular theories do particularly well when subjected to this form of experimental scrutiny. And what is more, a strong argument can be made that these mainstream theories of psychopathy not only lack empirical support but have essentially been falsified: they are refuted by and incongruent with the experimental evidence. When PCL psychopathic persons are tested in psychological studies, researchers have never been able to detect the sort of grave impairments of emotion and cognitive functions that are posited by these popular theories.

The Emotion Theory of Psychopathy: Basic Accounts and Experimental Work

From a historical point of view, the emotion theory of psychopathy is by far the most frequently invoked explanation of psychopathy among mainstream researchers, with its first iteration dating back to the earliest scientific writings. The first version of the theory appears already in Benjamin Rush's work from 1786,[21] whereafter the theory goes through numerous revisions in the nineteenth century before it is expressed in its most elaborate form by twentieth-century researchers like Ben Karpman,[22] David Henderson,[23] Hervey Cleckley,[24] Silvano Arieti,[25] David Lykken,[26] James Blair,[27] Christopher Patrick,[28] and others.[29] Although these authors are technically outlining different sub-versions of the emotion theory, the differences between them are of little consequence to the discussion in this chapter. What unites the various sub-versions is that they build on the same fundamental explanation: that what makes psychopathic persons abnormally different from the average person is that they have a severe lack of emotion.

A Summary of the Theoretical Framework

A central premise in the emotion theory of psychopathy is the assumption that a profound lack of emotion will fundamentally alter a person's psychology; or better, it will give rise to a type of psychology that is markedly different from a "normal" human psychology. Indeed, human beings are fundamentally emotional organisms. We are almost always in some form of emotional state, feeling joy and anger, jealousy and sadness, and so on.[30] It is rather trivial to observe that these emotional states are integral to how we see ourselves, how we behave, and how we otherwise interact with our environment. For instance, someone quits their job because they are *angered* by how the company treats its employees; a person might volunteer at the local shelter because they *sympathize* with the less fortunate of society; and two individuals form a long-lasting friendship because they *adore* each other. This is not to say that humans are always causally moved by emotions, but it seems obvious that emotions play a meaningful role in shaping our perspectives and behavioral inclinations.

What proponents of the emotion theory of psychopathy want to convince us is that a human mind deprived of these basic emotional states will eventually give rise to an abnormally different type of psychology—that is, abnormally different attitudes and behaviors. To fully appreciate this claim, it is helpful to first pay attention to the complexity of an emotional episode. For example, consider the feeling of fear. It begins with something that triggers your disposition to feel fear, such as a sudden encounter with an armed robber in the middle of the night. As the person points their gun at you, a string of processes begins to unfold in your body. Perhaps your heartbeat and breathing will slow as a cold sensation runs through your body. Soon the muscles in your upper corpus tighten, your hands become sweaty and jittery, and your facial expression turns from neutral to wide-eyed. As these processes come to realization, you may eventually be struck by a sudden involuntary inclination to run away from the robber. The processes I describe here are not necessarily universal across all people, but there is little doubt that when we go through the state of being fearful, multilayered complex changes occur both to our physiological and psychological states.

According to the emotion theory, then, what makes psychopathic individuals different is that they are largely deprived of some or all of these bodily sensations, and therefore also the psychological outcomes that are usually associated with them. When in a fearful situation, for instance, a psychopathic person's perspective is not affected by sensations like muscle tightening and chillness running down the spine; nor would they be struck with the involuntary behavioral tendencies that characterize emotional states. When they find themselves in emotion-triggering situations, psychopathic persons will have a relatively flattened and neutral reaction (or lack thereof). An illustrative way of describing this sort of emotional *neutrality* is to think of feelings (like fear, anger, joy, and so on) in terms of their more discrete psychological components. There is some consensus among researchers that feelings always have (at least) two different dimensions: namely a *valance* dimension, meaning that a feeling can be *negative*, *neutral*, or *positive*. But there is also an *arousal* dimension, such that we can speak of a feeling ranging from low to high activation.[31] A popular visualization of feelings as containing a valence and arousal dimension is the *circumplex model of affect* depicted in figure 6.1.

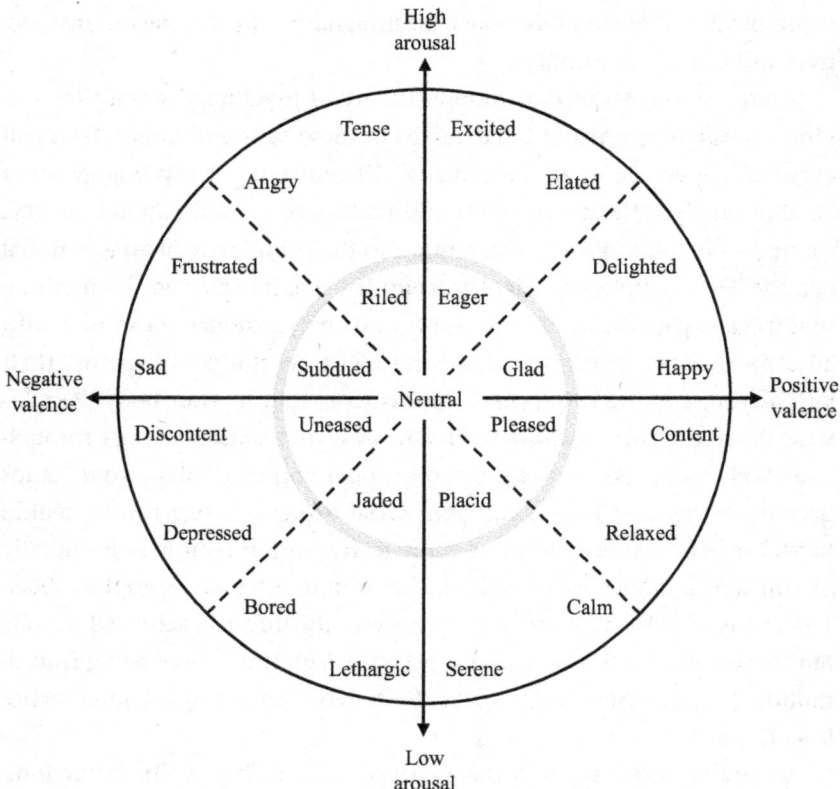

Figure 6.1
The circumplex model of affect: the affective states captured by each label indicate
the valence and to what degree the person is aroused. Adapted from Jonathan Posner
and colleagues (2005) and Lisa Feldman Barrett (2017).[32]

The circumplex model is probably an oversimplification of how emotions are construed,[33] but the model still helps us express more clearly what
might be psychologically different in psychopathic persons. According to
the emotion theory, psychopathy might be conceptualized as a condition
where emotional or affective states never really reach a degree of arousal that
significantly alters or *invades* the person's mental state. For instance, in his
version of the emotion theory, Hervey Cleckley speculates that psychopathic
persons might have *some* affect but that these states are always superficial
(i.e., less aroused).[34] Similarly, Robert Hare writes in *Without Conscience* that
psychopathic persons are characterized by a kind of "emotional poverty that

limits the range and depth of their feelings" and that careful observers can readily see that when they are emotionally moved, "little is going on below the surface."[35] More specifically, then, we might think of the emotion theory as proposing that psychopathic persons never have the types of feelings that are plotted onto the outer perimeters of the circumplex model, that they only have feelings of relatively near-neutral arousal and valence as indicated by the grayscale circle of figure 6.1. As Hervey Cleckley speculated, when it comes to these stronger feelings that move human affairs, psychopathic persons are simply "blind in this mode of awareness."[36]

An important qualification of the emotion theory is that psychopathic persons may learn to use words like fear or happy to describe their current psychological states—for instance, they can, of course, utter the sentence "the robber really frightened me"—but according to proponents of the emotion theory, psychopathic persons do not have any substantial bodily sensations associated with emotional words. As John Johns and Herbert Quay famously put it in a 1962 article: "The psychopath can thus be said to be one who knows the words but not the music; the denotative meaning of words and phrase may be intact, but the connotative emotional or motivational component is lost."[37]

To follow this line of thinking, the difference between psychopathic and ordinary non-psychopathic individuals is therefore thought to be quite remarkable. Ordinary people may experience many complex and uncanny bodily sensations when they are frightened (e.g., muscles tighten, hands sweat, cold sensation down the spine). And therefore, they connotatively associate the word *fear* with these fundamentally unpleasant sensations. However, according to the emotion theory of psychopathy, these bodily processes do not unfold to such a noticeable and invasive degree should a psychopathic person be in a similar fear-triggering situation. Thus, they are fundamentally unaware of what ordinary people mean when they use the word *fear*.[38]

The emotion theory of psychopathy evidently has an enormous intuitive appeal. I personally find it easy to imagine that should a person be deprived of emotion in this way, one obvious consequence would be a fundamentally different lived experience, and therefore they would presumably develop different perspectives about the world and human affairs. For example, as we covered in chapter 4, Hervey Cleckley was firmly convinced that if a person was deprived of feelings, it would cause major problems with understanding and caring about moral values,[39] a view that is expressed by almost all proponents of the emotion theory.[40] Some researchers have gone as far as arguing that psychopathic persons' emotional impairment leads to a profound carelessness and indifference, even about their own lives.[41] In a similar vein, David Lykken argued that an inability to experience fear makes such a person extremely dangerous insofar that they are to an extent undeterred by legal punishment.[42] James Blair has argued that the inability to experience strong negative feelings in general (e.g., distress, fear, and disgust) might cause all sorts of psychological disinhibitions, but the incapacity itself also causes developmental problems such as an inability to internalize and cultivate the sort of prosocial habits that grow from acknowledging other people's feelings and attitudes.[43] Altogether, it is perhaps this depiction of a cold, unemotional, and careless individual that has inspired so many pop-cultural (and mostly unrealistic, caricatured) portrayals of psychopathic villains, like the hitman Anton Chigurh (played by Javier Bardem) in Joel and Ethan Cohen's *No Country for Old Men* (2007) or the cannibalistic forensic psychiatrist Hannibal Lecter (played by Anthony Hopkins) in Jonathan Demme's *The Silence of the Lambs* (1991).

While speculations about the psychological consequence of lacking emotion is a curious intellectual exercise, we should remind ourselves that scientific theories are not only about intuitions and imaginations. The ultimate aim is to find out if there are evidence-based reasons for accepting the basic theoretical claim: that is, whether it is, in fact, the case that the individuals we assess with PCL psychopathy have a fundamental lack of emotions, that they *know the words, but not the music*.

Experimental Testing of the Emotion Theory

The emotion theory lends itself fairly well to scientific experimentation insofar that it makes relatively straightforward claims (i.e., what scientists often call "predictions"), which in turn allows for a similarly direct form of empirical testing. The main hypothesis goes something like this: if we measure emotions in stereotypical psychopathic persons, what we should find (i.e., the hypothetical prediction) is that these measurements are substantially different or muted when compared to ordinary non-psychopathic persons. A strong aspect of this hypothesis is that it clearly identifies not only what evidence would corroborate it (i.e., a lack of emotions) but also what evidence would *falsify* it. Namely, if we find that psychopathic persons' emotional capacities are similar or comparable to that of non-psychopathic persons, this would be incongruent with the explanation promoted by the theory. That is, in scientific parlor we say that such a discovery would falsify the theory; it would be reason to believe that the theory is wrong.

There are various ways scientists have tested the emotion theory; for example, in chapter 4, we surveyed evidence across dozens of experiments that have measured PCL psychopathic persons' ability to identify nonverbal emotional articulations (e.g., the facial expression of shame or fear). One could argue that if psychopathy is explained by a lack of affective sensations—the deprivation of the connotative association with emotion terms—this might lead to problems with recognizing emotions in other people. However, as we saw in chapter 4, no such deficiency has been detected in PCL psychopathic samples, who consistently perform just as well as non-psychopathic persons on these emotion recognition tasks. Of note, some sub-versions of the emotion theory—such as James Blair's popular theory—propose that psychopathy is only related to deficiencies in *some* emotion types, such as lacking the ability to experience fear, and that they have an intact capacity to experience other emotions (e.g., surprise and sadness). However, when my research group conducted our 2024 study where we reviewed all of this evidence,[44] we actually analyzed the data to see if there was support for such hypothesis. We found no such evidence: psychopathic persons' performance across all emotion types did not deviate in any meaningful way.

But perhaps these emotion recognition experiments are just not powerful enough to detect emotional deficiencies? It is at least conceivable that if a person has a deep emotional impairment like the one posited by the emotion theory of psychopathy, they could still learn clever ways to compensate for their incapacities such that they perform sufficiently well on these recognition tests. Maybe similar to how a colorblind person can learn that the middle light in a traffic signal is called *orange* and the top light is *red*, even if they have no connotative association with these color labels. Analogously, if a person has never experienced the emotion of fear, they might still be able to learn what a stereotypical fearful face looks like.

Therefore, perhaps a better and more powerful test of the emotion theory comes from so-called psychophysiological studies, which is a category of experiments that aim to pair behavioral tests with the measure of bodily, physiological processes. In the case of testing the emotion theory of psychopathy, these experiments usually involve a two-part procedure where (1) the participant performs a task that triggers or manipulates some form of emotional reaction in the participant, (2) whereafter the researcher attempts to measure changes in the bodily (physiological) systems associated with emotional states. As mentioned, when human beings are in an emotional state—for instance, when we feel joy or anger—these emotions are believed to be correlated with a number of intricate bodily processes. This includes changes to various anatomical systems such as the central nervous system (e.g., sensory activity), autonomic nervous system (e.g., organ, gland, and smooth muscle activity), and the somatic nervous system (e.g., skeletal muscle activity).[45] Taking advantage of this knowledge, and by deploying the right kind of technology, researchers can then measure and compare the bodily reactions in test participants and thereby make a more credible assessment of whether they are having an emotional reaction. If psychopathic persons have no physiological changes in these systems during an emotion-triggering task, this would corroborate the emotion theory of psychopathy. On the other hand, if there are salient and comparable changes to these physiological systems, this would be evidence against the theory.

There are various technologies that can facilitate this type of physiological measurement, including neuroimaging technology such as MRI (surveyed in chapter 5), as well as equipment that tracks activity like sweat gland secretion (*electrodermal* activity), heartrate (measured via *electrocardiography*), and muscle potentiation (via *electromyography*). In the following two subsections, I will review the research from these types of experiments by dividing them into two categories: those studies that use technology to measure physiological changes to the (1) central nervous system, and those that measure changes to the (2) autonomic nervous system (as of this writing, there is no research that measures changes in the somatic nervous system).

Experiments Measuring Central Nervous System Processes

One of the most common and reliable ways to measure changes in the central nervous system is to use neuroimaging technologies, which can give us an estimated assessment of real-time neurological activity inside the brain (i.e., the main organ in the central nervous system). As surveyed in chapter 5, there are various technologies and methods capable of measuring neurological activity, but the most sophisticated approach is functional MRI (fMRI). A typical fMRI study will collect data from the whole brain, but when testing the emotion theory of psychopathy, the key interest is whether there are different patterns of peak activity in the brain regions known to be functionally implicated in emotion processing.

Traditionally, psychopathy researchers have focused on activity levels deep inside the brain's limbic system—but more specifically in the amygdala region—as this part of the brain is widely believed to play a central role during emotion processes (MRI is one of the few technologies that can generate functional data from deep spatial locations in the brain). Although emotion researchers are still working on mapping the exact role of the amygdala— such as whether it is involved in all emotions or only some—a trove of evidence strongly suggests that there is a reliable increase in MRI signal coming from the amygdala when humans (and other mammalian species for that matter[46]) are exposed to emotional triggers.[47] For this reason, psychopathy researchers have hypothesized that the fMRI signal from the amygdala during a task-based activity should give some indication of the person's underlying emotional state. So, if they detect a comparatively muted resonance signal, this might suggest muted emotional reactions.

Drawing on the elaborate review study by Philip Deming and colleagues from 2022,[48] a total of 40 fMRI studies have measured brain activity in the amygdala region in PCL samples (see also my discussion in chapter 5). Of these studies, at least 36 experiments have used research methods that are designed to trigger some form of emotional reaction in participants (e.g., participants watching emotionally unpleasant images). These 36 experiments included 2,362 participants and yielded 163 effects. A total of 118 effects were nulls (72.39%), indicating no differences in activation patterns in the amygdala region during experimental tasks. Sixteen effects were positive (9.82%), indicating increased activity, and 29 effects were negative (17.79%), indicating relatively lower activity in PCL samples. Of all the 36 experiments, 16 (44.44%) yielded 100% nulls and nine (25.00%) experiments included some positive effects (i.e., relatively increased activity), suggesting that the results from 25 studies (69.44%) are fundamentally incongruent with the hypothesis of muted emotional reactions (i.e., nulls and positive effects go against the theory's main hypothesis). Across all 36 experiments, the proportion of nulls was associated with sample size, where the largest third studies had a null proportion of 81.58% (as opposed to 72.39% across all studies), suggesting a problem with low statistical power and prevalence of false-positive results (as was discussed in chapter 5).

Only three studies found 100% negative effects, making these the sole studies that provided evidence that can be reasonably interpreted as clearly corroborating the emotion theory (i.e., some studies find negative effects, but these might be inconclusive due to the presence of nulls and positive effects as well). The first study is an experiment by Kent Kiehl and colleagues from 2001[49] with a small sample size of 24 participants (but only eight PCL psychopathic persons), which found one negative effect. The second study is an experiment by Niels Birbaumer and colleagues from 2005,[50] also with a small sample size of 20 (10 PCL psychopathic persons), which found one negative effect. However, both of these studies were among the experiments that Philip Deming and colleagues identified as having mislabeled their results, where the fMRI peak coordinates were actually outside the amygdala region (i.e., 0% overlap). The third and final study is by Andrea Glenn and colleagues from 2009,[51] involving a small sample size of 17 (PCL scores not disclosed), which found five negative effects. But again, similar to the previous two studies, this experiment was also highlighted by the Deming group as being mislabeled, where peak coordinates only had a partial overlap (31.50%) with the amygdala region. In short, in the three studies consistently reporting negative results, the effects are dubious, essentially not detected in the amygdala region and therefore of uncertain inferential value.

Now, it should be briefly noted that there are many other brain regions implicated in emotion processes (not just the amygdala), and therefore to get a full overview of the evidence we should survey experimental data from these relevant regions as well. However, there is currently no indication in the review literature that results from these different brain regions are supportive of the emotion theory. For example, the most recent systematic review to provide a general overview of significant *and* null effects across neuroimaging experiments is a study by Stephanie Griffiths and Jarkko Jalava from 2017,[52] which maps the effects from 61 different experiments exploring 13 different brain regions (there are other review studies, but they do not provide an overview of null effects, thus generating a misleading representation of the evidence[53]). In Griffiths and Jalava's study, they demonstrate that there is no consistent pattern in fMRI signals to suggest reduced activity in any other brain regions usually implicated in emotion processing. Specifically, the authors stress that results across all MRI studies have been overly "inconsistent."[54] In addition, I analyzed evidence in chapter 5 from the prefrontal cortex regions, which are also understood to be involved during emotion processing (i.e., executive control of emotional impulses), but fMRI results from these studies are overwhelmingly null findings (84.65%), and significant effects were not only in contradictory directions (increased vs decreased activation) but they were also dispersed across different subregions in the prefrontal cortex.

In sum, there is currently no evidence from the neuroimaging research targeting the amygdala region—or any other relevant region—that appears to suggest clearly muted emotional reactions in PCL psychopathic persons. To the contrary, the results that have emerged from this paradigm are mostly incongruent with the hypothesis derived from the emotion theory.

Experiments Measuring Processes in the Autonomic Nervous System

Before the advent of neuroimaging technologies, psychologists commonly measured emotional processes by tracking changes in the autonomic nervous system, an anatomical network controlling vital bodily functions such as blood circulation (heart organ), respiratory activity (lungs and diaphragm), and temperature regulation (sweat glands) but also reflexes such as sneezing, coughing, frowning, and startle (smooth muscle activity). The rationale behind tracking changes in the autonomic nervous system is that emotions are evidently correlated with activations across this system; for example, when we are frightened, this might manifest in changes to heart rate (e.g., deceleration or acceleration), sweat gland activity (e.g., increased perspiration), and facial smooth muscles (e.g., eye startle).

Although researchers have used all sorts of technologies capable of measuring one or more elements of the autonomic nervous system, psychopathy researchers have mostly used the following three different technologies:

(1) *Electrodermal* technology is used to measure sweat gland activity (i.e., some emotions are correlated with perspiration). The technology consists of placing two electrodes on your hand/fingers to measure the electrical connection between the two points. When there is increased sweat gland activity, perspiration momentarily increases electric conductivity. The two most common readings from electrodermal technology are *phasic* (the relative change in activity across time) and *peak* levels (the highest level of conductivity during experimental task).

(2) *Electrocardiography* is used to measure changes in heart rate activity (i.e., different emotions are correlated with slowing/increasing heartbeat).

Researchers typically use a simple pulse reading instrument, such as strapping an arm- or wristband with a palpitation sensor onto the participant. The most common readings of heart rate activity are *resting state* (the lowest sustained heart rate between tasks), *phasic* (the relative change in heart rate across time and tasks), and *peak* levels (the highest pulse rate during experimental task).

(3) *Electromyography* is used to measure the activation of smooth muscles in the facial region (e.g., some emotions induce tensioning or easing of certain facial smooth muscles). The technology consists of placing electrodes above/under one or both eye sockets, which measure the electrical potentials coming from these muscles during stimulation (i.e., when muscles contract, you can measure an increased electric charge). The most common use of electromyography is to measure the *corrugator activity* (contraction of the large muscle that controls frowning/gaping) and *startle reflex* activity (i.e., contraction of muscles that control eye-blinking).

How exactly these three technologies are used can vary across experimental settings, but they are almost always deployed during a procedure where test participants are introduced to a range of emotion-triggering stimuli, whereafter the technologies generate data about physiological changes during the experiment (much similar to the procedures of an MRI experiment). Sometimes the emotional trigger is just one type of simple stimulus (e.g., the most common being emotion-laden pictures), where other studies are more elaborate and introduce a second form of stimulus such as a blast of loud white noise meant to trigger the startle reflex. The main incentive is to induce an emotional response that is potentially *readable* in the autonomic nervous system processes.

In table 6.1, I list all the peer-reviewed experiments published between 1980 and 2023 that have tested emotional responses in PCL samples using one of these three technologies. The list includes 27 experiments, involving a total of 1,849 participants, with an average sample size of 71 (an additional six studies[55] use these three technologies on PCL samples; however, these studies are not included here either because they did not clearly measure emotional reactions, were based on recycled data, and/or had incomplete reporting).

In the previous two chapters, I summarized the results from experiments by mapping the proportions of statistically significant results (positive and negative) and null findings. In table 6.1, I do something different, instead cataloging whether the experimental effects are indicative of either *supporting* or *disproving* the emotion theory, or whether the results are *inconclusive* (the following categorization is inspired by the common way of interpreting scientific data in relation to theory testing[56]):

- *Support* refers to when the experimental data is largely consistent with the hypothesis that psychopathy is explained by neutral or severely muted emotion processes. When a study supports the emotion theory, it means that most effects are statistically significant in the *negative* direction, signaling unusually muted reaction patterns compared to controls.

- *Disproof* refers to when the experimental data is largely incongruent with the hypothesis. When a study is disproving of the emotion theory, it means that a large majority of effects are either nulls or statistically significant in the *positive* direction, all in all showing a clear, non-muted pattern of physiological reactions to emotional stimulus.

- *Inconclusive* refers to when the experimental effects are so variable that there is no clear pattern to suggest if they are supporting or disproving the hypothesis. When a study is inconclusive, it means that effects make up an eclectic mix of null, positive, and negative effects.

Before I continue, I should first explain why I have decided to use these labels instead of simply displaying each experiment's number of statistically significant and null effects. To be sure, this latter approach would be the preferred way to proceed, because applying labels like *support* vs *disproof* vs *inconclusive* necessarily involves some degree of interpretation (and my personal interpretation might be biased). However, the problem is that the studies listed in table 6.1 have enormous diversity in how the researchers structure the data, how they analyze it, and lastly how and what effects they decide to report. For example, in their reporting of heart rate readings, one study might structure the data so that it represents the total average change in heart rate across the entire experiment, whereas another study might structure the data into blocks of averaged changes between different types of stimuli, and a third experiment may structure the data as the average change within an arbitrary time interval during the experiment. When researchers are dealing with continuous data in this form—where up to three different technologies give a constant flow of data throughout the entire experiment that often lasts between 30 and 60 minutes per participant—there will virtually be an

Table 6.1
Peer-reviewed psychophysiological studies involving PCL psychopathy samples

Study	Total sample	Study quality issues				Study results		
		FWER correct	Control confound	Exclusion criteria	Excluded middle	Skin conduct	Heart rate	Startle Reflex
Raine (1987)	36	No	No	No	No	Disproof	n/a	n/a
Patrick et al. (1989)	48	No	No	No	Yes	Disproof	Disproof	n/a
Ogloff & Wong (1990)	32	Yes	No	No	No	Support	Inconcl.	n/a
Arnett et al. (1993)	63	Yes	Yes	Yes	Yes	Disproof	Disproof	n/a
Patrick et al. (1993)[a]	54	No	No	No	No	Disproof	Disproof	Inconcl.
Blair et al. (1997)[a]	36	No	Yes	No	Yes	Inconcl.	n/a	n/a
Levenston et al. (2000)[a]	36	No	No	Yes	Yes	Disproof	Disproof	Disproof
Herpertz et al. (2001)[a]	66	Yes	Yes	Yes	Yes	Disproof	n/a	Inconcl.
Flor et al. (2002)[a]	21	Yes	No	Yes	No	Inconcl.	Inconcl.	Disproof
Sutton et al. (2002)[a]	172	No	No	Yes	No	Disproof	Disproof	Disproof
Pastor et al. (2003)[a]	48	No	No	Yes	No	Disproof	Disproof	Disproof
Vanman et al. (2003)	90	No	Yes	No	No	n/a	n/a	Disproof
Verona et al. (2004)	68	No	No	No	No	Inconcl.	Inconcl.	Disproof
Serafim et al. (2009)	110	No	Yes	Yes	No	n/a	Inconcl.	n/a
Newman et al. (2010)	125	No	No	Yes	No	n/a	n/a	Disproof
Vaidyanathan et al. (2011)[a]	61	No	No	No	Yes	n/a	n/a	Inconcl.

Study	N							
Baskin-Sommers et al. (2011a)	92	No	No	Yes	No	n/a	n/a	Disproof
Baskin-Sommers et al. (2011b)	87	No	No	No	Yes	n/a	n/a	Disproof
Gao et al. (2012)[a]	138	No	Yes	No	Yes	Disproof	Inconcl.	n/a
Rothemund et al. (2012)[a]	22	Yes	Yes	Yes	No	Support	Disproof	Inconcl.
Sadeh & Verona, (2012)[a]	63	No	No	Yes	No	n/a	n/a	Inconcl.
Anton et al. (2012)[a]	84	No	No	Yes	No	n/a	n/a	Disproof
Baskin-Sommers et al. (2013)	136	No	No	Yes	No	n/a	n/a	Disproof
Veit et al. (2013)	14	No	No	No	No	Inconcl.	n/a	n/a
Verona et al. (2013)[a]	48	No	Yes	No	Yes	n/a	n/a	Disproof
Loomans et al. (2015)[a]	53	Yes	Yes	No	Yes	n/a	n/a	Inconcl.
Degouis et al. (2023)[a]	46	Yes	No	Yes	No	Disproof	Disproof	n/a

Note. n/a = the study did not include this technology/measure. [a] The analysis was based on transformed data (e.g., log or standardized). Studies referenced in the table can be found in the endnotes.[57]

unlimited number of ways in which this continuous data can be structured and eventually analyzed (similar to how there are unlimited ways to cut a layered cake in different three-dimensional pieces). And when, in fact, we find such diversity across studies, it very easily becomes misleading to compare each study in terms of their proportion of significant vs null effects. Why? Because the experiments are not comparable at this very basic level. For instance, one study might have an enormous number of nulls because it structured the data a certain way and performed a specific set of analyses, which another study could also have done but simply did not. What is perhaps even more problematic is that when researchers have an open-ended number of ways to structure and analyze data, it increases the risk of reporting bias where some effects are left *unreported* in the scientific publication; this happens exactly because there is no strict consensus on how to structure and analyze data (unfortunately, a common problem in psychological research[58]).

The overview provided in table 6.1 thus indicates my personal interpretation of the effect *clusters* from each type of physiological measure, and how these results can be described as supportive, disproving, or inconclusive of the emotion theory (i.e., I call it an effect cluster because these physiological measures usually yield several and sometimes dozens of effects). Across these studies, there are 47 clusters of effects from skin conductance, heart rate, and startle reflex activity. I interpreted 30 (63.83%) of these clusters as indicating evidence disproving the emotion theory, 2 clusters (4.26%) were supportive, and 15 clusters (31.91%) were inconclusive. No study found evidence that was exclusively supportive of the emotion theory, where all studies that reported supportive evidence also reported at least one cluster of inconclusive or disproving effects.

The two studies that found supportive evidence were an experiment by James Ogloff and Stephen Wong from 1990[59] and an experiment by Yvonne Rothemund and colleagues from 2012.[60] The study by James Ogloff and Stephen Wong compared data between two small groups of offenders, one group of 16 people with PCL scores > 22 and one group of 16 people with PCL scores ≤ 22 (i.e., the researchers used the first non-revised version of the PCL scale). Test participants completed two different stimulus tasks, one where participants were subjected to a loud, highly unpleasant tone after a nine-point countdown, and a second task where participants could push a red button at the end of a countdown to prevent the onset of the unpleasant tone. The idea of this task is that the stimulus induces some form of anxious emotional response in participants as the countdown closes in on zero. During both tasks, the researchers monitored participants' skin conductance and heart rate activity. Results suggested that the high-scoring PCL group endured virtually no change in skin conductance measures, even as the countdown reached the end (in both tasks). However, the heart rate data were inconclusive. There was a statistically significant difference where the high-scoring PCL group presented lower activity across time intervals, but there was no group difference in heart rate resting state, and both groups show a clear spike or large increase in heart rate at the last countdown, indicating emotional intensification (an important caveat to this study is that they used an unconventional PCL cut-off score when grouping participants, which might be interpreted as introducing bias into the study).

The 2012 study by Yvonne Rothemund and colleagues used a different and more intricate experimental design referred to as a *classical* or *Pavlovian conditioning* study (inspired by Ivan Pavlov's classic study where he conditioned a dog to salivate when hearing the tone of a bell[61]). This type of experiment aims to test how efficiently a person learns that a preceding event (e.g., a picture of a neutral face) is a reliable cue of a subsequent unpleasant event (e.g., a burst of white noise). The experiment begins with presenting different stimuli, but when the cue (e.g., neutral face) is shown, it is followed by the unpleasant stimulus (e.g., noise). Early on in the experiment, participants will merely react to the unpleasant stimulus as a novelty, not noticing that there is a connection between the neutral face and the unpleasant probe. However, after a number of trials, participants will start observing or learn that the neutral face is a reliable cue of something unpleasant and should therefore start reacting to the cue as if it is something unpleasant (presumably in anticipation of the soon-to-come unpleasant stimulus). In their experiment, Rothemund and colleagues compared the responses from a very small sample consisting of a group of 11 PCL psychopathic persons (PCL-R mean 24.7) to a group of 11 healthy controls (no PCL score), and they found that the PCL group had largely unchanged skin conductance responses across all experimental conditions, consistent with the emotion theory. The electromyographic effects were inconclusive as there were no between-group differences in corrugator activity (frowning) to the noise probe, though data indicated a statistically significant yet small difference in startle reflex magnitude. There was no group difference in heart rate activity (an important caveat to this study is that the sample was nonrandomly selected, where the researchers hand-picked participants from a sample that was prescreened for psychopathy, selecting individuals for the PCL group that were known to have a factor 1 score > 8).

In sum, there is virtually no evidence in this research paradigm that is clearly consistent with the emotion theory of psychopathy, and most evidence (63.83%) suggests that the theory has been disproven. However, the prevalence of inconclusive results (31.91%) indicates that perhaps it is still too early to render the emotion theory falsified. Such an interpretation is not completely unreasonable, but there are various issues that weigh strongly against it.

For one, consider that only six (26.09%) out of the 27 studies corrected for the family-wise error rate (FWER, see chapter 4), suggesting that there is potentially a high prevalence of false-positive effects, given that many studies make dozens of post hoc analyses.

Secondly, only 10 studies (37.04%) tested and controlled for confounding variables. In studies like these, there is a high probability that statistically significant effects between groups are driven by extraneous factors such as comorbid psychiatric diagnoses, substance misuse, and so on. Why? Because we know that autonomic responses are affected by many different factors (e.g., some psychiatric medicines work as depressants, slowing the nervous system). And even among the studies that controlled for such confounding variables, almost none of them controlled for variables that are known to impair the autonomic nervous system. For example, only two studies controlled for anxiety and one study controlled for medicine usage; the rest of the studies controlled exclusively for age, intelligence (as measured by IQ test), education, gender, and/or race.

Thirdly, only 14 studies (51.85%) deployed theoretically informed exclusion criteria, where individuals are excluded from the study if it is suspected that their autonomic nervous system is compromised (e.g., psychotropic medicine, substance misuse, head trauma, and so on).

Fourth, a total of 10 studies (37.04%) tinkered with the group compositions by excluding the middle—that is, removing and/or deliberately not comparing performance from individuals who score in the middle range on the PCL assessment (see chapter 2 and 4). If these quality issues had been addressed by the researchers, it is entirely reasonable to expect that results would have been different—yielding many more null effects—and therefore not indicating inconclusive effects but instead disproving effects.

Lastly, a total of 15 studies (55.56%) were fully or partially based on what researchers refer to as "transformed" data, which is when scientists adjust their raw data in order to make it more suitable for certain types of analyses (e.g., a log-transformation[62] or standardized transformation[63]). It is not necessarily problematic to transform raw data for analytic purposes, but the practice of transforming psychological research data has in recent years been criticized as it can impact the calculation of p-values (the conventional metric for distinguishing between a null and statistically significant effect, see chapter 4).[64] In some studies in table 6.1, for instance, researchers reported that the raw data yielded 100% null effects, whereafter they still proceeded to analyze the transformed data, which then yielded several significant effects. In other studies, the researchers only report results from the transformed data, failing to disclose results on the raw data. And the decisions to transform the data were often insufficiently justified. Overall, this aspect of the research further detracts from its inferential value.

In my perspective, a much simpler way to interpret the prevalence of inconclusive results is that it is entirely unsurprising: it is simply what we should expect to find in a research paradigm like this. When researchers deploy methodologies with inbuilt analytic flexibilities—as the ones I have pointed out here, often referred to as *questionable research practices*[65]—and they use these flexible methods to study a subject matter where we know there is *some* variation in the general population (e.g., humans generally have different skin conductance, heart rate, and startle patterns), the analytic flexibilities will almost guarantee that some significant effects can be "beaten out" of the data. But let there be no mistake, when data are being carved up into an arbitrary tapestry of measurements, and then scrutinized in an open-ended number of ways, the effects that emerge as "statistically significant" are likely going to contain a substantial proportion of false positives. This is not a controversial perspective, but a widely accepted explanation of how psychological research (and other sciences for that matter[66]) often finds significant effects when we have good reasons to suspect that they are just random noise.[67]

So, perhaps a more fitting way to look at the prevalence of inconclusive results is that even under these formidable experimental and analytic flexibilities, researchers still predominately found null effects, generating results that are largely inconsistent with the emotion theory of psychopathy. With this in mind, it is curious to notice that since psychologists started paying attention to the problem of analytic flexibilities or questionable research practices, as they are usually called—a change that began to take shape around or after 2010[68]—the research on autonomic nervous system responses in PCL samples has stagnated. As indicated in table 6.1, while the most recent emotion-stimulus study on PCL psychopathy was published in 2023, this study was the first in almost a decade, suggesting that the research paradigm has effectively come to an end. A qualified guess why we are no longer seeing these studies appear with high frequency in academic journals is that when researchers are required to apply more rigorous analytic methods—a prerequisite that is increasingly enforced by journals and their reviewers—a probable outcome is that the experimental effects are overwhelmingly nulls. This would not only disincentivize researchers from submitting their work to a journal, but it would eventually disincentivize researchers from conducting these types of experiments in the first place.

The Cognitive Theory of Psychopathy: Basic Accounts and Experimental Work

Where the emotion theory of psychopathy can be traced back to some of the first scientific contributions in the field during the eighteenth and nineteenth centuries, the cognitive theory is an entirely contemporary addition. The first version of the cognitive theory was proposed by Ethan Gorenstein and Joseph Newman in 1980 (they called it the *response modulation hypothesis*),[69] and during the 1990s it swiftly gained traction in the research community as a novel explanation of psychopathy.[70] Since then it has gone through a couple of revisions and is broadly considered the only true rival to the emotion theory (although, as mentioned, some contemporary scientists see both theories as a dual-deficit explanation of psychopathy). Historically, what made the cognitive theory appealing to researchers was the realization that the emotion theory of psychopathy had clear shortcomings, namely the experimental discovery that PCL psychopathic persons have ordinary emotional reactions (as covered in the previous section). So, the cognitive theory emerged partly as an explanation of psychopathy that did not implicate a lack of emotion per se but rather involved cognitive deficits.[71] The basic tenet of the cognitive theory is that psychopathy can be explained as a deficit in attention and information processing, causing major problems with impulse control and behavioral modulation.

A Summary of the Theoretical Framework

A crucial aspect of the cognitive theory is the observation that human executive behaviors—that is, voluntary behaviors that arise from thinking and deliberation—are importantly linked to our ability to orient ourselves and process information about our immediate environment. On a daily basis, we make hundreds if not thousands of small executive decisions, where these are balanced according to our own personal needs and desires as well as the limitations and possibilities afforded by the world we inhabit. We get up early not because we necessarily have an innate desire to do so, but rather because of various external reasons—for instance, that the organization we work for demands that we arrive by a certain hour combined with the fact that we want to keep our job. Similarly, a person might go to a pub because they want to have fun with friends but also because they believe that friendships require some minimal social maintenance. And we choose to go to college not for its own sake but because we want to take up a specific occupation that requires a certain academic degree. Whenever we make decisions like these, it is complemented by an ongoing cognitive processing of a complex flow of information balanced with the interrogation of our own needs and desires, all upon which we eventually make and/or revise our decision-making.

Proponents of the cognitive theory of psychopathy believe that what makes psychopathic individuals different from ordinary people is that they have severe impairment in how they cognitively pay attention to and process this sort of complex information that precedes ordinary decision-making. More specifically, their cognitive problem can be reduced to an impairment in modulating goal-oriented behavior, a deficit in what researchers call *response modulation*. It is trivial to observe that people are ordinarily open to changing, adapting, or even canceling their behavior in accordance with how they scan and analyze their immediate environment. Many of our daily behaviors can be viewed as a process that is first initiated by an impulse (sometimes referred to as a *dominant response*), which may be modulated or canceled due to the subsequent processing of extraneous sensory information (sometimes referred to as a *modulated response*).

For example, consider a person walking down the street on a sunny day. As they are feeling increasingly warm, they decide to get an ice cream. We can think of this as a dominant response to the environmental affordances in the form of a goal-directed behavior: get ice cream. The person then walks into a café to quench their craving. However, as the person is about to walk up to the counter, they observe in their peripheral vision that there is a line of people inside the café and therefore that they are indeed about to *skip the line*. This processing of information triggers a disruption or cancellation of their initial goal-directed behavior to go straight to the counter, and instead they walk over to stand in line like everyone else. We can think of this moment as a response modulation, where a primary and dominant response (get ice cream) is canceled due to attention being automatically directed onto some extraneous or secondary information (the line of people). The primary goal-directed impulse is in this way modulated by the processing of peripheral secondary information, causing a clear change in the behavioral pattern.

According to the cognitive theory, psychopathy is thus best explained by an impaired response modulation—in other words, a deficit in the way human attention is ordinarily (and often automatic and effortlessly) changed from a focus on primary to secondary information cues. The outcome of such a deficit is that the person will almost always behave in accordance with the primary response or impulse, thus representing a poorly regulated and largely disinhibited behavior style. Needless to say, for this sort of impairment to explain the traits that are commonly associated with psychopathy—extreme carelessness, impulsivity, irresponsibility, and so on—the impairment would have to be quite substantial and therefore easily detectable. The main architects behind the theory, Mark Patterson and Joseph Newman argue in a 1993 paper that the impairments must be so severe that they essentially result in a failure "to use available information to anticipate risk," in a way that the deficit causes an unusual form of "curtailed reflectivity."[72]

What exactly is theorized to be the biological root cause of this sort of psychological deficit is still an open question. One proposal, from Arielle Baskin-Sommers and colleagues,[73] is that the deficit springs from an information processing *bottleneck* at the neurological level—hypothetically implicating a dysfunction in the prefrontal cortex—leading to a diminished capacity to process multiple sources of information during goal-directed behavior. Another neurological proposal, from Caroline Moul and colleagues,[74] suggests that the deficit is rooted in an impaired connectivity within the amygdala structure where signals caused by environmental triggers (e.g., the line you are about to skip) are not properly processed and thus never reach the level of cognitive awareness.[75]

The cognitive theory of psychopathy has garnered much support among mainstream researchers, often highlighted for how it can both account for the impulsive and disinhibited behaviors as well as the broader antisocial attitudes and personality characteristics conceptually associated with psychopathy. In the narrow scheme of things, the cognitive theory explains psychopathy as a basic inability to self-regulate due to an underlying attention deficit. However, proponents of the theory typically speculate that such an inability will have widespread consequences for the development of prosocial skills and attitudes. If a person has poor self-regulation, this will inevitably lead to fewer opportunities where the person can learn from their past mistakes—realize that they have done something wrong—simply because this form of perspective-taking is rarely formed at the cognitive level. As the person goes through their formative years in adolescence, this period will on average include fewer incidents where they are afforded with an opportunity to foster prosocial attitudes. The end product is a psychology that is fundamentally unaffected by the social, moral, and legal norms that generally shape or intervene with people's behavioral impulses. So, when clinicians report that psychopathic persons are callous and unemotional, according to the cognitive theory, this is not because they have an inability to process emotions. It is just that the social and moral issues that ordinarily trigger people's emotions are not interpreted by psychopathic persons as something that is emotionally salient, or that they simply fail to pay attention to salient information that would have otherwise triggered their emotions. In either case, they will come across as if they are carelessly disinhibited and deprived of any emotional concerns.

However, whether the cognitive theory is scientifically justified is not a question about its intuitive appeal and how well we think it fits with the conceptual descriptions of a stereotypical psychopathic person. Like with the emotion theory, the fundamental question is whether we have any evidence-based reasons for accepting it as a good explanation of psychopathy: whether the people we assess with PCL psychopathy also truly have severe attention deficits that manifest in obvious problems with impulse control and response modulation.

Experimental Testing of the Cognitive Theory: Evidence from Behavioral Studies

The cognitive theory explains psychopathy as a deficit of attention, which allegedly causes grave problems with response modulation and impulse control. From this explanation, we can infer that the main hypothesis entails the following prediction: If we subject PCL psychopathic persons to an experiment where we measure response modulation and impulse control, these types of behaviors will be unusual and greatly impaired compared to non-psychopathic persons. Similar to the emotion theory, we can also identify aspects that would readily falsify the cognitive theory, namely if we find that PCL psychopathic persons have equal or better response modulation performance compared to non-psychopathic persons.

The experiments that test for response modulation typically involve a behavioral task where participants are tested in their ability to shift attention from primary and obvious information cues to secondary information cues, where a successful shifting of attention results in some form of measurable behavioral change in the test participant. Two of the most popular experiments are (a) passive avoidance task and (b) go/no-go task discrimination.

Passive avoidance task: This type of experiment aims to test how quickly or effectively a person *learns* to modulate a dominant response when this behavior is peripherally associated with aversive consequences (i.e., the secondary information). For example, one version of this test has participants engage in a simple card gambling game where they do not know the rules in advance but must somehow learn how the game works by picking up on secondary information cues. First, participants are presented with two different types of cards. Second, to initiate the gamble, they must decide which card to pick/turn. The participants are unaware that the one type of card is associated with a monetary reward and the second type of card is a loss. Initially, the participants are puzzled as there appears to be no clear pattern to when they are rewarded vs punished, but eventually, they will *passively* learn the pattern and thus change their gambling behavior (i.e., avoid turning the types of cards that involve losing money). Sometimes, a study might enhance the primary response by running a couple of test trials where there is no clear pattern and participants are constantly winning the gamble, presumably making participants excited about the potential monetary gains from playing. In either case, the hypothesis is the same: if PCL psychopathy is associated with an attention deficit, this should be manifest in a substantially slower passive avoidance learning. Since psychopathic persons are so focused on winning, they are presumably overlooking cues when losing, thus on average needing more trials to shift attention and identify the reward-punishment pattern.

Go/no-go task: This experiment aims to test participants' ability to make speedy responses to primary information as well as swiftly modulating their dominant responses in the presence of peripheral, secondary information. For example, a popular version of this test has participants view a sequence of symbols displayed on a screen only a few seconds apart. Participants are instructed to press a button every time a specific target symbol occurs and avoid pressing the button when non-target symbols appear. During the trial, they are monetarily rewarded if they click the button when the target symbol is showing and punished when they click on non-target symbols. In this task, participants are allegedly locking in on the dominant response of identifying and clicking the button to gain a reward when the target symbol appears, a task that can be challenging as the symbols only show for a few seconds. But their performance also depends on the modulated response, which would be to avoid clicking the button when non-target symbols pop up on the screen (i.e., to avoid losing money). As with the passive avoidance task, the go/no-go test might be frontloaded with a test trial that aims to enhance the primary response by exposing participants to a disproportionate number of wins. But the hypothesis remains the same: if PCL psychopathic persons have an attention deficit and are then more profoundly locked in on the primary response, this should show in the data of them making a substantially higher number of errors where they push the button for the wrong symbols and then win less money during the trial.

Since the first outline of the cognitive theory by Ethan Gorenstein and Joseph Newman was published in 1980, more than 60 independent studies have tested PCL psychopathic persons' performance in these types of attention or response modulation experiments, and we therefore also have access to detailed systematic reviews and meta-analyses. The first review study was published in 1993 by Mark Patterson and Joseph Newman,[76] and the two latest review studies are from 2022 by Steven Gillespie and colleagues[77] and 2024 by Matthias Burghart and colleagues.[78] Though, the largest and most comprehensive review study is a meta-analysis by Sarah Smith and Scott Lilienfeld from 2015,[79] which synthesized the evidence from 56 experiments based on PCL samples, involving a total of 4,865 participants with an average sample size of 86 per study (their review also included studies involving non-PCL samples, but I have excluded these from the analysis here). Since the publication of Smith and Lilienfeld's meta-analysis in 2015, six additional studies have been published,[80] but since results from these studies are much similar to earlier findings, it is safe to assume the 2015 meta-analytic conclusions still hold. Here are three key takeaways from Smith and Lilienfeld's study, which I believe neatly sum up the evidence from this paradigm.

First, across the 56 experiments, the overall effect size is small, with an average r-value of $r = 0.19$. This means that even as there is a statistically significant effect, the correlation between PCL scores and test performance is so weak that it is debatable if the effect is indicating any *practically relevant* difference (Pearson's r-value is explained in chapter 4). One conventional way for psychologists to interpret an r-value is to think of it as expressing an explanatory relationship, here between test performance and PCL scores, which can be estimated by squaring the r-value (i.e., r^2).[81] Using this very basic heuristic, it would mean that the variance we find in *test performance* can only account for a dismal 3.61% of the variance in *PCL scores*. In other words, a person's test score is not going to reveal much about the person's PCL assessment score.

Second, the Smith-Lilienfeld review found that there was great heterogeneity across experiments, where effect sizes span from negative to positive r-values (i.e., from negative $r = -0.46$ to positive $r = 0.60$). That is, some experiments found that a PCL score was moderately associated with *increased* or *better* response modulation, where other studies found that it was moderately associated with *decreased* performance. In total, eight studies (14%) found that PCL psychopathic persons had better performance compared to controls. Such experimental results are clearly inconsistent with the generalizable claim that psychopathy is explained by attention and response modulation deficits.

Third, the meta-analysis found evidence of publication bias, where studies with larger effect sizes were more likely to be published than experiments with smaller effect sizes. The way the researchers investigated this was by asking their colleagues to share studies that they have not published and then compared them to the published literature (i.e., scientific journals sometimes prioritize publishing studies with large effects and may reject studies with nulls and/or small effects). After making this comparison, the researchers used a conventional method—called a *funnel plot*[82]—to generate a (very) rough estimate of what the average meta-analytic results would look like if such publication bias did not exist in the literature (i.e., if we assume that all conducted research in the field was submitted and published in academic journals, regardless of the nature of their results). By applying these methods, Smith and Lilienfeld found that the average effect size might drop with five percentage points, in this case to an effect size of approximately $r = 0.14$, thus nearing a negligible level of magnitude (i.e., an explanatory relationship of less than 1.96%). While there are many uncertainties involved in this type of bias analysis, it subtracts significantly from how confident we should be about the original effect size of $r = 0.19$ (which is still very small).

In sum, while the Smith-Lilienfeld study shows that there is evidence of some differences in response modulation in PCL samples, it is similarly evident that the magnitude of these differences is not consistent with what is predicted by the cognitive theory of psychopathy. Indeed, the weak correlations suggest that most PCL psychopathic persons in these experiments will have similar performance compared to non-psychopathic samples. For example, with an r-value of 0.19, it means that around 85% of all participants had similar task performances. It is largely for this reason that Smith and Lilienfeld concluded their review by stating that the cognitive theory does *not* appear to "provide a comprehensive account of the deficits of psychopathy."[83] Unsurprisingly, the 2024 review by Matthias Burghart and colleagues drew similar conclusions, stating that "psychopathy is *not* characterized by severe global EF [executive function] dysfunction and that EF impairments are *not* a key feature of this disorder" (original italics).[84]

While these meta-analytic results largely speak for themselves, I will make two additional observations that (to my knowledge) so far remain scarcely addressed in the literature.

The first observation is that effects from attention studies might not generalize onto real-life situations in the way that researchers often assume. Some researchers argue that if there is a small effect in experimental settings, this effect will also be present in any real-life situation. But I suspect this assumption is too hasty. Studies involving passive avoidance learning and go/no-go tasks generally test participants' ability to switch their attention when exposed to *subtle* secondary information in artificial low-stakes situations (e.g., passively noticing cues in trivial games where you can earn a few bucks). We can think of these cues as dispositional *triggers*, as a type of sensual impression that, if noticed, causes a response modulation. This is much like how a spider, if you see one on your bed pillow, might trigger your disposition to feel disgusted. However, what is often overlooked when generalizing findings from attention studies onto real-life situations is that the trigger conditions in real-life situations are almost always going to be much *stronger* compared to the ones in experimental conditions. But the way most human dispositions work is that their realization is proportionately related to the strength of the trigger. For example, if your response modulation is triggered by subtle information cues, it follows that it will be triggered much more reliably by stronger cues. This is roughly similar to how a person might be frightened by the sight of a small spider but will almost always be frightened by the sight of a large spider. So, where there might be a tiny difference in response modulation with subtle cues (i.e., the experimental situation), there might be no difference at all when comparing behavioral modulation with stronger cues (i.e., in real-life situations). The implication for the cognitive theory should be obvious: it posits that the attention deficit can explain a failure to engage in response modulation in morally and legally salient situations—that is, situations where the secondary information cues are undoubtedly much stronger. But if there are only small differences between groups in response modulation under subtle information cues, it would be irrational to think that these differences would remain the same or become larger under stronger attention triggers.

The second observation I would like to make is that even under the assumption that the meta-analytic results of the Smith-Lilienfeld study are reliable and generalize well onto real-life situations, it is still possible to question whether a small/weak effect size signals something that is specific to psychopathy and therefore capable of *explaining* the psychological traits we conceptually associate with it. It turns out that response modulation has been tested in many different non-psychopathic samples—mostly because this psychological function has been theorized to be implicated in other mental health diagnoses[85]—and in these studies it is common to discover small effect sizes across different samples. For instance, similar weak correlations have been detected with factors like IQ test score[86] and broadly defined psychiatric diagnoses such as attention-deficit/hyperactivity disorder[87] and internet gaming disorder.[88] However, since we do not conceptually associate low IQ and these other diagnostic labels with response modulation difficulties in morally and legally salient situations, it would be disanalogous to propose that all of a sudden this slight attention deficit can explain these key features in psychopathic persons. If a difficulty with response modulation is a common observation across much, much larger population groups defined by factors as broad as IQ and attention-deficit/hyperactivity disorder, it would be more reasonable to assume that the observed small effect in PCL samples is because these samples contain individuals with low IQ and comorbid disorders.

Concluding Remarks: Are Psychopathy Theories Falsified?

As mentioned in the beginning of the chapter, forensic practitioners use the PCL scales under the assumption that they offer a reliable way to measure psychopathy in justice-involved persons. And a fundamental justification behind these practices is that psychopathy is believed to be scientifically vetted—that is, that researchers have evidence-based explanations of why PCL psychopathic persons are different from non-psychopathic individuals. However, the evidence I have surveyed in this chapter heavily questions whether this basic criterion has been met, where it seems that the explanations researchers give are poorly supported by the evidence. Here, I will briefly argue that not only are the two mainstream explanations of psychopathy questionable, but more profoundly, the current evidence suggests that they are wrong or *falsified* explanations. If this argument is sound, it follows that when forensic practitioners are assessing individuals with the PCL scales, they do not have a good explanation of what they are doing, so to speak, aside from the fact that they are labeling and grouping people based on a list of arbitrary characteristics that make up the PCL item-checklist.

One of the most basic ground rules scientists follow when discussing the validity of scientific theories is the so-called *principle of falsifiability*. It is the idea that our theories must necessarily be formalized as explanations that are potentially defeasible, that they make predictions that can be proven wrong. When outlining scientific theories, researchers are encouraged to think carefully about what evidence will indeed be incongruent with their theory, thus expressing a readiness to accept that their theories might turn out to be wrong. Any experiment can therefore be seen as a potential step toward disproving their theory. The more incongruent evidence researchers discover, the more evidence they have that the theory is falsified, and the more irrational it would be to accept it as a viable explanation. Falsified theories are, by definition, poor explanations. They are imperfect attempts to account for the *why* of the *what*.[89]

What exact evidence (or amount of evidence) it takes for a theory to reach the critical threshold where scientists begin to think of the theory as falsified will almost always be a matter of scientific debate.[90] However, I think that a compelling argument can be made that both the emotion and cognitive theory of psychopathy have been falsified.

Why the Emotion Theory Is Falsified

A case against the emotion theory is relatively clear due to the overwhelming predominance of null effects across the numerous and diverse studies that measure neurological and autonomic responses to emotional stimuli. When exposed to emotional triggers, PCL psychopathic persons appear to respond in comparable ways to non-psychopathic controls, a pattern that is fundamentally incongruent with the hypothesis that psychopathy is explained by a lack of emotion or in any way having severely muted emotional dispositions. There were a few studies (in table 6.1) that found inconclusive evidence, but this can be accounted for by simply pointing out problems with their experimental design and analytic flexibility (i.e., questionable research practice). To argue that emotion theories have been falsified is therefore not unreasonable, but, more interestingly, nor does it run against what some psychopathy researchers have been arguing with increasing frequency during the past decade.

For example, one of the leading reasons researchers prefer the competing alternative to the emotion theory—namely the cognitive theory—is exactly because they find the emotion theory misleading due to its lack of evidential support.[91] As David Kosson and colleagues stated in a 2016 article, the "simple and widespread assumption that psychopaths are relatively emotionless is contradicted by the complexity of current research."[92] But there are also signs coming from within the research paradigm itself, including the curious observation that experimentation has virtually come to a full stop (e.g., only two classic stimulus studies have been published in the last decade; see table 6.1). And as argued by the lauded psychologist, Paul Meehl, psychological theories are rarely openly declared falsified; what often happens is that "experimenters lose interest" in a theory, and like old military generals, "they never die, they just slowly fade away."[93] The research paradigm testing the emotion theory is thus tainted by two heavy realities: the extensive prevalence of null effects combined with dwindling research interest. To me, this strongly suggests that there is no viable way forward, that the theory has been thoroughly falsified.

Why the Cognitive Theory Is Falsified

A case against the cognitive theory may start by pointing out that the theory itself hypothesizes that psychopathic persons have grave and deep problems with attention and response modulation. But since no such large attention deficits have ever been discovered in any experiment using PCL samples—the largest effect ever is just a moderate correlation—it should be uncontroversial to suggest that the theory has never amassed any substantial evidence to even tentatively corroborate it. Another way to look at this is that the small effect sizes that have emerged during decades of experimentation simply show that PCL psychopathic and non-psychopathic persons have similar response modulation capacity, an observation that is incongruent with the main claim of the cognitive theory.

For instance, a weak effect size of $r = 0.19$ suggests that whatever variance we find in response modulation test scores only accounts for 3.61% of the variance we find in PCL scores. To put this in perspective, consider that a PCL-R score ranges from 0 to 40 points. This means that response modulation test scores can only account for variance in the magnitude of 1.44 PCL-R scale points. These minute and barely measurable differences in attention tests simply cannot explain the traits that are conceptually associated with PCL psychopathy. And therefore, it is far more irrational to keep insisting that the theory is a potential explanation than it is to just reject it as a falsified explanation. Some researchers might argue that this would be analogous to the proverbial *throwing the baby out with the bathwater*, but the point is that at this stage, there is no *baby*; researchers have never detected any large attention differences, despite having steadily tried for the better part of three to four decades.

A final observation counting against the cognitive theory is that not only have researchers never found any large attention differences in PCL samples but they have failed to detect such differences under very favorable experimental conditions: testing attention and response modulation to subtle cues in trivial low-stakes scenarios. If PCL psychopathy is not associated with large differences under these favorable conditions, there will be little reason to think that there would be large differences in attention during real-life situations with proper moral and legal salience.

All in all, these observations can reasonably be interpreted as suggesting that the cognitive theory of psychopathy has been falsified.

7 A Dangerous Diagnosis: Why Psychopathy Assessments Are Ethically Problematic

Psychologists strive to benefit those with whom they work and take care to do no harm. In their professional actions, psychologists seek to safeguard the welfare and rights of those with whom they interact professionally and other affected persons . . . Because psychologists' scientific and professional judgments and actions may affect the lives of others, they are alert to and guard against personal, financial, social, organizational, or political factors that might lead to misuse of their influence.

—the American Psychological Association's *Ethical Principles* (2017)[1]

Both research literature and legal records suggest that PCL psychopathy assessments became an integrated part of mainstream forensic practices sometime during the 1990s. The PCL scales were promoted on the simple yet powerful promise that they could assist the legal system in identifying offenders with psychopathy, a largely mysterious but socially devastating personality disorder. The basic rationale for using the PCL scales is that psychopathic persons are dangerous social predators, responsible for most of the violent crime that plagues society. So, if our legal system becomes better at identifying these few individuals, and deals with them accordingly, all the better for the rest of society. During the 2000s, the PCL scales eventually became one of the most utilized assessment tools by forensic psychiatrists and psychologists, and they have since remained a central part of the information flow that supports decision-making processes in the legal system.

A central theme throughout this book has been the growing realization in recent years that there is little evidence to support this mainstream narrative. The individuals who are assessed with PCL psychopathy in the criminal justice system bear virtually no resemblance to the ordinary descriptions of them being vile and incorrigible social predators. Of course, this is not to say that there are no predatory people in our societies. But whoever these predators are, the research we have surveyed gives us little reason to think that they can be reliably identified using the PCL scales. Individuals who score high on the PCL scales are not the extraordinarily dangerous, chronically untreatable, and morally colorblind individuals that many researchers have consistently argued. Nor is there any compelling evidence to suggest that researchers have shown what makes them biologically and psychologically different from non-psychopathic persons.

The case I have been building over the preceding chapters thus challenges the rationale behind using PCL assessments to inform forensic practices. Recall that when a person is assessed or diagnosed with psychopathy, this information can have serious implications for that person's journey through the legal system. This includes (among many things) the person being viewed as "high risk," which may impact decisions about sentencing, placement, rehabilitation, parole, probation, and so on (see chapter 1). As I have shown, a primary concern about this practice is its lack of evidential justification. When legal decisions are informed by clinical expertise and knowledge, such guidance must be crucially anchored in the best available evidence, an approach also known as evidence-based practices.[2] It is not enough to simply state that psychopathic persons are dangerous social predators, and thus deal with them *as if* this is true. Instead, we demand of our legal institutions that whenever such decisions are made, it must be genuinely informed by credible evidence such as scientific research.

In acknowledging this disconnect between the scientific evidence and forensic practices, a small but ostensibly growing group of scholars is beginning to raise concerns over the role that psychopathy assessments currently play in the legal system. A recent example of this is a 2020 publication of a "statement of concerned experts" in the prominent academic journal *Psychology, Public Policy, and Law* by a group of 12 senior researchers and practitioners, led by forensic psychologist David DeMatteo. In this article, the authors voice strong reservations about the way the PCL-R is currently used to inform capital sentencing in the United States.[3] When a court is deciding whether an offender should be given a capital sentence (as opposed to life without parole), a decisive issue is whether the person is considered a chronic and imminent danger to fellow inmates and correctional staff. For the PCL-R to be informative, then, it must be shown that those who score high on the PCL-R also exert substantially higher risks of such institutional infractions. However, as DeMatteo and colleagues note (and as we saw in chapter 2), there is no compelling or decisive evidence to suggest that PCL psychopathic persons are linked to these unique risks. The authors therefore recommend that forensic practitioners should stop using the PCL-R to inform capital decisions (an opinion that ignited a considerable debate in the journal[4]).

Other examples of experts voicing reservation over the use of the psychopathy assessments are the numerous criticisms over how PCL psychopathic persons are sometimes excluded from rehabilitation programs, based on the belief that such individuals cannot benefit from these initiatives.[5] This debate was historically led by researchers like Jennifer Skeem, John Monahan, Randall Salekin, and others,[6] who have rightly pointed out that these pessimistic beliefs about rehabilitation have always lacked empirical substance. In addition, a handful of researchers and practitioners, including lauded scholars like Thomas Grisso, John Edens, and Jennifer Skeem,[7] have opposed the way PCL assessments have been used to inform the practice of transferring juvenile offenders to adult courts, based on the misconception that juvenile psychopathic persons are incapable of rehabilitation. Lastly, a growing number of researchers have flagged a broader concern about openly describing a defendant as "psychopathic" during court procedures, as surveys and mock trial experiments suggest that jurors and judges hold prejudicial misconceptions about psychopathy that may impact their judgments about culpability and sentencing.[8] Altogether, these are not the only criticisms of psychopathy assessments,[9] but they appear to be the themes that have attracted the most vocal discussion in recent years.

What all these critical voices have in common is that they are pointing out how current forensic practices fail to stay informed about and be guided by the best empirical research. As such, the criticisms relate to how the commitment to evidence-based practices is effectively downplayed, misconceived, or flat-out ignored, and sometimes even replaced by prejudicial misconceptions. This form of criticism is both serious and constructive, and at its core is an effort to set the record straight, to ensure that people assessed with PCL psychopathy are not treated unfairly in the legal system. Importantly, these criticisms are not saying that psychopathy assessments are *inherently* problematic. It is just that the practice itself is devoid of robust evidence that is needed to justify it.[10]

In this chapter, I shall add to these existing criticisms by arguing that the current use of PCL psychopathy assessments is much more problematic than what has so far been acknowledged and discussed in the mainstream literature. While scholars are right to point out that the empirical justification for using PCL assessments is fundamentally misconceived, there has so far been little conversation about whether it is at all appropriate for a forensic mental health practitioner—such as a psychiatrist or psychologist—to use their skills and expertise to conduct psychopathy assessments. In addressing this more basic question, I will argue that when forensic mental health professionals use the PCL scales (or any other type of psychopathy assessment), such practice will often be in direct violation of the *ethical codes* that guide their profession.

More specifically, I will argue that PCL psychopathy assessments may violate four core ethical commitments endorsed by forensic mental health professions, namely: *the principle of beneficence, the principle of nonmaleficence, the principle of respect for autonomy,* and *the principle of justice.* These ethical principles are the foundation of forensic ethical code of conduct, and they assert that a practitioner ought to (1) give primacy to the interest and well-being of patients, (2) under no circumstances inflict undue harm and/or compromise the well-being of patients, (3) refrain from acting against the will of patients, and (4) structure their practices in a way that is fair and equal to all patients. As soon as a practitioner conducts a PCL psychopathy assessment in a legal context, I shall argue, these principles will often be effectively violated. As a result, the only straightforward and guaranteed resolution to this ethical conundrum is for the field to consider an all-out *moratorium*, to immediately stop using the PCL scales.

Professional Ethical Obligations of Forensic Mental Health Practitioners

As outlined in chapter 1, PCL psychopathy assessments are usually administered as part of a broader procedure colloquially known as a *forensic evaluation*, where the main goal is to evaluate the justice-involved person in terms of their institutional risks, needs, and general responsivity toward rehabilitation goals.[11] The information that is generated during forensic evaluations is multifaceted (e.g., sociodemographic, substance use, and mental health) and typically stored in the person's forensic records—a file akin to a doctor's electronic medical records—which the legal institutions continuously update and use as an important source of information to support decisions about the person (the scope and quality of these reports can have significant variation across jurisdictions). An important observation here is that whenever a PCL assessment is conducted—whether as part of a forensic evaluation or not—the assessment itself is typically carried out by a forensic mental health practitioner. This generally implies that the practitioner is a person who holds an advanced degree and certification in a healthcare discipline such as psychiatry or psychology, and in this capacity belongs to a professional community that operates with clearly demarcated ethical obligations toward the people they work with.[12]

But what exactly does it mean to be *ethically obliged*, and what are the specific ethical obligations of forensic mental health practitioners? To be clear, in everyday life, an ethical obligation is usually understood as a formal or informal rule about a morally salient issue, in which the rule makes a distinction between what is considered the right/wrong thing to do (requirement) or not to do (prohibition). In short, there are obligations about what *you ought to do* and what you *ought not to do*. For example, in many cultures, it is generally believed to be morally wrong to tell a lie and morally right to keep a promise. In these social contexts, we are accustomed to saying that a person is ethically required to tell the truth and keep their promises, and that they are prohibited from lying. It is perhaps this characterization of ethics that most people think of when we use the term, and it also happens to be this broad characterization that has preoccupied philosophers for centuries, inspiring debates about whether there are universal ethical principles that ought to guide human behavior—an academic field often referred to as *normative* ethics.[13]

However, when we are addressing the ethical obligations of professional groups in forensic mental health, this conversation is importantly different in the sense that we are no longer discussing the person's everyday ethical obligations but rather a set of obligations that emerge from their role *as* a professional. That is, we can think of professional ethical obligations as a set of rules about what counts as appropriate and aspirational conduct, but also a set of rules that goes beyond (or is added to) what is expected of citizens' everyday ethical obligations. For example, when a person begins their workday as a forensic psychiatrist or psychologist, they immediately assume a wide array of obligations that they do not have when they are off duty as a mere civilian (more on this below).[14]

Professional ethical obligations are typically defined by the professional community itself, often communicated in the shape of a formal consensus document outlining explicit rules and guidelines, which the professional community demands that their members are both familiar with and follow (this document is often called an *ethics code*). While the purpose of these guidelines is multifaceted, it is first and foremost designed to offer: (1) broad aspirational goals, (2) a clear set of rules of conduct, and (3) a set of heuristics on best practices. Importantly, professional ethical codes also serve the purpose of reinforcing an implicit social contract that exists between society and the profession as an organized expert community. A contract where society recognizes the profession as an autonomous entity that is allowed to monopolize its knowledge base, where the broader society in exchange can expect a reliable quality of services. Thus, by setting a sustainable ethical framework for members of their expert community, and making sure that members follow them, the profession contributes to maintaining the mutual trust in and integrity of this social contract.[15]

In North America, forensic mental health services are usually provided by psychiatrists and psychologists (other professions include nurses and social workers), each belonging to their own professional community with their own set of distinct professional ethical obligations. However, while these professions might set many different requirements and expectations for their members, the actual differences between the ethical codes are in effect minimal. This is because there is a historical connection between how professional ethics were shaped in North American healthcare disciplines, which psychiatry and psychology either wholly or partially self-identify with. These healthcare professions have long subscribed to a worldview that places the well-being and rights of the individual person at the center of its professional-ethical orientations (note that many forensic psychologists do not work in health-related contexts, nor do they see themselves as healthcare professionals; however, it is almost always the case that those psychologists who conduct forensic evaluations and PCL assessments self-identify with such roles insofar that conducting a PCL assessment requires some form of clinical expertise).[16]

In their influential analysis of healthcare ethics, Thomas Beauchamp and James Childress have identified four overarching principles that run through most of the largest healthcare professions in North America and Europe (see below).[17] Historically, these principles started out as implicit assumptions, but since the *professionalization* of healthcare services in the twentieth century, they have figured as the backbone of the ethical codes we see in today's healthcare disciplines. The coordinated development of professional ethics codes was driven by many factors, but a crucial influence was a string of political reforms from the post–World War II era aimed at protecting citizens from disgraceful malpractices within these professions (e.g., the disclosure of the Tuskegee syphilis experiments in the United States influenced the *Belmont Report*, which outlined ethical requirements for scientific health research).[18]

Below is an abridged summary of each of the four principles identified by Thomas Beauchamp and James Childress, which offers a generalized picture of the professional roles of forensic mental health practitioners[19]:

The principle of beneficence: A healthcare professional must strive to give primacy to the interest and well-being of patients and must actively ensure that services are tailored and optimized toward such patient-centered ends.

The principle of nonmaleficence: A healthcare professional can under no circumstances inflict undue harm and/or compromise the well-being of patients, whether this is done through intentional or negligent behavior.

The principle of respect for autonomy: A healthcare professional must recognize and appreciate the patient as a dignified individual capable of making their own decisions, and proactively aim to enhance, as well as refrain from exerting restraint on this capacity.

The principle of justice: A healthcare professional ought to orient their practices in such a way that individuals are treated according to what is fair, appropriate, and bound by the assumption of social equality.

The influence of, or conformance with, these four principles in the various mental health professions is both broadly recognized and ubiquitous.[20] For example, the largest professional organization for forensic psychiatrists is the American Psychiatric Association, which has adopted the ethical code provided by the American Medical Association (AMA; like medical doctors, psychiatrists self-identify as healthcare *and* medical professionals).[21] This framework consists of nine regulatory principles, each of which exhibits obvious connotations to patient well-being and rights (see table 7.1), as well as

Table 7.1
Ethical guidelines in forensic psychiatry and psychology

American Medical Association (AMA)	American Psychological Association (APA)
Principle 1: A physician shall be dedicated to providing competent medical care, with compassion and respect for human dignity and rights.	*General Principles* (abbreviated text):
Principle 2: A physician shall uphold the standards of professionalism, be honest in all professional interactions, and strive to report physicians deficient in character or competence, or engaging in fraud or deception, to appropriate entities.	**Principle A—Beneficence and nonmaleficence:** Psychologists strive to benefit those with whom they work and take care to do no harm. In their professional actions, psychologists seek to safeguard the welfare and rights of those with whom they interact professionally and other affected persons, and the welfare of animal subjects of research.
Principle 3: A physician shall respect the law and also recognize a responsibility to seek changes in those requirements that are contrary to the best interests of the patient.	**Principle B—Fidelity and responsibility:** Psychologists establish relationships of trust with those with whom they work. They are aware of their professional and scientific responsibilities to society and to the specific communities in which they work.
Principle 4: A physician shall respect the rights of patients, colleagues, and other health professionals, and shall safeguard patient confidences and privacy within the constraints of the law.	**Principle C—Integrity:** Psychologists seek to promote accuracy, honesty, and truthfulness in the science, teaching, and practice of psychology.
Principle 5: A physician shall continue to study, apply, and advance scientific knowledge; maintain a commitment to medical education; make relevant information available to patients, colleagues, and the public; obtain consultation; and use the talents of other health professionals when indicated.	**Principle D—Justice:** Psychologists recognize that fairness and justice entitle all persons to access and benefit from the contributions of psychology and to equal quality in the processes, procedures, and services being conducted by psychologists.
	Principle E—Respect for people's rights and dignity: Psychologists respect the dignity and worth of all people, and the rights of individuals to privacy, confidentiality, and self-determination.
	Example of an ethical *standard*:

Principle 6: A physician shall, in the provision of appropriate patient care, except in emergencies, be free to choose whom to serve, with whom to associate, and the environment in which to provide medical care.

Principle 7: A physician shall recognize a responsibility to participate in activities contributing to the improvement of the community and the betterment of public health.

Principle 8: A physician shall, while caring for a patient, regard responsibility to the patient as paramount.

Principle 9: A physician shall support access to medical care for all people.

Standard 3.04 (a)—Avoiding Harm: Psychologists take reasonable steps to avoid harming their clients/patients, students, supervisees, research participants, organizational clients, and others with whom they work, and to minimize harm where it is foreseeable and unavoidable.

Notes. Text is adapted from the guidelines provided by the American Medical Association (AMA) and the American Psychological Association (APA).[22]

11 in-depth chapters on how these principles should be interpreted and how they relate to common themes in health care (e.g., consent, privacy, patient-relations, and research). Similarly, the ethical framework of the American Psychological Association (APA)—the largest professional community for forensic psychologists—is outlined in the compendium titled *Ethical Principles of Psychologists and Code of Conduct*.[23] This document includes guidelines on general aspirations of ideal practices (called *general principles*) as well as a set of mandatory and sanctionable rules that psychologists are strictly required to follow (called *standards*). The aspirational general principles (A–E in table 7.1) fully overlaps with (and adds to) the basic principles identified by Thomas Beauchamp and James Childress, and many of the ethical standards are formulated with direct or implicit reference to these overarching principles. Consider the specific requirement of *Standard 3.04—Avoiding Harm* (bottom of table 7.1), which, just like the principle of nonmaleficence, makes it unambiguously clear that professional psychologists cannot harm their clients or anyone with whom they work.

Another important way that professional ethics deviates from everyday normative ethics is how their respective rules are sanctioned when violated. In everyday life, if a person does something unethical, it may not have any real-life consequences (unless, of course, the unethical behavior also happens to be illegal). For example, if you break a promise to a friend or lie to a stranger in an online forum, there are typically no guaranteed repercussions (aside from a tainted personal relationship and a spotted conscience). However, the situation is significantly different in the context of professional ethics. When a forensic mental health practitioner conducts themselves in clear violation of their professional ethics, this might have serious ramifications for the person's future affiliation with that profession and, by extension, for their current and future employment as a working psychiatrist/psychologist. This is because their professional associations have assumed the adjudicative role of reviewing accusations and sanctioning violations of the ethics code. Thus, practitioners who fail to conduct themselves according to the ethics code may face different forms of sanctions from their professional community depending on the gravity of their misconduct, ranging from a reprimand to probation to an outward expulsion from the organization.[24] In cases where serious ethical misconduct has been sanctioned by the judicial arms of the profession, a working psychiatrist/psychologist might find it difficult to maintain a career within that profession.

So, returning to the initial question in this section: What does it mean for a forensic mental health practitioner to have an ethical obligation? A short answer to this question is that it means that whenever such a person is conducting themselves *as* a professional—that is, as a person operating in a role under the auspices of their profession—the individual is then expected to adhere to the ethical aspirations of that profession. This implies a commitment to various clearly demarcated guidelines and standards, which holds the interest and well-being of the patient to be paramount. Providing services that are germane to these ethical principles is a central element of maintaining a high level of professional integrity for the betterment of individuals and public health.

Why Psychopathy Assessments Violate Professional Ethical Obligations

It is from this background of professional ethics that we must now analyze the central question raised in this chapter: Are PCL psychopathy assessments reconcilable with the professional ethical obligations of forensic mental health practitioners? Here I defend a negative answer, which sees PCL assessments as fundamentally antagonistic to the ethical aspirations and rules in forensic psychiatry and psychology. More specifically, I believe that there is patent evidence to suggest that the way psychopathy assessments are ordinarily used in forensic settings is in direct violation of the four overarching principles that run through the ethics codes of these two disciplines, namely: (a) the principle of beneficence, (b) the principle of nonmaleficence, (c) the principle of respect for autonomy, and (d) the principle of justice. For the sake of brevity, I will limit my discussion to what I see as the most obvious and uncontroversial ways these principles are violated, but I foresee that these are not the only ways that PCL assessments are ethically problematic.[25]

(a) Violation of the Principle of Beneficence
The underlying rationale behind the principle of beneficence is the idea that a practitioner must be mindful of the basic intentions guiding their services as healthcare professionals. At its core is the view that the skills and knowledge they apply must always be intended to have some form of potential or de facto beneficial outcome. This rationale is manifest in the way that healthcare professionals typically characterize their professional roles as

that of healers or helpers. This role does not imply that every little thing that a practitioner does throughout their workday must have a measurable health-related benefit. The principle of beneficence is not an articulation of a micromanagement imperative but rather a foundational aspiration about the overall end product that healthcare professionals aim to provide: they strive to heal, help, and benefit the people they work with.

As a foundational objective, the principle of beneficence is referenced in various implicit and explicit ways throughout the ethics codes of psychiatrists and psychologists. For example, the AMA's nine principles of medical ethics—in particular, principles 7 and 8—invoke clear language about beneficial intentions, such as physicians must work toward the "improvement" of public health and that care of and responsibility to the patients are "paramount." And in chapter 1 of their ethical code, the AMA writes that a healthcare worker must always "advocate for their patients' welfare."[26] Similarly, the APA emphasizes beneficial intentions in their aspirational *General Principle A*, but they also invoke this language throughout their sanctionable ethical standards, including how they bar practitioners from providing unbeneficial services. For example, *Standard 10.10* prescribes that practitioners are obligated to disengage or terminate non-beneficial therapeutic services.[27]

So, how might we characterize the underlying intentions behind PCL psychopathy assessments? To be clear, the assessment tools themselves are made to assist the practitioner in "measuring" psychopathy. According to the PCL-R manual, this is "strictly speaking" what it was designed to do.[28] And, of course, this functionality says little about what might be the specific intention behind using them. Just like the function of a pair of scissors is to cut, people can use them with different intentions—for instance, when a doctor uses them during surgery (beneficial) or when they are used as a weapon (harmful). However, when we analyze the intentions behind the ordinary use of PCL assessments, what immediately stands out is that the practice appears to be entirely devoid of any conceivable consideration of benefiting the person being assessed.

For example, in chapter 1, I outlined some of the best documented applications of PCL assessments, which are revealing for the intentions that guide their use. Namely, the assessment itself is primarily used to inform decisions about: (1) pre-sentencing and sentencing, (2) correctional placement, (3) capital punishment, (4) dangerous and long-term offender status, (5) sexually violent predator status, (6) juvenile transfer, (7) treatment and rehabilitation,

(8) parole, and (9) correctional and probation supervision and management. More importantly, across these contexts, a PCL assessment is used as a tool for supporting specifically *restrictive* forensic measures. For instance, when a person is assessed with psychopathy, such as scoring above 30 on the PCL-R, this information supports various decisions including (but not limited to) unmitigated sentencing, placement in a maximum-security institution, receiving capital punishment, exclusion from rehabilitation programs, the possible denial of parole, heightened probation supervision, and so forth.

In none of these documented uses can we convincingly argue that the intention behind conducting a PCL assessment is to benefit the person being assessed, unless, of course, we are willing to accept the dubious (or absurd?) claim that imprisonment or enhanced probation supervision is beneficial to that individual's life. On a more serious note, the way we determine what is beneficial in the context of psychiatric and psychological practices rests largely on whether the service itself is reconcilable with the basic role of the professional as a healer and helper. That a PCL assessment is sometimes used—as highlighted by critics[29]—as a discriminatory factor in providing healthcare services (i.e., exclusion from rehabilitation programs) speaks volumes about the primary intentions of these assessments: it is aggravating information meant to enhance legal restriction, and not about care and help.

It is important to stress that while the common uses of PCL assessments are scarcely driven by beneficial intentions, this does not mean that the assessments could never be used to inform beneficial practices. What matters is the purpose behind its use. It is interesting here to note that some scholars have worked on developing treatment and rehabilitation programs that are specifically tailored to individuals who score high on the PCL-R.[30] In the context of such a treatment program, the use of the PCL-R could be seen as beneficial to the patient insofar that it is used to select the most beneficial treatment/rehabilitation plan. However, initiatives like these remain a matter of experimental and academic interest, and there is so far little evidence that these care-oriented applications of PCL assessments are broadly endorsed across legal institutions.[31] In addition, practitioners should note that if a PCL assessment score is recorded in a person's forensic files, practitioners cannot guarantee that this information is exclusively being used for targeted treatment purposes, as the information may be used later on to inform other decisions where the practitioner is not part of the decision-making process (e.g., placement or parole). In other words, while there are potentially beneficial uses of the PCL assessments, it can never be guaranteed that all types of uses are beneficial; and what is more, practitioners have overwhelming reasons to believe that most common uses of the PCL scales lead to non-beneficial outcomes for the person receiving the assessment.

(b) Violation of the Principle of Nonmaleficence

Of all the four basic principles, the principle of nonmaleficence is perhaps the most intuitive in terms of what exactly it prescribes. What makes it easy to interpret is its "prohibitory" definition, clearly stating what a practitioner universally *cannot do* (as opposed to prescribing an open-ended range of general aspirations). The principle of nonmaleficence is historically rooted in the culturally well-known *Hippocratic Oath* (from ancient history), which includes the promise "to abstain from doing harm"—a dictum that was explicitly invoked in Western healthcare disciplines around the nineteenth century under the Latinized phrase *primum non nocere*, "first, do no harm." In short, whatever forensic mental health practitioners do in their daily professional roles, they must always abstain from practices that have known or potentially harmful effects on persons they work with and/or whom are the target of their services.

The nonmaleficence principle is also one of the more explicitly invoked principles throughout the ethics codes in psychiatry and psychology. For instance, the AMA writes in the preamble to its code that all its ethical rules are "developed primarily for the benefit of the patient," and, as mentioned earlier, in principle 8 it is stated that the care and responsibility toward patients is "paramount." Further, in chapter 1 of AMA's ethics code, it is stated that the relationship between a healthcare worker and a patient is "fundamentally a moral activity that arises from the imperative to care for patients and to alleviate suffering."[32] Similarly, the APA includes the principle of nonmaleficence in *General Principle A*, thus characterizing nonmaleficence as a basic aspiration of the profession. The principle also appears later in the sanctionable *Standard 3.04—Avoiding Harm* (see table 7.1).[33]

An important qualification of the principle of nonmaleficence is that it must not be narrowly interpreted to imply that a practitioner must avoid all forms of harmful acts. Within healthcare professions, there is room for clinical services to involve elements of harm, but only as long as they are necessary to achieve an overall non-harmful and beneficial service for the patient. For example, similar to a medical surgeon who necessarily has to cut through skin tissue before removing a tumor, a forensic psychiatrist might have to prescribe medicine with unpleasant side effects or engage in a certain type of behavioral therapy that stirs up negative feelings. But just as the surgeon's interaction eventually heals the patient from a fatal disease, if the psychiatric intervention is conclusively non-harmful and beneficial to the patient (e.g., overall improvement of mental health), the practice will still be reconcilable with the principle of nonmaleficence (though, a practitioner must still strive to minimize harmful elements and terminate services with zero prospects of benefits).

So, how does a PCL assessment measure up in this harm-benefit analysis? In some cases, a high PCL score might not have any consequence for the person being assessed. But when it does, the primary effect is always some form of legal restriction, such as an amplification of legal sanctions and decisions, ranging from enhanced correctional supervision to denial of parole to increased chances of capital sentencing. Therefore, the harm-benefit analysis essentially boils down to how we must characterize these restrictive consequences. Answering this question runs the risk of sounding foolishly circular, as it quickly turns into an exercise of articulating something that is self-explainable and perhaps trivially obvious. However, it seems uncontroversial to assume that, from the perspective of a person who is subjected to enhanced legal sanctions due to a high PCL score, such decisions are almost always viewed as undesirable due to the pain and suffering that these judicial constraints are designed to cause (i.e., there is an element of harm in how legal sanctions are intended as punishment, retribution, and so on).

This characterization might have some stipulative elements, but it happens to correspond with how legal sanctions are generally thought of as a means of deterrence. Indeed, legal sanctions are typically seen as a sociopolitical instrument designed to encourage socially desirable behaviors. By making it extremely undesirable (i.e., harmful) to be the subject of legal sanctions, our society aims to deter people from illegal activity. Hence, being on the receiving end of a legal sanction will ordinarily be seen as harmful, and having that sanction *enhanced* is, of course, even more undesirable. Thus, when a PCL assessment is used to enhance legal sanctions, we can view it as an instrument that cofacilitates these harmful consequences. And therefore, their ordinary use is in direct violation of the ethical obligation to do no harm.

(c) Violation of the Principle of Respect for Autonomy

Forensic mental health services always involve people. This includes patients enrolled in therapy programs, undergoing psychological testing, or when subjects participate in research. During these activities, the principle of respect for autonomy implies that these individuals—be they patients or research participants—must be regarded not as mere entities in the control of the healthcare professional, but instead as people with self-governance capacities ("auto" means *self*, "nomos" means *ruling*, i.e., ruling oneself). It is a perspective that addresses the individual person as a collaborator or "co-player" in the practitioner's decision-making processes. Characterizing the individual as an entity that deserves respect runs parallel to how constitutional and liberal democracies in North America and Europe have traditionally conceived individual rights and freedoms as a central pillar of the social fabric (e.g., as formalized in the *Bill of Rights* in the United States and the *Canadian Charter of Rights and Freedoms* in Canada).

There are many explicit ways in which the principle of respect for autonomy is expressed in healthcare ethics codes, as the principle has historically served as a key inspiration to healthcare professions to formalize their ethics codes in the first place. For instance, virtually all of AMA's nine principles can be read as a direct or indirect reference to patient autonomy, where principle 1 is a direct iteration of it: "Respect for human rights and dignity." Further, in chapter 1 of the AMA code it is stated that the relationship between healthcare workers and patients is a "collaborative effort" and a "mutually respectful alliance."[34] Similarly, the principled respect for autonomy surfaces throughout the APA's ethics code, with direct reference evoked in *General Principle E*, emphasizing the respect for patient "dignity" and "self-determination" (table 7.1). The principle also figures in various places within the sanctionable standards—for example, *Standard 3.06—Conflict of Interest* prohibits psychologists from serving interests that conflict with those of the individuals in their care, and *Standard 3.08—Exploitative Relationships* bars psychologists from taking advantage of or exploiting individuals they work with.[35]

Perhaps the most well-known ethical obligation that the principle of respect for autonomy has historically inspired is the requirement of psychiatrists and psychologists to obtain *informed consent* from people who are subject to their care, research, and other services (the requirement of informed consent has also become a legal standard in North America and Europe).[36]

The concept of informed consent, and the practice of obtaining it, generally implies that four primary conditions have been met:[37]

1. Disclosure: The practitioner has sufficiently communicated to the patient relevant information about their services, including potential and de facto risks and benefits.
2. Understanding: The information about the services has been sufficiently understood by the patient.
3. Voluntariness: The patient's decision to consent or refuse must be voluntary and free of coercion/manipulation.
4. Competence: The person who is consenting to healthcare services has intact decision-making capacities, such as being an adult with ordinary cognitive capabilities.

The requirement to obtain informed consent is expressed in chapter 2 of the AMA's ethics code, where they write that "Patients have the right to receive information and ask questions about recommended treatments so that they can make well-considered decisions about care."[38] The requirement also figures in the APA's ethics code in *Standard 3.10—Informed Consent.*[39] In short, the obligation to obtain informed consent refers to how a psychiatrist or psychologist must ensure that the person with whom they work has been given a clear opportunity to deliberate and voluntarily decide whether they want to either *authorize* or *refuse* to partake in the practitioner's services.

With this standard definition of informed consent, there are at least two apparent reasons why psychiatrists and psychologists should be wary of the prospects of obtaining valid informed consent for a PCL psychopathy assessment: first, given that psychopathy assessments (in most cases) have no conceivable benefits to the person being assessed, and are associated with potential and obvious harms, these realities suggest that any reasonable person (if sufficiently informed) would never voluntarily consent to it. The problem here is obvious: even in a situation where a healthcare professional has obtained informed consent for a PCL assessment, they should still be fundamentally suspicious about whether the patient has properly understood what the potential consequences of the assessment are. Since there are no conceivable benefits associated with the assessment, it suggests that they have altogether failed to understand its proper consequences and purposes (i.e., violating condition 2 above).

Second, given that PCL assessments are often part of a larger forensic evaluation, which eventually feeds into the decision-processes across legal institutions such as a parole board or probation management, there will always be a risk that a person's refusal to undergo a PCL assessment will reflect negatively on their overall evaluation (e.g., a parole board member might find it suspicious that a person refused the assessment). This context therefore generates an implicit incentive for the patient to collaborate, which in this case is better interpreted as an element of coercion in that the person knows that refusal will have negative repercussions. Indeed, as has been acknowledged by other scholars,[40] when a person is viewed as a *non-collaborator* in a forensic setting—for example, by refusing forensic mental health services and assessments—this may negatively impact the overall risk management and rehabilitation assessments, which in turn can have various undesirable consequences for that person when this information is used to support decisions. In such a context, a practitioner will (or should) always have some form of looming doubt that if they obtain consent for a PCL assessment, this may not have been fully voluntary (i.e., violating condition 3 above).

In short, if a patient gives their informed consent to undergo a PCL assessment, it will remain virtually impossible for a practitioner to ascertain whether the conditions for informed consent have been met: First, if an individual consents to a PCL assessment, it implies that the information about its consequences was not sufficiently understood (condition 2), and, secondly, it is never clear if the decision was voluntary and generally free of coercion (condition 3). Should a practitioner choose to ignore these ground facts and proceed with conducting the assessment, it would generally amount to violating the principle of respect for autonomy.

A curious nuance to this conclusion is that the APA's *Standard 3.10* on informed consent provides some wiggle room for performing legally mandated assessments without obtaining informed consent, namely when an assessment is either legally required or court-ordered (note that the AMA does *not* leave room for this). It must be noted though, that PCL assessments are generally not legally mandated in this way. A court might issue an order for a person to be evaluated by a psychologist, but it is typically the practitioner themself who decides what sort of assessment tools to use (e.g., they can choose a less problematic type of assessment). Some jurisdictions have mandated PCL psychopathy assessment for dangerous offender and sexual violent predator status hearings, but these evaluations tend to be conducted by psychiatrists due to their (higher) authority as medical practitioners (and they are strictly required to obtain consent in such a case). Furthermore, when a court orders an assessment of a person, it is not the psychologist who has been ordered to perform an assessment; it is the assessment itself that has been given a legal mandate. Despite these caveats, it still implies that when psychologists (but not psychiatrists) are commissioned by the legal system to conduct a legally mandated psychopathy assessment, the requirement of informed consent is effectively eased.

However, even as there might be cases where a psychologist is on ethically stable ground with regard to informed consent, this does not release forensic psychologists from their broader ethical obligation to align their practices with other requirements implied by the principle of respect for autonomy. An interesting paradox here is that forensic psychologists still have the obligation to evade exploitative relations (*Standard 3.08*), which arguably precludes services where a psychologist solicits assessments that are harmful to the person, in exchange for a monetary payment for their services from a third party like a legal institution. Breaking this standard would be another way of violating the ethical obligation to respect a person's autonomy.

Finally, it is important to stress that this situation is a paradox and *not a dilemma*; that is, it is *not* a situation where there are no good solutions. Should a psychologist be asked to perform a legally mandated PCL assessment—which they have good reasons to believe violate their ethical commitments—the APA ethics code requires them to resolve this paradox by refraining from providing the service. This is stated clearly by the APA in a special commentary publication on legal vs ethics conflicts, where they write: "In cases in which a conflict between legal and professional obligations occurs, forensic practitioners make known their commitment to the EPPCC [Ethics Principles of Psychologists and Code of Conduct] and the Guidelines and take steps to achieve an appropriate resolution consistent with the EPPCC and the Guidelines."[41]

(d) Violation of the Principle of Justice

Healthcare professionals do not work in a vacuum but see their profession as one with a genuine social responsibility and purpose. Their commitment to the principle of justice encapsulates this view, as it characterizes healthcare skills and services as a social good that must be distributed to members of society in an equal and fair manner (notice that the use of the noun "justice" in this principle is not understood as a strictly legal term, but instead as a concept about social fairness). Healthcare services are, for obvious reasons, central to every citizen's prosperity and happiness, and it is upon this observation that practitioners assume the responsibility of justly administering these essential services, both at the broader social level (i.e., equal distribution) and at the individual level (i.e., nondiscriminatory delivery). This implies, among many things, that healthcare practitioners work to ensure that similar cases are provided with similar treatment and, correlatively, that there is no unfair discrimination between people in their care. As Tom Beauchamp puts it: "Like cases should be treated alike."[42]

We find the principle of justice expressed throughout the ethics codes of psychiatry and psychology. For example, the AMA makes it clear in principles 1 and 7 that professionals must provide "competent" medical care with the respect for individuals' "dignity and rights," further recognizing a responsibility to improve "community" and "public health" (table 7.1). And in chapter 1 of the code, AMA writes that practitioners must not "discriminate against a prospective patient" based on "social characteristics that are not clinically relevant."[43] Similarly, the APA writes in *General Principle D* that every person who is subject to psychological services has a right to "equal quality" of care (table 7.1). This main aspiration also figures in several standards in the code, such as *Standard 3.01—Unfair Discrimination*, which states that psychologists must not engage in "unfair discrimination" throughout their "work-related activities."[44]

Perhaps one of the clearest, downstream implications of the principle of justice is the way healthcare professions have uniformly committed themselves to having their practices be guided by scientific knowledge, namely as captured by the guidelines for evidence-based practices (which I have mentioned on several occasions in this book).[45] Where evidence-based practices are often talked about as a *quality* criterion—that services must be supported by reliable, empirical evidence to enhance efficacy—this commitment also happens to be an ethical criterion. As Kenneth Goodman puts it: "Ethical practice is, in large part, scientifically sound practice."[46] This view shows up in the way that references to scientific evidence are continuously evoked throughout the ethics codes: for instance, in the APA's *Standard 9.02—Use of Assessments*, which states that services that rely on assessments tools must be guided by the "research on or evidence of the usefulness and proper application of the techniques."[47]

While evidence-based practices are generally consistent with the four basic ethical principles, it is above all a way of promoting and safeguarding fairness in healthcare services: scientific evidence does not discriminate—at least not if it is conducted according to proper scientific standards—so when services are based on generalizable scientific knowledge, we reduce the risk of personal biases and arbitrary decisions unjustly impacting services. Conversely, when professionals engage in practices that lack scientific evidence altogether, this is another way of saying that people in their care have been serviced in an inequitable, unjust way.[48]

With this definition of the principle of justice, and its link to evidence-based practices, it follows that there is a fundamental ethical tension when forensic practitioners conduct PCL psychopathy assessments. Indeed, the recurring theme throughout this book has been how there is a profound lack of evidence to support common claims about PCL psychopathy. Where psychopathic persons are typically described as unconscionable social predators, research tells us that this claim is not anchored in empirical facts. However, since PCL psychopathic persons are still being dealt with *as if* they have all of these socially undesirable proclivities—as if they are extremely dangerous, untreatable, morally colorblind, and so forth—this is effectively violating the commitment to evidence-based practices. And it conclusively amounts to unfair discrimination of the individuals who have been labeled by the assessment. In sum, when the legal system is restricting individuals based (wholly or partially) on their PCL assessment, they are doing something that is lacking scientific support, and psychiatrists and psychologists are therefore ethically complicit when they act as a facilitator of such discriminatory practices by conducting the assessments (I mention in passing this type of unfairness is sometimes referred to under the broader category of *epistemic injustice*[49]).

Building a Constructive Dialogue: The Four Most Common Objections

I have here presented an overarching argument for how the common use of PCL psychopathy assessments appears to violate the four basic ethical principles outlined in the work of Thomas Beauchamp and James Childress—principles that have inspired many of the ethical guidelines and sanctionable standards we find in forensic psychiatry and psychology. In short, I argued that in all the common ways that psychopathy assessments are used in our legal system, it is reasonable to suspect that (1) these assessments have no benefit to the person being evaluated, (2) that there are potential and de facto harms associated with the assessments, (3) that no reasonable person would ever voluntarily consent to it, and (4) that the assessments lead to unjust discriminatory practices insofar that the way they are used to support decision-making lacks empirical evidence. If this analysis is sound, it follows that forensic practitioners ought to immediately stop using PCL assessments as part of their services.

This argument is not intended to be controversial. It is just four straight-forward ways that PCL assessments might end up violating professional ethical obligations. Pointing out ethical problems with certain forensic services is a necessary part of maintaining the quality and integrity of the profession. My experience is also that researchers and practitioners generally acknowledge that PCL psychopathy assessments are ethically problematic, even if they do not agree with every single aspect of what I outlined above. And for what it is worth, we are currently seeing these concerns being addressed in peer-reviewed academic writings, perhaps indicating that a consensus is starting to take shape.[50]

However, this is not to say that there is no need for further dialogue about the issue, nor am I saying that there are no important caveats or counterarguments against the view I have expressed above. In an aim to anticipate and be forthcoming about such a dialogue, I shall here briefly rehash four of the most common objections I have come across over the years of discussing this issue with scholars, practitioners, and university students. These objections are important to grapple with, as they help us nuance the discussion. But, as I will argue, these objections nevertheless fail to seriously challenge the overarching conclusion that PCL assessments are ethically untenable.[51]

Objection #1: My Client Is the Legal Institution, Not the Patient

By far the most common "push-back" I receive from forensic experts at academic conferences—both in North America and in Europe—is the view that my argument is fundamentally misrepresenting the proper scope and target of a practitioner's ethical obligations. The counterargument typically goes something like this: when a forensic practitioner enters the legal system and conducts a PCL assessment, their ethical obligations only pertain to their client who pays their bill, in this case, the legal institutions, as represented by a court, parole board, prison, district attorney, and so on. In essence, forensic practitioners hold no ethical responsibility toward the person they are assessing. That there are or may be negative downstream consequences of a PCL psychopathy assessment, such as it contributing to capital sentencing, denial of parole, or juvenile transfer, is really of no ethical concern to the practitioners themselves.

This counterargument has often taken me by surprise, mostly because I think it builds on a narrow and essentially wrong interpretation of the ethics code. Nowhere in the AMA or the APA's ethical codes is it stated that practitioners are relieved of their ordinary ethical responsibilities toward people once they enter the legal system or are otherwise hired by a third-party entity. Quite the contrary, it is repeatedly and explicitly stated throughout the ethics codes that a practitioner's ethical responsibility is primarily and/or equally toward the person who is subject to their services, regardless of whether they are working *for* that person or whoever is paying the bill.[52] Every person with whom a healthcare professional works with *as* a healthcare professional is considered a dignified person in their care and is thus entitled to ethically responsible conduct from the practitioner. For instance, of all the ethical obligations healthcare professionals are committed to, at least one thing ought to be unmistakably clear: under no circumstances can they knowingly participate in or facilitate harmful and non-beneficial practices, regardless of who they are working for or who is the target of their services. As we have seen, these obligations are unambiguously stated in the professional ethics codes, such as the APA's *Standard 3.04—Avoiding Harm* (table 7.1). To neglect these imperatives is, by definition, unethical practice.

While there might be circumstances where forensic psychiatrists or psychologists only hold obligations toward the legal institution that hired them, these instances are very limited, mostly pertaining to cases where there are no other implicated people—for example, where there are no persons who could be directly impacted by the expertise that the psychiatrist or psychologist bring *as* a professional. For example, such a situation might include when they are serving as expert witness a court or in an advisory role about an abstract topic (e.g., the efficacy of psychotropic drugs or behavioral therapy). But cases like this are very different from when a professional enters a clinical context and performs a PCL assessment of a person. In this context, whomever is being assessed must always be seen as properly deserving of respect and dignity, as a person toward whom the professional holds ethical obligations.

In addition, there is plenty of precedence from landmark cases of ethical misconduct where the AMA and APA have made it clear that a practitioner's responsibility is not nullified just because they work for a third-party entity such as a legal institution. This includes how the AMA has prohibited its members from participating in the legal system's facilitation of the death penalty,[53] and how the APA has barred its members from developing interrogation and torture programs on the behalf of a state actor like the US government.[54] While these types of professional misconduct are importantly different from conducting PCL assessments, the main reason why death penalty assistance and torture program consultations are prohibited is exactly because practitioners are not relieved from their ethical responsibility just because they work for a third-party entity. In both cases, the practices therefore violate fundamental professional ethical principles, perhaps most obviously the principle of beneficence and nonmaleficence.

Objection #2: I'm Obliged to Service a Legal Institution

Another common counterargument is the view that forensic practitioners are justified in conducting PCL assessments because of an imminent responsibility to assist the legal system in the administration of justice. What I have often heard practitioners say is that they believe they are essentially in a dilemma-like predicament where they recognize their ethical responsibilities toward the person they are working *with*, but that they also have professional obligations toward the legal institutions they are working *for*. A forensic psychologist may be commissioned to conduct a PCL assessment by a court or a parole board, and in this situation, the argument goes, the legal institution will effectively trump their ethical obligations toward the individual person.

Like the previous *Objection #1*, this counterargument also rests on a misunderstanding of where a practitioner's primary obligations belong. In fact, their responsibility is exactly the opposite: should a practitioner find themselves in a situation where their ethical responsibilities seemingly conflict with the requirements of a legal institution, they are then expected to resolve the situation by defaulting or adhering to the ethics code. For example, the AMA writes in Principle 6 that a physician is "free to choose whom to serve," and they elaborate in the *Preamble* to the code that "ethical responsibilities should supersede legal duties."[55] Further, the APA writes in *Standard 1.02—Conflicts Between Ethics and Law, Regulations, or Other Governing Legal Authority*: "If psychologists' ethical responsibilities conflict with law, regulations, or other governing legal authority, psychologists clarify the nature of the conflict, make known their commitment to the Ethics Code, and take reasonable steps to resolve the conflict consistent with the General Principles and Ethical Standards of the Ethics Code."[56]

This is not to say that there are no exceptions to this rule. However, these exceptions only pertain to cases where there are legal statutes that make the practitioner *criminally* liable should they choose to ignore them. A common example of this is when a healthcare practitioner is in possession of confidential information that their patient is at risk of harming another person(s), a predicament known as *Tarasoff liability*.[57] In this situation, the practitioner has an ethical responsibility toward protecting client privacy and confidentiality; however, they are also legally obligated to report to authorities that other people are in imminent danger. The professional consensus is, then, that a practitioner is relieved of their ethical duties and therefore ought to comply with their legal responsibilities (to not incriminate themselves). Though, when it comes to legally mandated PCL assessments, a practitioner is never criminally liable in this way.

Objection #3: I Have a Duty to Society to Conduct Psychopathy Assessments

A third common objection is the belief that forensic practitioners have a duty to society to report when they have a psychopathic person in their care, or that they are somehow obliged to conduct psychopathy assessments as a form of security precaution. The idea is that there are social and public health benefits associated with identifying psychopathic persons (e.g., it minimizes the risk of violence), and by assessing and reporting this information to authorities, there is a way in which the subsequent personal harm to the patient (e.g., enhanced legal confinement) is *outweighed* by the overall social benefits (I have heard many versions of this counterargument, but it always builds on the notion that social responsibilities outweigh professional ethical obligations toward individuals).

The basic idea behind this objection has some merit as healthcare professionals sometimes must make decisions that prioritize the health and well-being of the many over the few (sometimes referred to as *public health* decisions/interventions[58]). When they do so, there is some consensus that the justification usually has to meet various criteria, such as (a) that the pursued good for the many *far outweighs* the harms to the individual, (b) that harming the individual is the *only way* to achieve this, and (c) that the social good is *certain* to be fulfilled.[59] Illustrative examples of these kinds of decisions include when individuals with a highly infectious and lethal disease are forcibly quarantined in order to prevent community-wide contagion, or when individuals with serious mental illness are forcibly committed due to immediate danger to themselves and others. In such unique scenarios, healthcare professionals may be justified in assisting authorities in the effort to spare the life of the many on the minimal cost to the individual.[60]

However, as has been the repetitive theme throughout this book, there is no reliable evidence to suggest that individuals assessed with PCL psychopathy exert such a threat to public health and well-being. Therefore, PCL assessments cannot be justified under these conditions—that is, there are no evidence-based reasons for believing that subjecting PCL psychopathic persons to enhanced legal restriction will bring any large-scale public health benefits.

Objection #4: Forensic Practitioners Have Uniquely Different Ethical Obligations

A final objection that I frequently encounter is based on the claim that *forensic* professionals operate with a different and more nuanced set of ethical obligations compared to *ordinary* psychiatrists and psychologists. In essence, the counterargument is that the common ethics codes like those of the AMA and APA either have no bearing on forensic practitioners or that practitioners somehow are permitted to occasionally curtail or "pause" these ethical obligations once they practice inside the legal system. So, where it might be the case that PCL assessments appear to violate the principles outlined in the AMA and APA ethics codes, this is really of no concern to forensic professionals.

The main idea behind this counterargument—that forensic professionals are bound by uniquely different ethical obligations—has also been laid out in the scholarly literature, perhaps most famously by Paul Appelbaum in a 1997 article.[61] In this paper, Appelbaum proposes that forensic mental health practitioners should work toward outlining an alternative ethics, citing precedence in the medical field where ethical obligations seem to deviate between institutional contexts. Of particular interest to Appelbaum is when medical professionals engage in experimental *research*: for example, testing the efficacy of a new form of pharmaceutical treatment. In such a context,

Appelbaum argues, it is no longer clear that physicians are adhering to their ordinary ethical obligation to prioritize the well-being of their patients (e.g., the test medicine is randomly prescribed and might not be beneficial).[62] So, if there are contexts where medical professionals can curtail their ordinary ethical obligations or somehow deviate from its requirements, then it might not be that difficult to imagine forensic practices being bound by alternative ethics as well.

However, there are several reasons why I do not find this type of counterargument convincing. The first and most pressing reason is that it is simply inaccurate to think that forensic practitioners are bound by uniquely different ethical obligations. Any professional who self-identifies as a psychiatrist or psychologist, and thus operates under the auspices of professional organizations like the American Psychiatric Association and the American Psychological Association will be fully expected to adhere to these organizations' ethical codes and guidelines. This goes for all subspecialities, including its forensic practitioners. This fact is ubiquitous and clearly stated in these organizations' bylaws and policies,[63] and it is generally not something that scholars, nor practitioners take issue with. As stated by Thomas Grisso, the APA ethics code "covers all applications of psychology in every aspect of society."[64] There might be smaller professional organizations (such as colleges affiliated with APA and AMA, or separate certification boards) that operate with their own ethics codes. But these codes do not meaningfully deviate from the four basic principles I outlined in this chapter, nor have any of these organizations given up on the widely acknowledged best practices that are derived from these principles (e.g., evidence-based practices and commitment to informed consent).[65]

Second, where some people might have been swayed by the argument given by Paul Appelbaum—and thus personally believe that forensic practitioners can ignore ethics codes—it pays to notice that such personal beliefs are not legitimate excuses that somehow relieve them of their obligations. When scholars like Applebaum write an academic article proposing a rethinking of forensic ethics, this is merely an intellectual contribution to an academic conversation. But this does not mean that the professional community agrees with his view. In fact, such proposals have been uniformly dismissed, evident by the long-held commitment among forensic practitioners of adhering to the AMA and APA professional codes (even Appelbaum appears to have changed his view since 1997[66]).

Third, to think that medical ethics deviates across contexts—like Appelbaum's article suggests is the case in medical research—is to seriously misrepresent the reality in these contexts. For example, in medical research, scientists must still adhere to extremely strict ethics protocols, which altogether epitomize the four general ethics principles considered in the chapter. Above all, medical research must obtain informed consent from participants and be conducted in a manner that minimizes harm. And while an experimental treatment cannot be guaranteed to have any immediate benefits to the participant, the basic motivation behind the research is still to enhance medical knowledge—something that clearly fits with the general aspirations of healthcare professionals.

Concluding Remarks: A Case for a Moratorium

In recent years, there has been a growing conversation in the field of forensic psychology and psychiatry about the ethics of using PCL assessments to inform legal decision-making. This chapter adds to this dialogue, arguing that the clinical practice of conducting PCL psychopathy assessments is in violation of the professional ethical code of forensic psychiatry and psychology, the two main professions that are commonly tasked with conducting these assessments in the criminal justice system. I advanced this argument on the basis of four premises, namely: (1) a PCL assessment is usually nonbeneficial to the person being assessed, (2) a PCL assessment might generate harmful downstream consequences, (3) no rational person would ever voluntarily consent to receiving a PCL assessment, and (4) that the inferences that are commonly derived from PCL assessments lack empirical support. Since these premises all appear to be true, I believe that it then follows that psychiatrists and psychologists will almost always be violating their ethical code when conducting PCL assessments. One qualification to this argument is that there might be unique instances where a PCL assessment is ethically justified (e.g., in treatment settings where the PCL score is not recorded in the person's forensic files), but such cases are presumably rare and certainly fall outside the mainstream use of PCL assessments, which has been the focus of this book.

It is important to stress that this argument appears to be particularly resilient to counterarguments. What makes the argument resilient is that for the conclusion to be true—that PCL assessments violate professional ethical obligations—all four premises need not to be true. Moreover, even if *one* or *some* of these premises are true, or some of them are *sometimes* true, it would still follow that the practice of conducting PCL assessments is ethically problematic. For instance, even if it was only the case that one of the premises are true (e.g., if we agree that PCL assessments can only lead to harmful consequences), or that only some of them are *sometimes* true (e.g., if we agree that sometimes PCL assessments cause harm), it will still follow that the practice is ethically untenable. And still, I believe that a strong case can be made that all four premises might consistently be true in most legal contexts.

If the argument provided in this chapter is sound, it appears to follow that—from a professional-ethical standpoint—the disciplines of forensic psychiatry and psychology ought to recommend an immediate moratorium of PCL psychopathy assessments in legal contexts (such a moratorium would not necessarily apply to research-based settings, where it is not clear that these assessments are harmful, unjust, and so on).[67] Such a recommendation has the potential to stir controversy among longtime advocates of PCL assessments. However, to researchers and practitioners who prioritize the efficacy of legal decision-making, and care less about what toolbox they apply to promote these ends, the call for a moratorium should be welcomed as a necessary step in the right direction. This is because a recommendation of a moratorium is completely in line with the principle of evidence-based practices. In other words, there will be two upsides of a moratorium: first, the field would end the use of an assessment tool that is ethically dubious. Second, the field would stop using a tool that has few (if any) evidence-based justifications.

Indeed, from the viewpoint of evidence-based practices, it still remains thoroughly unclear what exactly practitioners—and the legal system as a whole—stand to *lose* from a moratorium on PCL assessments. As documented throughout this book, there is currently no clear evidence to suggest that all the common claims made about PCL psychopathic persons are true. So, when PCL assessments are used to inform legal and clinical decisions, research shows that the information practitioners rely on is overly unreliable and essentially misleading. Hence, the PCL scales are oftentimes not actually informing anything in the strict meaning of the word, but, in the worst-case scenario, they are a powerful source of bias and confusion. Thus, ending the use of PCL assessments will not necessarily subtract from the efficacy of legal decisions, and it might even enhance them as it frees up resources that can be better spent (recall that it can take hours to complete one PCL assessment, and a justice-involved person might be assessed several times).

Furthermore, practitioners and decision-makers should note that the justifications for using the PCL scales in legal settings are not only poorly justified but also relatively overstated. The tools were traditionally promoted as if they could provide unique information for legal decision-making purposes. However, as I have uncovered throughout this book, empirical research has consistently demonstrated that the PCL scales have no such unique information to provide. On the most generous interpretation of the research, the forensic use of the PCL scales only appears to have one type of evidence-based justification, namely when it is used in risk assessment contexts. For example, a case can be made that there is a weak to moderate link between PCL scores and criminal recidivism (though, it has important caveats; see chapter 2). However, in such risk assessment contexts, the PCL scales are still reliably *outperformed* by competing risk assessment tools.[68] And while there evidently are better and more efficient ways to perform risk assessment, an important part of this observation is that forensic practitioners are already trained in using these alternative and better risk assessment strategies. Thus, if the use of the PCL scales in these contexts would be phased out following a moratorium, it does not entail that legal institutions (and private consultants) must invest resources into education and/or restructuring of existing services.

In sum, recommending a moratorium on PCL assessments in legal settings is therefore not only in line with professional ethical conduct; it appears to be in line with the broader aspirations of working forensic psychiatrists and psychologists to deliver optimal, evidence-based services.

8 Is Psychopathy a "Zombie" Idea?
Explaining the Lack of Scientific Progress

A zombie idea is a view that's been thoroughly refuted by a mountain of empirical evidence but nonetheless refuses to die, being continually reanimated by our deeply held beliefs . . . quietly biding their time in peer-reviewed papers and textbooks, waiting to infect another generation of unsuspecting psychological scientists."
—Lisa Feldman Barrett in *Zombie Ideas* (2019)[1]

In this book, I have scrutinized the most central, mainstream ideas that researchers and forensic practitioners usually invoke when they speak about psychopathy. The overarching theme that emerges across the chapters is that the conventional depiction of psychopathy as an empirically well-validated disorder is seriously misleading. Not only are the common prognostic claims about extreme dangerousness, untreatability, and moral incapacities poorly supported by the available evidence, but there are currently no evidence-based reasons for believing that psychopathy has discrete biological causes, nor are there compelling reasons to believe that scientists can theoretically explain what distinguishes PCL psychopathic persons from other people. That is, aside from the fact that these individuals have been clinically assessed and labeled according to the PCL item-checklist.

In this chapter, I will switch the focus away from the experimental work and its forensic implications, and instead try to address a comparatively much harder question: Why is it that there is such an enormous disconnect between what scientists and clinicians *say* about psychopathy and how PCL psychopathic persons *really are* when we study them in experimental settings? Or to put it differently: Why is it that there has never been any *scientific progress* in corroborating some of the most basic claims about psychopathy?

Before I explore potential answers to this question, I first want to point out how extraordinary it is that I can even ask this very question. Usually, when scientists fail to find empirical support for a specific idea, this typically leads to a change in how scientists collectively think about the idea itself. At first, the lack of evidence may lead to a slowly growing skepticism about the idea, which then later leads to revisions of earlier theoretical frameworks. And if the experimental evidence keeps falling short, the idea itself might eventually become entirely ignored, discarded, or better, replaced by competing ideas. This sort of tit-for-tat process is one of the most foundational trademarks of a scientific field, commonly referred to as the *self-correction* of science, where ideas are systematically tested and only the ones with evidential support earn the endorsement of scientists.[2] Scientific fields therefore rarely find themselves in a situation where there is a fundamental disconnect between its central ideas and the available evidence. Because it is exactly these types of unsupported ideas that scientific work is supposed to prevent us from endorsing.

What is particularly odd about psychopathy research is that not only has this self-correction process never taken place in mainstream circles—for instance, there is a clear similarity between Benjamin Rush's early theory to that of Hervey Cleckley's theory, which in turn is endorsed by many of today's researchers—but there have also never been any good, evidence-based reasons for accepting the many common claims about psychopathy in the first place. As I have demonstrated throughout this book, experimental evidence has essentially always been lacking. And what makes it even more strange is that this track record of scientific futility appears to have had little impact on the intellectual appeal of the idea and the field's scientific status: psychopathy is today a widely regarded area of study in mainstream scientific circles, researchers continue to publish their work in prestigious academic journals, and university students are systematically exposed to research on psychopathy in academic seminars and textbooks (as I covered in detail in chapter 1).

So, how did the field wind up in this truly peculiar situation where there is a long-lasting disconnect between what researchers say about psychopathy and what the empirical evidence is showing?

It is my impression that there are many different and reasonable answers to this question. Perhaps the most common answer I have encountered is that the lack of progress is a testimony to how difficult it is to study a personality disorder like psychopathy. This perspective is frequently expressed in scientific review articles and academic books, where authors often round up their work by acknowledging the lack of "hard" evidence and recommending new and more innovative directions for future research.[3] Sometimes scientists will emphasize that many of the methods and technologies we use to investigate psychopathy might not be powerful enough to detect their psychological differences. For example, task-based neuroimaging studies sure are fascinating, but we can legitimately question if these techniques are mature enough to study something as complex as a personality disorder.[4] According to this answer, then, the failure to corroborate common claims about psychopathy is not an indicator that these claims are fundamentally wrongheaded or false; it is an indicator of how scientists are operating within the confines of limited means. Perhaps it is similar to searching for extraterrestrial lifeforms in our galaxy with a pair of binoculars. The point being that if this is really the case, psychopathy research is then excused for its obvious lack of scientific progress.

I think this explanation gets many things right, but I am not sure I find it particularly compelling. There are many reasons for this, but the most pressing one is that it runs the risk of greatly misrepresenting how much good research there is on PCL psychopathic persons. Although I have expressed concerns about the quality of many studies throughout this book, it would be wrong to think that all research has methodological shortcomings. There are hundreds of studies led by prominent behavioral scientists with formidable research capabilities, and in many cases their methods are truly innovative and well executed. And what is more, these researchers are not searching for the metaphorical *needle in the haystack*, but they are testing extraordinary hypotheses that should be relatively straightforward to measure. If common claims about psychopathy are true—for instance, that they lack empathy or that they are extremely impulsive—I then fail to see why we would not have been able to measure it using the methods and technologies currently at our disposal. Testing hypotheses about such psychological extremities should be something that contemporary behavioral scientists are able to do. So, I do not think the lack of progress in psychopathy research can be sufficiently explained by the general difficulty of studying it.

A second popular explanation for the lack of evidence in the field is that the *clinical assessment tools* researchers use for sampling purposes may be fundamentally flawed, such that they are not reliably identifying "truly" psychopathic persons to include in their studies. This sort of explanation has resurfaced many times in the research history, and it has given rise to a debate about which assessment tools are best suited to identify so-called real or truly psychopathic individuals.[5] According to this explanation, the lack of research progress I have documented in this book can thus be boiled down to a problem with the PCL scales, such that when we conduct studies on samples that have been selected with these scales, we are effectively not studying a "clean" sample of psychopathic persons. Of course, what ends up happening when scientists are studying misrepresentative samples is that they keep getting mixed results. It would be similar to a situation where scientists are studying a specific disease where they are unaware that their sample is contaminated with people who are not carrying that specific disease (e.g., due to misdiagnosing). Such studies are likely to yield all sorts of puzzling results, making scientific progress unlikely.

Again, I think there are important elements in this explanation, but I am not sure how compelling it is. While sample contamination is a major issue in mental health research,[6] there are at least four reasons why I do not think it sufficiently explains the lack of evidence in psychopathy studies. For one, without getting into too much detail, in the experiments we have explored throughout this book, scientists often incorporate several strategies to test for sample contamination, such as running elaborate moderator, regression, or post hoc exploratory analyses of the raw data. Second, there is a decades-long broad consensus among leading researchers and clinicians that the PCL scales are the best tools to identify the psychopathic patient stereotype. And it is this stereotype that researchers make claims about, not some other, different stereotype. Third, if we look at the research conducted on samples selected using competing assessment tools, the experimental results still fail to clearly corroborate common claims about psychopathy, suggesting that samples that are alleged to be less contaminated still yield no better evidence.[7] A fourth and final reason is that we must always be suspicious when behavioral scientists explain the lack of evidence by pointing to (unproven) problems with sample selection, because this type of explanation can almost always be invoked when evidence is lacking: we can always claim that the sample was misrepresentative if we did not find the evidence we expected. For that reason, it is also a type of explanation that conveniently gives biased and wrongheaded ideas an "out," allowing for the perpetuation of ideas that would otherwise have been deemed falsified. Furthermore, notice that researchers rarely think there is a problem with assessment tools if their theoretical predictions were corroborated; the contamination-explanation is typically used to explain *false negatives*, not false positives. This, of course, makes the concern about sample contamination inherently suspicious.

In this chapter, I will explore a third and radically different explanation of the lack of scientific progress in the field, and also an explanation that has yet to be broadly discussed in the contemporary mainstream literature. I will consider the possibility that perhaps the lack of progress is rooted in psychopathy being an example of what scientists colloquially refer to as a "zombie" idea.[8] A zombie idea can be defined as an idea that is intuitively appealing to researchers but, unbeknownst to them, is fundamentally misconceived, with no corresponding basis in reality. In essence, I will explore whether there are good reasons to be skeptical about the "reality" of psychopathy as a personality disorder, which simply means to doubt whether the idea itself is fact-based: to doubt whether there are any individuals whose psychology truly corresponds with the ordinary description of psychopathy (i.e., social predator, morally colorblind, extremely impulsive, unemotional, and so on). If no such individuals truly exist, that would directly imply that psychopathy is not real or (as philosophers would say) that its reality is that of a *fiction*. However, the fact that scientists are currently studying psychopathy, and thus presumably find the idea intuitively appealing—dealing with it *as if* it is a fact-based idea despite the lack of evidence—thus makes it a textbook example of a zombie idea.

However, before I consider the zombie explanation in greater detail, I shall first say something about what it takes or better, how difficult it is to show that a particular idea is also a *zombie* idea.

Zombie Ideas and How to Identify Them

When scientists are unknowingly caught up in studying a zombie idea—for instance, consider historical examples like *phrenology*, *luminiferous aether*, or the *geocentric* view of the universe—what ends up happening is that generations of researchers keep being attracted by the idea presumably because of its intuitive appeal. Like zombies, the idea "bites" and "infects" new generations of researchers. However, since the idea has no corresponding basis in reality—like zombies, the idea is not "alive"—researchers will continuously fail to find any clear and convincing evidence to support common claims about the idea, even as they relentlessly execute one experiment after the other. Yet, upcoming generations will keep repeating the cycle, somehow collectively overlooking or ignoring all the tell-tale signs that their scientific field is making little or no progress. To me, whether we like to admit it or not, this looks a lot like what has been going on in psychopathy research, especially in the past three to four decades where the number of scientific publications grew with exponential speed. I am, of course, not ignorant about the fact that some researchers will find it controversial and be strongly opposed to framing psychopathy as a zombie idea. But it should be relatively *uncontroversial* to suggest that this type of explanation can account for the lack of scientific progress that I have documented throughout this book.

Now, one major challenge with this explanation is the difficulty in *proving* that psychopathy is not real. To say that psychopathy is a zombie idea is exactly to propose that it is not real, that the idea is essentially pure fiction. However, to make such an argument in a sufficiently convincing way, we would basically have to *prove a negative*, which requires an extraordinary amount of interrelated high-quality evidence, which I do not think is currently available in the field.[9] Another major challenge with the zombie explanation is that it can very quickly be accused of making a logical fallacy. For example, it might seem tempting to argue that since we have not found any positive evidence of psychopathy, it must somehow mean that psychopathy is not real. But making this argument would be equivalent to making the informal logical fallacy of *appeal to ignorance*. Surely, the absence of evidence rarely proves a negative.[10]

To avoid wrestling with these two major challenges, I will not attempt to prove or argue that psychopathy is a zombie idea. Instead, I will aim to provide reasons why scientists should be skeptical about psychopathy and establish why such skepticism about the idea cannot be ruled out and might even be a rationally appealing position at this stage in the research history (where we benefit from "20/20 hindsight").

To build this case, I will proceed in ways that scientists usually do when they distinguish between good and bad scientific ideas. That is, when they distinguish between ideas they provisionally accept as legitimate targets of experimental work vs ideas they are skeptical about and thus deem less deserving of scientific inquiry.[11] In making this distinction, it is conventionally believed that skepticism is warranted if at least two conditions can be established:

Condition 1: Skepticism is warranted if there is no robust experimental evidence to support the idea. For example, if scientists have thoroughly tested an idea but continuously failed to find good evidence to support it, this absence of evidence is enough reason for skepticism. This is because scientific ideas are generally weighed in terms of the empirical support they can muster. When ideas lack empirical evidence, they risk falling prey to a brutal form of collegial skepticism expressed in the common saying: that which is asserted without evidence can be dismissed without evidence (sometimes referred to as *Hitchens' razor*, a play on the more famous *Occam's razor*).

Condition 2: Skepticism is ordinarily warranted if the idea contains elements that make it rationally unappealing. For example, if it can be convincingly pointed out that there are problems with the idea that undermine why we should believe it, that is, rationally accept it, this is sufficient grounds for skepticism. This is because when an idea is rationally unappealing, it detracts from the very role ideas play in scientific inquiries. Ideas are, after all, expressions of human rationality, and when something is deemed rationally unappealing, it should by definition warrant skeptical attitudes (especially in the absence of corroborative empirical evidence).

Thus, I shall premise the conversation in the remainder of this chapter on these two conditions. That is, if it can be successfully demonstrated that the idea of psychopathy fails on these two conditions, then I think it should be uncontroversial to suggest that we should also be skeptical about whether there is anything in reality that corresponds with what is entailed by the idea—whether there are real persons that exhibit everything that is ordinarily entailed by the idea of psychopathy.

For the sake of brevity, I shall assume that I have successfully established *Condition 1* in chapters 2–6, where I have shown that there is no clear scientific evidence to suggest that the individuals we identify as clinically psychopathic—those who score high on the PCL assessment—fit the conceptual and theoretical accounts researchers invoke with the term psychopathy. Thus, in the following pages, I will focus primarily on *Condition 2*, exploring elements that might render the idea of psychopathy rationally unappealing. To make this case, I will proceed by showing that the idea of psychopathy faces two fundamental problems when it is weighed against conventional standards that scientists use to evaluate scientific ideas. I will close the chapter by considering two common objections to my skeptical position.

Problem #1: Uncritical Acceptance of Abductive Triggers

The first problematic element embedded in the idea of psychopathy relates to how scientific inquiries ordinarily begin and how researchers generally justify their efforts to explore an idea. Not everything that can be conceived of necessarily qualifies as a proper subject of scientific scrutiny. There are certain ground rules that scientists follow when deciding what factual observations to direct their focus and resources at. For example, contemporary scientists generally do not spend resources testing whether planet Earth is a flat disk. A major reason why scientists snub an idea like this is that they do not think it passes some minimal rational requirements of being worthy of their time. In this section, I consider whether this might also be the case for the idea of psychopathy, that it too similarly fails to pass some minimal standard requirements for initiating and continuing scientific inquiry.

To make my point, we must take a step back and appreciate more carefully what science *is*. At its most basic level, science can be described as an *empirical* approach to generating knowledge, meaning that whatever scientists are studying, and whatever they claim to have scientific knowledge about, will always be based on some form of measurable facts.[12] In this way, scientists are often engaged in a type of discovery process where they are aiming to make sense of specific observations—that is, they are trying to understand why certain measurable facts are the way they are. As described in chapter 6, science might be characterized as seeking a *why* of a *what* (where the *what* is the observed facts and the *why* is the explanation of said facts). This process of seeking explanations for specific factual observations almost always begins with an act of what the philosopher of science Charles S. Peirce crowned *abductive reasoning*,[13] where scientists draw their attention to some salient aspect of reality (i.e., facts), of which they believe they have an insufficient understanding. And it is at this initial stage of the inquiry process that scientists begin to form potential theories that account for the observed facts (theories that are then tested experimentally at a later stage in the scientific process).

We can therefore think of this scientific process as beginning with a factual observation that surprises us, or rather *triggers* our curiosity, and from there on we initiate the intellectual pursuit of wanting to understand this observable fact.[14] For instance, to use a historical cliché, many people have heard the tale of Isaac Newton sitting under an apple tree and being struck by a falling apple, which then triggered him to ask the question: Why are objects moving downward and not upward? More realistically, Newton was probably inspired by a comet that passed by the Earth in the year 1680. This comet had a long bright tail (of dust and gas) that appeared to indicate that it was moving in a curved path (which comets do if they are in a strong gravitational field). This observation might have inspired Newton to ask the following question: Why is the comet moving in a curve and not a straight line? The central thing to notice here is not so much what exactly gave rise to these thoughts, and what thoughts subsequently led Newton to his theory of gravitation. For our purpose, the central thing to notice is that the entire process began with something that triggered Newton's curiosity and made him realize that there was an aspect of reality he did not fully understand: the comet's curved path (i.e., in this case, the comet's path *is* the abductive trigger).

Obviously, not everyone was triggered by the comet in the same way as Newton. Some people who observed it might have been surprised by other facts, such as the comet's brightness in the evening sky or its eventual disappearance after several days in plain sight. But to Newton, it was the curved path that caught his attention. What exactly it takes for people to have their curiosity triggered in this specific way presumably depends on their overall background knowledge and attitudes about the world.[15] If a person did not find the comet's curved path surprising, if they did not pick up on that detail, it is probably because their background knowledge did not allow for that detail to inspire any form of surprise. Maybe because to them, a straight or curved tail would be all the same. Without the appropriate background knowledge there will be no abductive trigger, no reason for surprise. Consequently, there will be no occasion where the person shifts into the "scientific mode of thinking," where they approach the issue methodologically to formalize and test explanations for the observed reality: seeking an empirically informed *why* of the *what*.

We can use this example to pinpoint more closely how the scientific study of psychopathy began, and therefore also how the reasoning about the idea was shaped. Historically, what must have occurred is that scientists— physicians, psychiatrists, psychologists, and others—were surprised by an encounter with a specific patient stereotype. That is, a type of patient that perplexed them and inspired the need to explain why these people are the way they are—specifically, to explain what makes these individuals psychologically different from ordinary people. This abduction-triggering occasion would not have occurred, of course, if there was nothing unusual about these patients. In other words, the background knowledge of the scientists who initiated psychopathy research must have been in such a state where they could not square their observation—about the patient stereotype—with what they collectively believed to be *normal* human psychology. The patient stereotype was, to these early scientists, an extraordinarily surprising and curiosity-triggering encounter.

With this perspective in mind, it is therefore interesting to observe that throughout history, psychopathy researchers have disagreed enormously on how to describe this stereotypical psychopathic patient. For example, when we read some of the earliest accounts of psychopathy—by Benjamin Rush, Philippe Pinel, Jean-Étienne Dominique Esquirol, James C. Prichard, Cesare Lombroso, Henry Maudsley, Emil Kraepelin, and others—they can reasonably be accused of not even describing the same type of patient.[16] As documented in chapter 1, this sort of confusion and disagreement persisted well into the modern era. In the 1970s and 1980s, one of the largest debates in the field was on how to *describe* a stereotypical psychopath, where Aubrey Lewis summed up the debacle in 1974 by calling psychopathy the "most elusive category" in psychiatry.[17] It was not before the 1990s that psychopathy researchers came to some form of consensus on what facts they aspired to explain. Namely, consensus took shape around Hervey Cleckley's description outlined in his iconic book *The Mask of Sanity*, a consensus that was later reshaped into the stereotype described in the PCL (1980) and the PCL-R (1991) manual by Robert Hare and his colleagues.

What this history reveals is that when researchers were describing different patient stereotypes in the years before the 1990s, they were, by definition, speaking about different abductive triggers. They were referring to different types of observations that had kindled surprise and curiosity, observations that they needed and wanted to better understand. These different patient descriptions are different kinds of *what* in need of a *why*. Of course, there was some meaningful overlap between these types of *what*, but scholars who study the history of psychopathy agree that the differences were often large. As Robert Hare mentions in a 2007 article, it was almost as if experts in the 1970s were talking about different disorders.[18]

When seeing the history of psychopathy research through this lens, it is then important to note that the patient stereotype Hervey Cleckley is

talking about in *The Mask of Sanity* was largely brushed aside by his contemporaries as something that was entirely unworthy of their intellectual
time and effort. In a personal correspondence with Robert Hare in the 1970s,
Cleckley describes himself as a "voice crying in the wilderness,"[19] and we can
read in the foreword to the fifth edition of *The Mask of Sanity*, published in
1976, that the author is disappointed about the status of his lifelong work,
regretfully admitting that it has had virtually no impact on contemporary
psychiatry. To Cleckley, the notion of a psychopathic person had become the
"forgotten man of psychiatry."[20] This lack of impact cannot be explained by
Cleckley's contemporaries being unaware of his conception of psychopathy
because, by the 1960s, Cleckley was already one of the most famous psychiatrists in North America, being the coauthor of the best-selling book on multiple personality disorder, *The Three Faces of Eve*, published in 1957, which in
the same year was adopted into an Oscar-winning movie. A few years later,
in the year 1964, *The Mask of Sanity* was published in its fourth edition. In
short, there is nothing to suggest that scientists and professionals alike were
unaware of what Cleckley had to say about psychopathy.

Because of these historical facts, it is only reasonable to assume that in
the 1980s it was considered an entirely rational and mainstream position to
be skeptical about Cleckley's ideas, perhaps seeing them as something unde-
serving of scientific scrutiny. Of course, not everyone agreed. For instance,
David Lykken and Robert Hare were some of his most prolific supporters in
the 1970s,[21] and a small community of psychopathy researchers and clinicians honored Cleckley's work on psychopathy after his passing in 1984.[22]
However, according to mainstream psychiatrists and psychologists in the
1980s, the stereotype described in Cleckley's book—which contains detailed
vignettes of 15 different patients—did not inspire any curiosity. One way
to interpret this is that to mainstream scientists, these stereotypes were not
considered appropriate abductive triggers.[23]

There may be various reasons why researchers did not find Cleckley's
stereotype scientifically interesting. For example, maybe Cleckley's contemporaries just saw his stereotype as something that could be explained by other
simpler and more commonsense ideas. There are at least two good reasons for
assuming this: first, many of the patients described in *The Mask of Sanity* had
obvious comorbidities such as severe addiction and tertiary syphilis (which
can cause all sorts of aggravated psychiatric symptoms). Second, it is possible that mainstream scientists simply did not see Cleckley's psychopathic

stereotype as something that needed an explanation, but rather as an entirely ordinary instantiation of normal human psychology (I explore these two possibilities in more detail in the next subsection). Perhaps this is why there was vocal opposition to Cleckley's idea of psychopathy: for instance, when it was ridiculed by people like Hans Eysenck, describing the idea as a *white elephant* and advising his colleagues to reject it as a proper theme of scientific inquiry (to be fair, Eysenck's criticism was not only directed at Cleckley but ostensibly directed at the entire idea of psychopathy).[24]

However, as we know, this status quo began to change in the 1980s, and by the mid-1990s, there are clear signs that the patient stereotype described by Cleckley and the PCL-R manual was no longer ignored by contemporary scientists. It is difficult to say what exactly made hundreds of scientists change their minds, but prevailing disagreements and skepticism about psychopathy were becoming increasingly rare in academic journals during that period, and the disorder was beginning to make its way into psychiatric and forensic practices with breathtaking speed.[25] In my perspective, a combination of various influences may have contributed to this swift change in attitude.

First, one reason (and I think, the most obvious one) might be that it was the empirical research that changed the minds of skeptics. As I have documented throughout this book, there are at least a handful of early experimental results that appear to suggest that Hervey Cleckley's conception of psychopathy was vindicated. For example, his theory predicts that psychopathic persons would have deep difficulties with moral reasoning, and this is what was concluded in James Blair's famous 1995 experiment testing psychopathic persons' ability to distinguish between moral vs conventional issues.[26] Similarly, some early researchers reported that psychopathy was strongly linked to criminal recidivism, like a 1995 prediction study by Ralph Serin and Nancy Amos.[27] There is also the study by Marnie Rice and colleagues from 1992 that suggested that psychopathic persons were not only untreatable but that they get worse when subjected to rehabilitation programs.[28] Of course, early studies like these are no longer compelling due to their obvious methodological shortcomings and the collective failure to replicate their results. But it is easy to imagine that they must have had a significant impact on scientists who were skeptical about psychopathy. And one might add that this would certainly be an appropriate way to adjust one's views on psychopathy insofar that it would be driven by empirical evidence. To give an example of how profound the change was, consider that a scholar like Ronald Blackburn characterizes psychopathy as a "myth" in the year 1988,[29] but by the mid-1990s, he is becoming increasingly involved in the research paradigm,[30] and psychopathy is discussed in detail in his popular university textbook *The Psychology of Criminal Conduct* (first edition in 1993, latest reprint in 2008).[31]

Second, in 1994, the American Psychiatric Association included (for the first time) a direct reference to psychopathy in the fourth edition of their influential *Diagnostic and Statistical Manual of Mental Disorders*. Although the fourth edition attracted criticism for not properly defining psychopathy—essentially designating psychopathy as a synonym for antisocial personality disorder (see chapter 1)—the fact that the term psychopathy was now included in an authoritative, "official" resource implies that researchers may have felt beholden to treat it as an idea worthy of scientific pursuit. An often-forgotten part of these developments is that before the publication of the fourth edition of the diagnostic manual, the American Psychiatric Association actually commissioned Robert Hare and his colleagues to participate in a workgroup meant to (among other things) align the definition of antisocial personality disorder more closely with how psychopathy is defined in the PCL-R manual.[32] Again, this suggests that the shifting attitudes among scientists and clinicians might have been propelled by institutional influence as well.

Third, scientists and clinicians may have changed their minds because they were increasingly swayed by scores of anecdotal descriptions of stereotypical psychopathic patients surfacing in the clinical literature. These accounts are packed with first-hand encounters with patients who are described as lacking emotion, empathy, and conscience, as if it is a matter of indisputable fact. For instance, J. Reid Meloy's book from 1988, *The Psychopathic Mind*, draws on his personal experiences from working in clinical psychiatry.[33] And Robert Hare's bestselling book from 1993, *Without Conscience*, includes many examples from within the prison system.[34] One can imagine that some researchers might have been wondering that if there was some truth to these anecdotal accounts then it is certainly worthy of scientific inquiry. Today, we are much more skeptical about the inferential value of anecdotal evidence like this—because they are mostly based on the clinician's personal impressions, which might be biased and misleading in myriad ways—but in the 1980s and 1990s, anecdotal evidence might have been more accepted and seen as genuinely informative, and therefore perceived as scientifically more relevant than it is today.

Fourth, what may also have changed researchers' minds is that popular culture was increasingly capitalizing on the idea of psychopathy.[35] For instance, one of the most popular novels of the 1990s is Bret Easton Ellis's *American Psycho*, published in 1991, which was turned into a movie in 2000, starring a young Christian Bale as the paranoid serial killer Patrick Bateman. One of the most successful movies at the box office in the 1990s was Jonathan Demme's *The Silence of the Lambs* from 1991, starring Anthony Hopkins as Hannibal Lecter, the cannibalistic murderer. Other popular films from that period exploring the theme of psychopathy include Stanley Kubrick's *A Clockwork Orange* (1971), Terrence Malick's *Badlands* (1973), Martin Scorsese's *Taxi Driver* (1976), and Oliver Stone's *Wallstreet* (1987). This increasing focus on psychopathy in pop-cultural productions may have made it difficult to resist the temptation to study psychopathy if you were a forensic psychiatrist or psychologist. Not only would a cultural fascination have generated a demand for expert knowledge, but many scientists would probably agree that it is much easier to receive grant money to study an idea that interests the broader public. In addition, it is difficult to ignore the possibility that pop culture may have shaped the minds of upcoming generations of scientists. So, if you grew up watching fascinating movies about psychopathy in the 1980s and 1990s, this might precondition you into thinking that the idea legitimately deserves scientific scrutiny. And, perhaps, such preconditioning might nudge you into picking psychopathy as a research topic, first as a graduate student and later as a career scientist.

Other influences could have been the explosion of literature about serial killers in the 1980s, which, as Eric Hickey has documented, really took off after the live broadcasting of the 1979 trial of the infamous serial killer Ted Bundy, which captivated the mainstream media and millions of people.[36] In addition, North America went through a period of a dramatic increase in crime rates in the 1980s, which later ignited the so-called tough-on-crime movement.[37] With this movement, it is fair to speculate that funding for crime-related research may have increased, including forensic psychiatry and psychology, where psychopathy research would have been an obvious way to justify the relevance of these disciplines. Another curious observation related to the tough-on-crime movement is how the media and politicians (and some researchers) spun off ideas like the *super predator*, which was largely a synonym for juvenile psychopathy.[38]

My point in highlighting these observations is that it is likely that there were many different incentives for scientists to engage in the study of psychopathy. But regardless of what may have been the main influences, something must have changed the opinion of scientists and clinicians in the 1990s, such that the idea of psychopathy was no longer being actively ignored or stamped as rationally unappealing, instead seen as something that warranted scientific scrutiny. However, what is crucial to notice about this development is that the idea of psychopathy itself never changed. The switch that we see among researchers regards a change of mind and their perspectives, not a change in ideas and its evidence. With the turn of the decade in 1990, the psychopathic stereotype described by Hervey Cleckley and the PCL manual is increasingly being seen by scientists as a proper curiosity-triggering aspect of reality—a proper abductive trigger—where before the 1990s it was not. But what if scientists and clinicians were simply mistaken in making this switch in perspective? I do not find this speculation far-fetched, as many of the influences I mentioned above are entirely unrelated to the scientific reasoning and discovery process.

To fully understand the consequence of *wrongfully* designating an observation as a proper abductive trigger, we must appreciate what sort of intellectual efforts are set in motion with this change in attitude. As mentioned earlier, what usually happens is that scientists will begin to generate novel ideas and theories meant to account for and make sense of the observed reality. However, if the observed reality is not a "proper" abductive trigger—for instance, because we already have a different but true or reasonable understanding of what we observe—scientists are bound to create and formulate explanations that are fundamentally misleading. The outcome being that their ideas do not *correspond* to anything in reality, let alone the observed reality they intend to explain (i.e., the most basic aspect of a zombie idea).

To illustrate this, consider a scenario where a patient shows up in a psychiatric clinic displaying a set of severely debilitating symptoms. One psychiatrist finds the patient entirely unremarkable, stating that we have good reasons to think that the patient has withdrawal symptoms related to substance misuse. Another psychiatrist at the clinic disagrees and finds the patient curiosity-triggering, and immediately begins formulating a theory about a novel personality disorder meant to account for the symptoms observed in the patient. In this scenario, only one of the two psychiatrists can be right.

And if we imagine that it is the former account of substance misuse that ends up being the true account, it then follows that the second psychiatrist—due to wrongly perceiving the patient as an abductive trigger—was led astray by their own perspective, thus creating a thoroughly misleading account of the initial observations.

We only need to tweak this imaginary example a little bit to see its implications for psychopathy research. In this field, researchers are preoccupied with making sense of or explaining the patient stereotype laid out by Hervey Cleckley and subsequently in the PCL-R manual. These efforts include the assumption that the patient stereotype suffers from a personality disorder, and various theories have been formulated to explain the psychological symptoms and their causes, generating a wide array of claims and prognostic predictions. However, what if these intellectual efforts were fundamentally misguided in similar ways as the psychiatrist in the example above, who misinterprets withdrawal symptoms for a mysterious personality disorder?

It should be uncontroversial to suggest that when scientists are mistaken about abductive triggers, it has the potential to lead them on a metaphorical wild goose chase. But the question I pursue in this chapter is whether we have reason to believe that psychopathy research falls into this category, that the field has been established on a hollow foundation, wrongfully assuming that the patient stereotype is a proper abductive trigger. One way to probe this suspicion is to look at the actual accounts of the patient stereotype (i.e., the abductive trigger) described in Hervey Cleckley or Robert Hare's work—which both were elemental for the development of the PCL-R manual, as well as the current research paradigm—and then ask ourselves if we (today) still think that these descriptions indicate something extraordinary, something that ought to trigger our curiosity.

To be sure, there are elements of Cleckley's and Hare's descriptions that I think *should* trigger our curiosity—that is, *if* these elements were indeed true. For example, *if* the notion of extreme dangerousness, untreatability, moral colorblindness, and lack of emotions and/or poor impulse control were truly measured in the patient stereotype, then I would be the first to admit that it would be a curiosity-triggering observation. Consider, for instance, how Cleckley in *The Mask of Sanity* compares psychopathic persons to the glass flowers exhibited at the Harvard Museum of Natural History (created by Leopold and Rudolf Blaschka), which are so craftily made that it is almost impossible to tell that they are not real organic material. Using this example, Cleckley conveys the idea that his patients have a near-perfect resemblance to that of ordinary people—just like the Blaschka-glass flowers resemble real flowers—but that the emotional life of psychopathic persons is somehow like glass, hardened and inanimate, "They are blind to this mode of awareness," Cleckley states.[39] If this characterization was true, if psychopathic persons truly were emotionally blind, that would certainly be a curiosity-triggering fact. But the point is that traits like these have never been corroborated in experimental work and might therefore be better characterized as misleading *impressionistic* accounts. Cleckley probably believed that his characterization was true, but that does not mean that he could not have been mistaken.

The consequence of seeding doubt about these clinical impressions is far-reaching. Consider what happens if we remove from the Cleckley-Hare stereotype the traits of shallow emotions, extreme impulsivity, and moral colorblindness (elements that have been thoroughly challenged by empirical research). What we are left with is a description of a person that bears little resemblance to the idea of psychopathy—of course, a person that may have problems adapting to society's expectations of proper behavior—but nevertheless an individual without any clear psychological abnormalities along the lines that psychopathy researchers ordinarily claim. And, from a scientific point of view, such a person might even constitute a rather mundane observation—an observation that we can question whether it even warrants further scientific inquiry, let alone setting in motion an entire scientific field to study it.

Many of the curiosity-triggering qualities of the Cleckley-Hare conception of psychopathy appear to hinge on a handful of basic impressions, such as the notion that these individuals are emotionless and therefore entirely colorblind to the values of human affairs. It was this quality that inspired Cleckley's metaphor of psychopathic persons wearing a *mask*—that is, a veneer of normality that conceals that they are cold, amoral, disinterested people. And this metaphor is the centerpiece in Hare's global bestseller *Without Conscience*, where the reader is introduced to this psychological duality, where psychopathic persons look normal but behave in vile and abhorrent ways without the slightest sense of guilt and remorse. Altogether, this characterization runs through the definitions of the item-checklist of the PCL-R manual, most obviously in item 5, *conning/manipulative*, item 6, *lack of remorse or guilt*, item 7, *shallow affect,* and item 8, *callous/lack of empathy* (see figure 1.2 in chapter 1).

To be clear, with these observations, I am not making a statement about which stereotype of psychopathy might be the most appropriate, nor am I criticizing the scholarly work of Cleckley and Hare. What I am pointing out is that when we are assuming that the Cleckley-Hare psychopathic stereotype is a proper abductive trigger, this is in and of itself potentially problematic. What is perhaps needed in the field is a deeper and genuine conversation about what exactly it is that is so extraordinary about the patient stereotype—those who score high on the PCL scales—that we necessarily need sophisticated science to make sense of these individuals. That there appears to be ample room for such debate is a problem that significantly detracts from the rational appeal of the idea and could furthermore be taken as a sufficient reason to be skeptical about its reality.

On a more personal note, and with the obvious advantage of hindsight, I remain overly conservative about this issue and suspect that all of the qualities that have historically been presented by researchers as the true curiosity-triggering elements of psychopathy might have been nothing but inflated and misleading impressionistic views. Notably, this would not be the first time such an interpretation has dawned on Hervey Cleckley's work. For example, it has previously been alleged that his influential work on *multiple personality disorder* (today called *dissociative identity disorder*) generated proper confusion in the field due to a highly misleading and impressionistic account of one of his patients.[40] One particularly salient example is how his book, *The Three Faces of Eve* from 1957 (coauthored with Corbett Thigpen),[41] narrates a flip rendition of his patient "Eve," who later published several books that cast serious doubt on how Cleckley characterized her disabilities.[42]

So, with all these considerations in mind, I believe scientists would be wise in maintaining a skeptical attitude about the reality of the psychopathy stereotype until proven otherwise.

Problem #2: Redundancy and the Principle of Simplicity

The second problem that might make psychopathy a rationally unappealing idea has to do with how the idea itself could be seen as *redundant,* that it is used to explain something that we already have a sufficiently good and accurate understanding of. More specifically, my suspicion is that the psychopathic stereotype outlined in the historical and mainstream literature does not signify anything abnormal or truly puzzling—something in need of separate theorizing—but can instead be accounted for with established theories in the behavioral sciences.[43] For example, the individuals traditionally deemed "psychopathic," and labeled with the elusive personality disorder "psychopathy," might be nothing but mere variations of normal human psychology or perhaps suffering from something different than psychopathy. If this suspicion is warranted, it would make psychopathy a redundant idea; and such ideas are, by definition, rationally unappealing insofar that scientists typically follow an *epistemic rule-of-thumb* where a parsimony of ideas is preferred over a redundant plurality of competing ideas.

To fully appreciate the suspicion about redundancy, we must again take a step back and focus on some of the most basic procedures in a scientific inquiry. I outlined in the previous section how all scientific endeavors begin with the encounter of an abductive trigger, defined as a factual observation that one or more scientists think requires scrutiny. After this encounter, the next step is to formulate a conceivable explanation of this specific observation; we must somehow assemble our collective intellectual capabilities and imagine a *why* of the *what*, equivalent to Isaac Newton formulating his theory of gravity to account for the comet's curved path. When we work on formulating a theory in this way, scientists generally agree that there are certain quality standards that any theory must live up to. They do not believe that every explanation "goes," so to speak—that all explanations are equally appealing—but will generally only pursue, test, and study those theories that meet conventional quality standards (and when scientists ignore these standards, it usually leads them astray[44]).

Throughout history, scientists have proposed many different types of quality standards that are meant to, from a purely rational point of view, demarcate good theories from bad theories, separating theories scientists believe are worthy of further scrutiny from those that they deem less worthy. For example, some of the most well-known, mainstream quality standards are those of *falsifiability*, *explanatoriness*, *depth*, and *simplicity*, which have today become almost synonymous with how science operates.[45]

1. *Falsifiability:* A theory must be formulated as a defeasible explanation, meaning that scientists must be able to identify what type of evidence would undermine it. If a theory is not falsifiable, it cannot be shown to be wrong, and therefore it is unfit for scientific experimentation.

2. *Explanatoriness:* A theory must be able to explain all relevant facts about the observation the scientists aim to understand. If a theory cannot account for all the central facts, it is, by definition, an incomplete explanation.

3. *Depth:* A theory must be able to explain the observed facts without raising further questions. If a theory causes perplexity and confusion, it is therefore counterproductive to the very act of trying to enhance our understanding of the relevant facts.

4. *Simplicity:* A theory must be able to explain the observed facts in a manner that is not overly convoluted compared to alternative, competing explanations. If an explanation is unnecessarily complicated compared to its alternatives, scientists will deem the theory unappealing due to its redundancy.

In this section, I am limiting the conversation to a discussion of simplicity (even as there might be issues related to the other three standards). The quality standard of simplicity—also known as the *principle of simplicity* or the *law of parsimony*—is one of the oldest and most central doctrines of a working scientist, sometimes described in university textbooks as *Occam's razor* (attributed to the work of William of Occam in the fourteenth century[46]). At its core is a commitment to always prefer simpler explanations over complicated ones, but for a working contemporary scientist, it often boils down to ensuring that whenever we aim to understand and explain a specific aspect of reality, we must first consult the possibility that other scientists might already have outlined a simpler and therefore better explanation. Or, as Paul Meehl famously summarized the principle to behavioral scientists: "Do not concoct theories to explain facts already explained by an ensconced theory."[47] In short, if there are already reasonable and established ways of explaining whatever facts have triggered our curiosity, then we should resist the temptation to indulge in additional (and thus potentially redundant) theory formulation.

If we circle back to our previous example above with the two psychiatrists debating how to diagnose a particular patient, we will easily recognize that it is exactly this type of misstep of ignoring the principle of simplicity that the second psychiatrist makes. That is, the first psychiatrist makes the most rational decision insofar that they explain the observed symptoms (i.e., facts) by referencing established theories about substance misuse withdrawal. The second psychiatrist ignores this knowledge and embarks on a hopeless intellectual pursuit, concocting an explanation about a mysterious personality disorder, an explanation that is comparatively much more complicated than the established theory about substance misuse. There are, of course, cases where established theories provide insufficient explanations—which would be a violation of the principle of *explanatoriness* outlined above—and in these cases scientists would generally be justified in embarking on novel theorizing. Nevertheless, when simpler, established theories provide sufficiently qualified explanations, scientists should be rationally compelled by the principle of simplicity to favor this explanation.

With this example in mind, it may be fair to think of psychopathy as being caught up in a similar conundrum about redundancy, as if the idea is an overly complicated way of explaining something that common psychological theories are already capable of explaining. For instance, maybe the classic psychopathic stereotype is already sufficiently understood as a variation of normal human personality? In recent years, some researchers have indeed proposed this as a viable possibility. For example, the prominent researcher Scott Lilienfeld wrote (just before his premature death in 2020) that scientists "have learned that psychopathy is not independent of the general personality realm," where he argues that "psychopathy appears [to] be a constellation or configuration of several well-established personality dimensions" (sic., *my addition*).[48] Moreover, Lilienfeld speculates that perhaps psychopathy is not a uniquely discrete condition as has traditionally been assumed:

> The burgeoning evidence of close empirical linkages between psychopathy and general personality . . . raises the important question of whether psychopathy is as distinctive an entity as many of us, myself included, have long assumed. Rather than a unique condition warranting unique etiological models . . . might psychopathy instead be better conceptualized as a constellation or configuration of traits . . . that have long been familiar to personality psychologists . . . ? And if so, should the discipline of psychopathy be merged into the broader discipline of trait psychology?[49]

Lilienfeld is not alone in making these observations, that the idea of psychopathy aims to make sense of something that behavioral scientists already have a sufficient understanding of.[50] However, it is not entirely clear to me that other researchers would be willing to characterize the idea itself as *redundant*, as Lilienfeld is articulating in the quote above (even Lilienfeld himself appears to have gone back and forth on the issue[51]). Nevertheless, it is reasonable to conclude that if the psychopathic stereotype can be fully explained by invoking ordinary psychological theories about normal personality, this would also render the idea itself redundant.

Notice that many psychopathy researchers appear to disagree with this viewpoint, and I have often heard verbal responses (but seen very few written responses) stating that personality psychology *cannot* account for the psychopathic stereotype exactly because it is a personality disorder. That is, the traits that are associated with psychopathy are *abnormal* instantiations of human psychology and personality. For example, while it may be normal for humans to have varying degrees of empathy,[52] researchers may claim that with psychopathic persons this capacity is abnormally lacking or severely muted. Similarly, where humans generally have diverging ways of reasoning about moral issues,[53] psychopathy researchers might respond that psychopathic persons are morally colorblind, that they are abnormally incapable of comprehending moral values. Since these varying forms of abnormalities are not directly accounted for within ordinary psychological theorizing, researchers might hold that the idea of psychopathy is warranted as an attempt to explain these abnormalities. The problem with such potential responses, though, is that they are not anchored in facts. That is, to the best of our knowledge, researchers have never detected a person who is completely deprived of empathic capacities, nor have they documented a case of moral colorblindness. And for this reason, scientists have no clear motivation to indulge in speculations about such abnormalities.

To further demonstrate how psychopathy as an idea might be considered redundant, I shall now go back to where the modern study of psychopathy started—with Hervey Cleckley's *The Mask of Sanity*, fifth edition published in 1976—and analyze if there might have been simpler, established theories that could have accounted for the stereotype Cleckley is theorizing about. *The Mask of Sanity* is broadly celebrated for containing the most detailed descriptions of the psychopathic patient stereotype ever to be published, and it therefore allows us to analyze whether the idea of psychopathy was "concocted," to use Paul Meehl's term, in light of already established and simpler explanations. Cleckley's book contains a total of 15 patient vignettes—which are based on Cleckley's own clinical work at a hospital in the US state of Georgia—each listed with a different pseudonym (e.g., "Tom"). On average, each vignette contains around 10 pages of information, totaling 168 pages in the 1988-reprint of the fifth edition. Just for comparison, consider that modern resources generally avoid including real clinical vignettes, for instance, James Blair and colleagues' 2005 book, *The Psychopath*, only has a couple of pages describing a fictional account of the stereotype,[54] and Jacqueline Helfgott's 2019 book, *No Remorse*, is limited to describing abstract general traits (just as I did in chapter 1).

Overall, it is my impression that it should be fairly uncontroversial to say that all of the 15 vignettes included in *The Mask of Sanity* can be fully accounted for by established theories in psychology, that there is nothing mystifying and strange about these cases that would warrant how Cleckley initiates his theorizing about psychopathy. More specifically, Cleckley's fallacy is that he overlooks that his patients' symptoms and behaviors could have been explained either by (1) invoking theories from personality psychology (as proposed by Lilienfeld above), or (2) by invoking theories about other mental health diagnoses. In this sense, I believe that all 15 vignettes may be categorized into three different groups, a categorization that I think most contemporary readers would immediately recognize:

a) *Normal personality (non-criminal):* Patients that are utterly normal but may not have behaved in ways that complied with the social norms of upper- or middle-class life in the state of Georgia during the post–World War II era (five cases: Roberta, Pierre, Anna, Milt, and Stanley).

b) *Normal personality (criminal):* Patients who are mostly petty criminals often with unsavory personalities, but arguably with no clear indications of psychological abnormalities (two cases: Tom and Gregory).

c) *Other mental health diagnoses:* Patients who have a long history of severe problems with addiction, post-traumatic stress disorder (PTSD), homelessness, clinical depression, tertiary syphilis, suicide attempts and threats, and so on (eight cases: Max, Arnold, George, Frank, Jack, Chester, Walter, and Joe).

In short, a compelling argument can be made that every single example Cleckley writes about is not a proper abductive trigger that justifies additional scientific theorizing about a disorder like psychopathy. Here I give a brief commentary on each of the three categories.

Category (a)—Normal personality (non-criminal): As a contemporary reader of *The Mask of Sanity*, it is difficult to read through these five vignettes without also thinking that the reason Cleckley finds these patients so extraordinary is because he holds all sorts of unconscious biases, such as preconceptions about gender roles and what a "normal" social life should look like. There are multiple examples in Cleckley's writings that gives rise to a suspicion about bias, but it is especially clear in his write-up of the two female cases (Roberta and Anna). For example, Cleckley is continuously expressing a subtle distaste for the two women's decision to have a non-monogamous sex life. He also appears to attribute moral fault to the women for not living up to the strict expectations of the socio-conservative lifestyle afforded by their parents in the segregated state of Georgia. It gets particularly odd when Cleckley describes how Roberta's parents forced her to see a gynecologist to check her hymen and for pregnancy, where Cleckley could be accused of ignoring the more pressing issue, that Roberta appears to have a mutually destructive relationship with her parents.[55] In the case of Anna, Cleckley dedicates several pages to detailing how Anna embarrassed her parents as a teen by fooling around with several boys at her school, an incident that led to the parents sending Anna off to a "fine boarding school in a distant state."[56] During her time away, Anna gets involved with minor shenanigans and rule-breaking (e.g., smoking), which all in all seems to be extraordinarily trivial, especially for a contemporary reader. Later in life, Anna enters a marriage that eventually fails—which Cleckley solely blames on Anna—after which the rest of the description centers around describing what would probably strike a modern reader as a moralizing distaste for Anna's bisexual proclivities, and her decision to simply live an unmarried life.

In general, none of the five cases in category (a) involve any major criminal behavior, nor any behaviors that a contemporary reader would see as a sign of psychological abnormality. The accounts mostly circle around trifling mischiefs, confrontations with family members as a youth, and problems internal to the person's marriage (e.g., domestic disturbance and violence). Or, as Cleckley puts it in the case of Milt, most accounts just show "signs of maladjustment."[57] Occasionally, these cases even mention information that is directly inconsistent with the idea of psychopathy, such as the case of Stanley who clearly has a very high-tempered personality (i.e., not a lack of emotion). But overall, it is safe to say that these five cases would not attract the attention of behavioral scientists had Cleckley published his work today.

Category (b)—Normal personality (criminal): This is the smallest category with only two vignettes (Tom and Gregory), both of which involve a history of antisocial problems as a child and adolescent that slowly evolve into more serious forms of crimes during adulthood (mostly property crimes). For instance, Tom is a school drop-out who drifts around in town, going from job to job and with changing social relationships, tallying up a record of dozens of encounters with law enforcement. But again, there are signs of strange biases when you read through the cases. Cleckley appears to show deep distaste for how Tom is "desultorily promiscuous" and that he marries a "girl who had achieved considerable local recognition as a prostitute and as one whose fee was moderate." Cleckley emphasizes that Tom "had previously shared her offerings during an evening (on a commercial basis) with friends or with brief acquaintances among whom he found himself."[58] In the case of Tom and Gregory, Cleckley never mentions any information about comorbidities or substance misuse, though he occasionally hints at it, which will probably make contemporary readers suspicious about whether he is withholding a type of information that could explain behaviors like property theft (e.g., to support substance dependent lifestyle).

Nevertheless, as a contemporary reader, a case can be made that Cleckley makes the fallacy of thinking that these two cases should trigger our curiosity as behavioral scientists. Indeed, it could be argued that Tom and Gregory are trivial textbook cases of juveniles getting progressively caught up in an antisocial lifestyle, which never evolves into anything serious enough to warrant speculations about a personality disorder like psychopathy.

Category (c)—Other mental health diagnoses: These eight vignettes amount to more than 50% of the observations that Cleckley bases his theory of psychopathy on. While all eight individuals have a long history of antisocial behaviors—and occasionally display extremely erratic impulses—they all have evidence of multiple mental health issues. Six out of the eight cases involve war combatant veterans (Max, George, Frank, Jack, Chester, and Walter) who are suffering from disabling comorbidities like PTSD, depression, addiction, and other illnesses like tertiary syphilis (the latter causes extraordinary psychiatric symptoms). The remaining two cases (Arnold and Joe) have endured extreme, decades-long alcohol addiction. When reading *The Mask of Sanity*, it can be a particularly strange experience when you realize that Cleckley is fully recognizing these comorbidities, even talking about them in minute detail, whereafter you begin to wonder if Cleckley is simply blinded (by tunnel vision) to these obvious and simpler explanations for his patients' behaviors.

For example, consider his patient Arnold, whom the hospital continuously allows probation, but who always ends up heavily intoxicated on the streets. One time the staff found him wandering half-naked in the rain, clearly delusional and "cursing in aimless violence" (Cleckley's book is a tapestry of such examples).[59] In all these cases, it is hardly disputable, nor should it be controversial, to state that these individuals' behavior can be sufficiently explained by the presence of PTSD, addiction, and so forth. One might go as far as calling them textbook examples of such diagnoses. Ironically, at one point in the book, Cleckley appears to notice that there might be a problem with overlooking the correlation between alcoholism and his patients, writing that "nonalcoholic psychopaths are not so rare as the preceding accounts might lead one to believe."[60] But if it is not so rare, why is most of the book still slated toward discussing these addiction cases? My personal view is that if you removed these cases, *The Mask of Sanity* would have been an utter bore, with little realistic chance of publication.

Obviously, many contemporary psychopathy researchers might take issue with the vignettes in *The Mask of Sanity*, and they may fairly point out that today there are better case examples to rely on as a researcher. However, the type of skepticism I am voicing in these paragraphs about *simplicity* is not a conversation about which case example is the best and most illustrative. The contention I am making is that *all* the stereotypes researchers are discussing in the literature will always be caught in the same scientific predicament as Cleckley's examples, where the attempt to theoretically account for the stereotype fail to meet the quality standard of *simplicity*: that their concrete clinical examples can be explained either by (1) invoking theories from personality psychology or (2) by invoking theories about other mental health diagnoses (or a combination hereof). In short, that the idea of psychopathy is fundamentally a redundant attempt to account for something that we already have a good understanding of.

Common Objections

In the preceding pages, I have showcased how there are reasons to be skeptical about the reality of psychopathy. This perspective is based on a two-fold observation: first, that researchers have yet to find any convincing experimental evidence of the existence of truly psychopathic individuals. Second, that there are two potential problems with the idea of psychopathy, which, from a scientific viewpoint, makes the idea rationally unappealing, namely: (a) that the field lacks a clear and obvious abductive-triggering observation in need of an explanation and (b) that the idea of psychopathy appears to be redundant. Altogether, this suggests that we have reasons to suspect that psychopathy is not real—that there are no individuals whose psychology truly corresponds with the ordinary description of psychopathy—which in turn would explain the lack of scientific progress in the field of psychopathy research.

I acknowledge that the perspective I have expressed in this chapter is speculative, and for that reason alone, my view is bound to be controversial, which in turn makes it a justified target of criticism. To forego such criticism, I will here survey two objections that I have frequently encountered when speaking to experts in the field and students at my home university, objections that I think that skeptics (like myself) must be prepared to answer.

Objection #1: The Existence of Serial Killers Proves that Psychopathy Is Real

The first objection I have in mind comes in many different variations, but it basically attempts to refute any skepticism by appealing to examples of alleged real-life psychopathic persons such as the existence of serial killers. Usually, it goes something like this:

> To claim that psychopathy is not real is to ignore basic observational facts, such as the existence of serial killers. These individuals obviously exhibit all the traits commonly associated with psychopathy. This includes their basic incapacity to empathize with and care for other human beings. The work by psychopathy researchers is an attempt to make sense of these salient fact-based observations, which arguably cannot be squared with what is considered "normal" human psychology. Hence, psychopathy as personality disorder is our best attempt to explain how these individuals deviate from the rest of the population.

I do not mean to make light of this objection, as it expresses a perspective that I personally accepted for many years, first as a graduate student and later as an aspiring researcher. And while I do admit that it carries enormous intuitive appeal, I have now come to believe that this appeal is vastly overrated. Here I summarize why I think that.

First, as a scientist, if you find this objection convincing, you must also recognize that the objection is not based on empirical facts obtained during scientific studies. To invoke examples about serial killers—and, correlatively, make sweeping claims about their psychology—is to draw inference from largely anecdotal information, something that psychologists are generally discouraged to do. The problem with anecdotal evidence is that we do not *know* that these individuals truly lack the capacity to empathize or that they otherwise match the conceptual definition of psychopathy (e.g., moral colorblindness or extreme impulsivity) because we have never subjected them to any controlled scientific experimentation. To claim otherwise is to resort to little more than guesswork.

Second, there is not much evidence to suggest that infamous serial killers—like Ted Bundy, John Gacy, Edmund Kemper, Jeffrey Dahmer, or Gary Ridgway—actually match mainstream conceptions of psychopathy. For example, a telling study by Eric Hickey and colleagues from 2018 analyzed extensive information about these five serial killer cases and found that only one of them scored above the clinical threshold on the PCL-R scale (i.e., Ted Bundy, who scored 34 out of 40). This suggests that the serial-killer objection has minimal generalizing appeal, as it shows that serial killers are psychologically much more heterogenous than assumed by the counterargument above.[61]

Third, a common response to the second point above is that even though many serial killers would not be deemed clinically psychopathic—such as a PCL-R score above 30—there are still well-known cases like Ted Bundy's that meet this threshold. And therefore, we have examples of real-life cases of a psychopathic person. However, there are plenty of reasons to be suspicious about how well a person like Ted Bundy matches the scientific conception of psychopathy, or better, that we need a concept like psychopathy to explain what makes him different from other people. For instance, on most accounts, Bundy was not an unemotional person, where anecdotal evidence predominately describes him as a highly aggressive person, with proneness to emotional outbursts (e.g., evident in his extremely violent crimes, some of which seemed driven more by rage than cold calculation). But perhaps my biggest issue with the Bundy example is that he arguably exhibits signs of comorbidities that might offer a simpler and therefore better explanation of his horrific behaviors and odd personality. This includes evidence of extreme alcohol intoxication during the offenses he was found guilty of, evidence of necrophiliac urges, and signs of delusional behavior. It is at least conceivable that Bundy's behavior is best (i.e., simpler) explained by a comorbid constellation of addiction, paraphilia, and cognitive disorder, rather than a scientifically elusive personality disorder called psychopathy. Having studied many serial killer cases, my impression is that this uniformly holds true across all cases, that psychopathy is never an exhaustive and/or even an informative explanation.[62]

Fourth, the serial-killer objection appears to overlook that it is not so much a defense of psychopathy research as it is conceding a huge shortcoming. The field of psychopathy research was never initiated as an attempt to make sense of serial killers, nor does the field intend to do so. The general claim is that psychopathic persons make up about 1% of the general population, which amounts to around 4,000,000 people in North America alone.[63] Compare this to the presence of serial killers, where some scholars estimate that there are currently fewer than 200 in the United States, of which only some of them might meet the clinical criteria for psychopathy.[64] Thus, as a scientist, if you find the serial-killer objection convincing, you must then concede that while the idea of psychopathy might help us make sense of a tiny part of the human population, its initial promise of explaining most of the violent crime in our society is poorly substantiated. You cannot have it both ways.

It is largely because of these four reasons that I have started to see the serial-killer objection as less than convincing. And as a scientist, I cannot come up with good reasons for why we should keep studying the idea of psychopathy if our evidence of its existence merely boils down to anecdotal evidence from a handful of serial killer cases. Not only is such anecdotal evidence never enough to establish and justify a scientific field, but such a research paradigm is essentially a non-starter, as we might never be able to study these serial killer cases under controlled experimental settings (i.e., because of methodological and ethical restrictions, as well as "red tape" bureaucratic rules).

Objection #2: There is Some Evidence of Psychopathy in the Research Record

A second objection that I often hear when I voice my speculations to university students or at academic conferences is that *if* psychopathy is not real, why have researchers found *some* evidence of psychopathy in their experiments. As we saw throughout chapters 2–6, there is a small portion of experiments that do find some statistically significant effects consistent with the idea of psychopathy. Or to put it differently, it is not all null findings.

I find this objection important, but as an antidote to skepticism, it is much less convincing than the first objection. The main reason for this is that it is based on a fallacious assumption: it wrongly assumes that if we are studying an idea in psychological experiments that is not real, then we should find absolutely no evidence of it, that all effects should be nulls. But this is demonstrably a wrong assumption.

The different types of experiments we have reviewed in this book all involve enough methodological and analytical flexibility that researchers are likely to find a variety of false-positive effects—that is, statistically significant effects that appear to indicate *some* form of unusual differences between test participants, but where these differences might be aberrations and/or not truly generalizable at the population level (i.e., this is what researchers call *false-positive* effects). We know this because scientists have studied what they call *null fields* (e.g., homeopathic treatments[65]), which are scientific paradigms that are investigating a specific hypothesis where we are very certain that there should be no evidence to support it, but where experiments nonetheless find some effects to suggest otherwise.

Perhaps the most famous example in the field of psychology is a cohort of experiments reported in a 2011 article titled "Feeling the Future," where the author purported to have found evidence of so-called *precognition*, a phenomenon where a person's current behavior is retroactively influenced by a future event (essentially, predicting the future because of having extrasensory access to information passed down from the future).[66] Of course, the conclusion from these experiments has been widely rejected and ridiculed by scientists. However, the effects reported in the experiments are not fraudulent; what is seemingly causing these effects is that the research methods, data structuring, and statistical analyses all contain enough flexibility to *force out* statistically significant, yet false-positive effects.[67]

As I have argued throughout this book, I think we have good reasons to conclude that this is also what is going on in many psychopathy experiments. One of the more telling examples of this was the evidence from empathy studies, where our 2024-review study showed that the proportion of nulls was 84.79% and 91.28% across two different paradigms. But when we sorted studies according to their analytic flexibility, those studies with the most rigorous methods yielded a higher proportion of nulls, around 95%, exactly the number you would expect from a null field, as our statistical methods conventionally tolerate a 5% false-positive rate (i.e., the alpha-level of 5% in a p-value calculation; see chapter 4).

In short, the response to the objection above is that whatever significant effects that have been found across psychopathy studies, these cannot reasonably be seen as supporting the idea of psychopathy. In fact, the extent of these significant effects is exactly what we should expect given the flexibility of the methods that are deployed.

Concluding Remarks: Psychopathy as a Zombie Idea

In this chapter, I have provided reasons why I believe we should be skeptical about psychopathy. Importantly, I did not aim to *prove* that psychopathy is not real, but instead I pointed out that there are aspects about the idea of psychopathy that might be problematic in a scientific context. First, we can be skeptical about whether the impressionistic accounts of the psychopathic stereotype outlined by clinicians like Hervey Cleckley and others are something that ought to trigger our curiosity into a scientific mode of thinking. Second, the mainstream idea of psychopathy might turn out to be redundant, insofar as the clinical stereotype can be explained by other already established theories in behavioral science. Altogether, these are sufficient reasons for why we should be skeptical about the reality of psychopathy, but it also suggests that psychopathy could be a zombie idea: an idea that is intuitively appealing but has no basis in reality. This final point may therefore serve as an explanation for why there has been little if any scientific progress in the field.

What I find particularly appealing about describing psychopathy as a zombie idea is first and foremost that it provides a good explanation for the lack of scientific progress. I take this view to be utterly uncontroversial because when scientists study zombie ideas—as has happened many times in history—their empirical work always comes out short. This is arguably what makes the scientific method so extraordinary: *in the long run, you cannot amass compelling scientific evidence for that which does not exist.* Thus, what makes the zombie-idea explanation potentially controversial is perhaps more related to whether researchers and clinicians think it is a wrongheaded explanation. But this should not necessarily be a point of controversy, because one crucial aspect that makes the zombie-idea explanation reasonable is exactly that it is a *falsifiable* explanation. Indeed, to those who find my speculations wrong or even controversial, there is a straightforward way to show that psychopathy is *not* a zombie idea: simply find and present a real-life example of a *truly psychopathic* person, a person who meets all the common descriptors of stereotypical psychopathy (e.g., moral colorblindness, lack of emotions, poor impulse control, and so on), and a person where *there could not possibly be a simpler or more established explanation* of that person's behavior and personality other than the idea of psychopathy. This would be the most direct way to overcome skepticism about psychopathy. If psychopathic persons are *real*—and, for instance, make up around 1% of the entire population as commonly claimed—it should also be a relatively easy thing to do.

In the meantime, researchers interested in psychopathy might begin exploring it *as* a zombie idea, as it could be helpful to future mental health researchers to better understand the social dynamics that drive scientific paradigms researching zombie ideas. Arguably, it is still poorly understood how zombie-fields emerge. However, I imagine that it involves some form of cognitive bias and dissonance where a large proportion of scientists fail to realize that there are sufficiently good reasons to stop investigating the idea and therefore keep perpetuating the research paradigm.

In an ideal world, what should have happened when psychopathy researchers realized that the stereotype they are studying is inconsistent with the idea of psychopathy is that it should have kindled another abductive trigger, inspiring the following question: Why is it we are not finding evidence to support our ideas? As I have shown throughout this book, all the claims made by mainstream researchers could have been reasonably questioned already in the 1990s, but most certainly during the 2000s, and therefore—in the eternal clarity of hindsight—researchers should have interrogated their own motivations for the continuation of the research. Of course, this did not happen, which is exactly why the paradigm has ended up where it is today, where an increasing number of researchers are beginning to question why there is a lack of progress in the field.

In relation to this latter observation, it must be stressed that I am not the first person to raise skepticism about psychopathy. As mentioned, it was indeed a mainstream position all the way into the 1980s to reject the idea altogether. But there have also been more recent attempts to reiterate these early skeptical attitudes. Perhaps the first broad criticism of psychopathy is a 2015 book by Jarkko Jalava, Stephanie Griffiths, and Mike Maraun, titled *The Myth of the Born Criminal*. In this book, the authors focus their criticism of the notion that psychopathy is a biological condition, showing that not only is there little evidence to support this claim but there is also little reason to think that it is a good hypothesis.[68] Furthermore, and as mentioned, the prominent psychologist, Scott Lilienfeld, has suggested that stereotypical psychopathic persons might be explained by ordinary theories in psychology as opposed to a theory about a disorder in the classical sense.[59]

The problem is, however, that every time these sorts of deeper criticisms have emerged in scientific publications, they appear to have been largely ignored by scholars from within the research paradigm itself. For example, Jarkko Jalava and colleagues' book is cited only 41 times (Google Scholar), and it does not have a single citation from any of the leading mainstream researchers, which to me indicates that their work has mustered little recognition. Such an observation might hint at how zombie-fields form; they take shape when critical discourse dwindles, generating sanitized echo chambers free of skeptical voices. A useful illustration of how skepticism has been largely ineffective can be found in the early days of psychopathy research.

One of the most cited experiments on Hervey Cleckley's notion of psychopathy is a study by David Lykken from 1957 (2,014 citations according to Google Scholar).[70] This experiment is frequently cited by today's psychopathy researchers (346 citations between 2018 and 2023) for being the first study to find supportive evidence of psychopathy. However, it is literally never mentioned that the experiment faced a searing and perfectly legitimate criticism by Roy Persons and Carol Persons a few years after its publication in 1965.[71] This commentary only has 13 citations (Google Scholar), and it has never been cited by any mainstream psychopathy researcher. In their criticism of Lykken's experiment, the authors mention methodological problems similar to the ones I have stressed in earlier chapters, highlighting issues with methodological flexibilities, the selective exclusion of experimental data, the use of questionable statistical analyses, and finally criticizing Lykken's unjustified, sweeping conclusions. Shortly after its publication, Lykken wrote a brief response published in the same journal, brushing aside the criticism, and insinuating that it was motivated as a *red herring* (i.e., a deliberate attempt to distract from the relevancy of psychopathy research).[72] All in all, I think contemporary readers would recognize that Lykken's response lacks the sort of intellectual seriousness that we usually expect from an academic dialogue.

The problem is that Roy Persons and Carol Persons were right and David Lykken was wrong. The criticisms they raised already back in 1965 rest on the exact same type of observations that contemporary psychologists recognize as grave problems in the behavioral sciences: problems that are ordinarily called questionable research practices, which, in turn, are often invoked to explain the ensuing *replication crisis* that is sweeping through the behavioral sciences.[73] It is tempting to look at David Lykken's experiment—and Persons and Persons' criticism—as a high-profile example of an opportunity that was lost, where a lesson could have been learned.

For all the problems I have pointed out in the research paradigm, one glowing optimistic note shines brighter than the pessimism: the strength of the scientific method is exactly that if we study an idea that has no basis in reality—an idea that is not real—we will continuously fail to find convincing evidence of its existence. Thus, the shortcomings that I am pointing out relate to a human problem where we scientists fail to accept the proper message told by our own experimental evidence. When we refuse to face such evidence and draw proper conclusions from it, tragedy is potentially born, such as when these misguided ideas make it into the legal system under the false pretense of science, where they have real negative impact on human lives. This is the thorny situation that has been born out of psychopathy research, and as researchers we are principally responsible for causing but also cleaning up the mess. The first step we must take to end this tragedy is to cultivate skepticism and critical debates in our field.

Conclusion: The Fall of a Dangerous Diagnosis

It's not what you don't know that gets you into trouble. It's what you know for sure that just isn't so.

—Anonymous

The idea of psychopathy has an enormous intuitive appeal. What could possibly better explain crime and social turmoil other than an endemic personality disorder characterized by a profound lack of empathy and sense of moral responsibility?

This book was written primarily as a summary of the empirical research that has investigated the idea of psychopathy as defined by the *Hare Psychopathy Checklist* (PCL). This summary revealed that the most common claims about psychopathy are both seriously misleading and unsupported by scientific evidence. In addition, I argued for a moratorium on the forensic use of PCL assessments, premised on the observation that this practice appears to violate the professional-ethical codes in psychiatry and psychology. My recommendation of a forensic moratorium is especially urgent given that legal institutions and practitioners already have access to alternative assessment strategies that reliably outperform the PCL scales in their ability to inform decision-making. In short, these institutions have nothing to lose by changing their procedures, and they may even have a lot to gain by reallocating resources (freed by a moratorium) to more effective evaluation strategies.

The recommendation of a moratorium on psychopathy assessments in the legal system will strike some readers as an overreaction, as there appears to be some specific contexts where these assessments are potentially informative to decision-makers. For example, one of the most common uses of the PCL scales is for risk assessment purposes, and, as I showed in chapter 2, this is one of the contexts where the PCL scales have a relatively robust evidence base. However, as I also pointed out in chapter 2, just because an assessment tool is capable of risk prediction, it does not automatically follow that it is justified under the principle of evidence-based practices. This principle prescribes that practices must be guided by the *best available* evidence, and for all we know, the PCL scales are far from the best tools available to inform risk-related decisions. And not only are the PCL scales reliably outperformed by other strategies but we also have a well-documented understanding of how these assessments may introduce bias as decision-makers often associate the label of "psychopathy" with inflated risk levels.

In the final chapter 8, I explored the possibility that perhaps psychopathy is a "zombie" idea—that psychopathy has no basis in reality—which I proposed could explain the lack of scientific progress in the field. More empirical research is needed to sufficiently corroborate such a view (i.e., proving a negative), and the view is perhaps destined to attract vocal opposition among long-time advocates of psychopathy research. However, it is similarly possible that many stakeholders will welcome this viewpoint, as it resonates with a more general development in mental health research where categorical diagnostic entities are increasingly seen as problematic. In recent years, the use of labels and categories in both research and clinical procedures is gradually being phased out, especially in the realm of *personality disorders*, where consensus is shifting toward viewing such disabilities as multimodal and multidimensional entities without clear-cut boundaries.[1] It is currently unclear how this switch in perspective will impact the field of psychopathy research, as well as the current reference to psychopathy in official diagnostic manuals such as the *Diagnostic and Statistical Manual of Mental Disorders*. And it is even more unclear how it is going to affect the forensic use of PCL psychopathy assessments, as these are not always used for diagnostic purposes (e.g., risk assessment, supervision management, or rehabilitation).

One of the most important observations made in this book is that psychopathy research appears to have an engrained problem with so-called questionable research practices, where experimenters introduce bias and flexibility into the research process, which dramatically undermines the integrity of the scientific output.[2] Furthermore, some review studies are marred by reporting bias—a problem that is particularly obvious in neuroimaging research—where the experimental results are presented in a much more positive light than it deserves, often by resorting to rhetorical overinterpretation of results or not fully disclosing the true extent of null effects. In the behavioral sciences, the problem of questionable research practices is typically framed as a purely academic issue, but with psychopathy research, it may have had real-life consequences due the legal influence of the research. More work is still needed to fully understand just how "questionable" psychopathy research is, and what role these practices played in the revival of psychopathy research in the academic world in 1980s and its growing application in the criminal justice system through the 1990s.

The observation that psychopathy research is entangled in these dubious practices will not come as a surprise to people familiar with developments in the behavioral sciences. For the past 10–15 years, these disciplines have gone through a renaissance where researchers have worked hard to improve and rethink the way they conduct experiments and disseminate their results, such as innovating methods, analyses, theories, and reporting practices (a movement sometimes referred to as *open science*).[3] This process began with a number of stunning discoveries. For instance, that scientists anonymously admit to scientific spin[4] and that experimental results are rarely "reproduced" in subsequent replication studies—a finding that eventually morphed into what we today call the *replication crisis*.[5] Behavioral scientists have now come to the realization that questionable research practices are endemic to their disciplines and are presumably a major reason behind the replication crisis itself. This historical context suggests that we cannot genuinely discuss psychopathy research without also recognizing that the paradigm rose to prominence during a period (1990 to 2010) when scientists were much more relaxed and/or unaware of the methodological shortcomings of their work. However, where other research paradigms have rebuilt their credibility by acknowledging these problems and embracing constructive resolutions, this sort of reflection and change has been comparatively less pronounced in mainstream psychopathy research. The future of the field may therefore very well depend on how researchers collectively respond to these challenges.[6]

I feel obligated to both acknowledge and underscore that the message conveyed throughout this book is not my own original creation. Many of the chapters simply rehash the important work from recent review studies and meta-analyses published in peer-reviewed academic journals. Even the message in chapter 8 that expresses skepticism about the reality of psychopathy (which might be seen as the most controversial part of the book) has been voiced by multiple researchers before me.[7] In fact, there is a way in which the current book can be seen as little more than compiling evidence to substantiate a perspective about psychopathy that was dominant in the decades prior to the 1990s, an era where psychopathy was widely seen as an elusive *wastebasket diagnosis* with no foundation in reality. Back then, such a view was based mostly on armchair speculation. What is different today—and what I partially aimed to demonstrate with this book—is that we now have plenty of evidence to support it. Psychopathy research is no longer conducted from a standpoint of uncertainty and theoretical speculation; researchers who are pursuing the idea today are doing so *despite* the overwhelming prevalence of null results and falsifying evidence. Therefore, the time is arguably due to have a constructive conversation about how sustainable the future looks for researchers (and graduate students) interested in the idea of psychopathy.

Even though this book builds on observations made by scholars before me, it still contains a plot line that is novel and markedly different from the mainstream depiction of psychopathy research. It is a plot line that draws a major contrast between how there was a *rise* in the popularity of researching the disorder in the 1990s and 2000s, where we today are facing what looks more and more like an imminent *fall*. Over the course of roughly one decade beginning in the mid-1990s, the idea of psychopathy (re)gained traction in scientific circles and swiftly made its way into the legal system. This rise was not without its critics. Already in 1983, the forensic psychologist and legal scholar James Wulach warned of the ethical complications that practitioners would be facing if the diagnosis were to be embraced by our legal institutions.[8] And as these institutions increasingly adopted the PCL scales into decision-making protocols, researchers continuously called for caution as they noted that the science was incomplete and often misinterpreted by its end users.[9] Still, psychopathy research kept rising in popularity, its legal influence kept growing with breathtaking speed, and by the mid-2000s psychopathy was frequently referred to as one of the most important innovations in forensic psychology.

Today, we can look back at this history with the proverbial clarity of hindsight. The rise in popularity was not driven by a corresponding sophistication of scientific knowledge, but presumably driven more by extra-scientific factors such as questionable research methods, bias, scientific spin, and perhaps even social demands afforded by tough-on-crime policies. And it is exactly because the scientific evidence was weak and poorly construed that the tides are now changing. As mentioned in chapter 8, one of the truly fascinating powers of the scientific method is that if we study overstated or misconceived ideas, we will eventually fail to find compelling evidence to support them. So, this is where we are today: the common claims about psychopathy have been thoroughly challenged by empirical research, and for that reason, a fall from scientific grace is taking shape on the horizon. This is not to say that psychopathy research might never gain a "second wind," but the current extent of null results in the field leaves little room for optimism.

Faced with these realities, I believe a strong case can be made that the use of PCL psychopathy assessments in forensic decision-making has become a liability to our legal system and collective sense of justice. One of the cornerstones of the criminal justice system is the right we have to a fair and equitable process. The legal implementation of the PCL scales muddies this process by introducing bias and therefore an opportunity for unjustified disparity in decision-making. As covered in chapter 1, there are many ways a PCL assessment can impact these legal decisions, but when it does there is a high risk that it aggravates legal decisions for all the arbitrary reasons embedded in the conception of psychopathy.

At the beginning of the book, I sketched the common understanding of *the problem of psychopathy* as a conversation that emerged among forensic researchers and practitioners in the 1990s. As they saw it, the problem of psychopathy regarded how our society should effectively respond to the presence of psychopathic persons in our communities. This way of describing the disorder as an alarming social problem is no longer valid. The more "realistic" problem of psychopathy, it seems, is the actual practices that have emerged from taking claims about the disorder seriously. Although the exact number is unknown, it is quite possible that hundreds of thousands of individuals undergo a PCL psychopathy assessment every year across North American jurisdictions. Unfortunately, we do not have good insight into how often and to what extent these assessments impact decision-making. For instance, we do not yet know the answer to important questions like: How many juvenile offenders have been transferred to adult courts because of a high PCL score? How many people were given a capital sentence instead of life without parole? How many were refused bail and remain in detention, and how many have had their chances of parole ruined because of a high PCL score? Sadly, the answer to these questions is most certainly not *zero*. This is the reason why I have chosen to describe psychopathy as a *dangerous diagnosis*. It will be up to future researchers to ascertain just how dangerous a psychopathy assessment really is.

People who find the overall message in this book compelling, and who agree that there is a pressing and legitimate case for ending the use of PCL assessments, might wonder how this development is going to take place. It is tempting to think that all it takes is for the scientific community to *change their minds* about psychopathy—that if enough researchers reject the idea of psychopathy, it will then generate a trickle-down effect into the criminal justice system, thus compelling decision-makers to end these assessment practices. I think that this anticipation is far too naïve, mostly for two different reasons.

First, expecting an automatic change in practices naively overlooks the intricacies of how the legal system operates as a complex network of various institutional bodies that are largely independent of developments in the scientific community. The problem is that PCL assessments have already become an integrated part of the complex flow of information that legal decision-makers rely on. These practices were never directly shaped by scientists, and nor will they be directly changed by scientists. To facilitate change, our legal system will rely on activists who take it upon themselves to challenge the use of the PCL scales. This means that precedence must be established in courtrooms, for instance, by experts testifying that these assessments provide little reliable and useful information. It also involves an active and reflective conversation inside legal institutions like correctional facilities, parole boards, and juvenile detention centers, where decision-makers must come to the realization that their resources are better spent on other evaluation strategies. This process is likely going to be a long and bumpy journey, but it is a process that is necessary to end the bias and harm that surrounds the use of PCL assessments.

Second, the view that a change in scientists' attitudes about psychopathy will promote change in the legal system fails to acknowledge the possibility that the idea of psychopathy might have sprung from deeply held intuitions about crime. For instance, the intuition that much of the criminal behavior we observe in society can be explained with reference to psychological abnormality, that the majority of violent crime is caused by a few psychologically "rotten apples." This intuition is likely to persist even if scientists abandon the idea of psychopathy and may therefore continue to generate ideas similar to the current conception of psychopathy. And with each such new idea, there is a chance that it will make its way into the legal system in similar ways as psychopathy has.[10]

This latter point is not armchair speculation but is already taking shape. During the last two decades, there has been a steady uptick in different research programs exploring the link between crime and abnormal personality constructs. For example, one program investigates what is referred to as *callous and unemotional traits*,[11] and another paradigm goes under the name *The Dark Triad*.[12] Similarly, some researchers are now studying "psychopathic traits" in ordinary healthy persons.[13] My suspicion is that these paradigms are built on the same intuitions that undergird the field of psychopathy research, namely that crime can be explained by psychological abnormality. However, if we are serious about the lessons learned from centuries of psychopathy research, we should be skeptical about these newer iterations. They too may similarly collapse under the pressure of genuine scientific scrutiny. But before they do, there is a chance that they will make their way into the legal system, thus repeating yet another tragic cycle where good and honest scientific intentions give rise to harmful practices.

To ensure that these ideas do not prematurely influence our legal system, I recommend that we wrestle more directly with the research tradition itself, including the underlying assumption that science about personality "traits" or "constructs" can readily inform legal decision-making. Indeed, even if researchers one day discovered a strong link between a personality construct and criminal behaviors, we can still question if there is a reliable and responsible way to utilize such knowledge in the legal system. For a starter, one of the most conventional forms of wisdom to emerge from the discipline of criminology is that criminal behavior is heavily context dependent. People get caught up in a criminal lifestyle for a myriad of reasons, often largely independent of their personality.[14] So, will it not always be doubtful if scientific knowledge about personality can genuinely inform decision-making about the individual person?

Second, some researchers might claim to have discovered valid ways to measure a particular personality construct, but that does not mean that their methods can be easily implemented in legal institutions where a great diversity in education and competence exists among practitioners tasked with conducting these assessments.[15] So, is it reasonable to assume that a tool with proven reliability in controlled laboratory settings will have similar reliability in real-life forensic settings?

Third, the entire premise of asserting a predictive and/or explanatory link between a personality construct and criminal behavior is fundamentally problematic as it deviates strongly from mainstream views in contemporary studies of personality. Researchers in this domain—a domain separate from psychopathy studies and forensic psychology in general—have long given up on the idea that personality traits are discrete and stable dispositions that can strongly predict and explain behaviors across contexts.[16] But if experts acknowledge that personality is an imperfect predictor of *basic* human behaviors, why do some forensic psychologists keep insisting that personality can predict and explain something as complex as *criminal* behavior?

Besides wrestling with these basic questions, we must also critically discuss the deeper intuition that gives birth to ideas like psychopathy. When we study the history of psychopathy research, it appears that the idea itself springs from a belief that criminal behaviors—and especially instrumental remorseless violence—are somehow signs of the person being psychological abnormal. I suspect this is why researchers eventually concocted elaborate theories about a mysterious personality disorder called "psychopathy." It was an attempt to explain something they found deeply abnormal. As Benjamin Rush put it in 1786, the symptoms associated with the disorder were "contrary to nature."[17] However, this underlying intuition about remorseless violence being a sign of psychological abnormality does not stand up to scrutiny. It might be admirable when we feel empathy toward one another, and when we acknowledge a wider moral responsibility to all members of our species. But, according to much anthropological and sociological research, prosocial behaviors like empathy and remorse appear to be heavily dependent on their social context, and therefore are improper indications of what is *normal* vs *abnormal*. As numerous scholars have pointed out, perhaps the most fascinating and terrifying aspect about human violence and destruction is the ease at which *ordinary* people readily engage in it.[18]

Thus, a reasonable alternative perspective is to see the underlying intuition in question—and by extension the idea of psychopathy—as a byproduct of a bias that prevails in our society, a society where most people are rarely exposed to any grave form of human nastiness. If you are never on the receiving end of violent behavior, you might be inclined to interpret it as something truly puzzling when you are confronted with it. Perhaps this attitude is especially true for those individuals who eventually end up studying ideas like psychopathy at prestigious universities. First as socioeconomically privileged undergraduates, later as carefully selected PhD students, and finally as scholars that are groomed in the image of the bias itself. This perspective is not meant as a polemical accusation. Rather, it is an acknowledgement of how cognitive biases are known to make their way into the behavioral sciences, causing a peculiar form of shared theory-induced blindness in the scientific discourse. So, if the idea of psychopathy is rooted in such biased and blinding intuitions, this would not only explain the profound lack of scientific progress in the field but it would also explain why psychopathy can be described as a *zombie* idea that is difficult to get rid of.

Notes

Introduction

1. Kent A. Kiehl and Morris B. Hoffman, "The criminal psychopath: History, neuroscience, treatment, and economics," *Jurimetrics* 51 (2011): 355–397.

2. Bear F. Braumoeller, *Only the dead: The persistence of war in the modern age* (New York: Oxford University Press, 2019); "War and peace," 2023, Our World in Data, https://ourworldindata.org/war-and-peace.

3. David Livingstone Smith, *Less than human: Why we demean, enslave, and exterminate others* (New York: St. Martin's Press, 2011).

4. "Refugee data finder," UNHCR—The UN Refugee Agency, accessed January 24, 2024, https://www.unhcr.org/refugee-statistics/.

5. "Crime statistics in the U.S.," Statista, accessed on June 1, 2023, https://www.statista.com/topics/2153/crime-in-the-united-states/.

6. U.S. Government Accountability Office, *Costs of crime: Experts report challenges estimating costs and suggest improvements to better inform policy decisions* (2017), accessed September 8, 2022, https://www.gao.gov/products/gao-17-732.

7. Robert Hare, *Without conscience: The disturbing world of the psychopaths among us* (New York: The Guilford Press, 1993), xi.

8. Hare, *Without conscience*; Jeremy Coid et al., "Prevalence and correlates of psychopathic traits in the household population of Great Britain," *International Journal of Law and Psychiatry* 32, no. 2 (2009): 65–73.

9. Stephane A. De Brito et al., "Psychopathy," *Nature Reviews Disease Primers* 7, no. 1 (2021): 49; Dennis E. Reidy and Katherine W. Bogen, "Public health considerations in psychopathy," in *The complexity of psychopathy*, ed. Jennifer E. Vitale (Cham: Springer Nature, 2022), 611–635.

10. Kiehl and Hoffman, "Criminal psychopath."

11. G. T. Harris, T. A. Skilling, and M. E. Rice, "The construct of psychopathy," *Crime and Justice* 28 (2001): 197–264; M. DeLisi, "Psychopathy and crime are inextricably linked," in *Routledge international handbook of psychopathy and crime*, ed. M. DeLisi (New York: Routledge, 2018), 3–12; S. Porter, M. Woodworth, and P. Black, "Psychopathy and aggression," in *Handbook of psychopathy*, ed. C. Patrick (New York: The Guilford Press, 2018), 611–634; Bryanna Fox and Matt DeLisi, "Psychopathic killers: A meta-analytic review of the psychopathy-homicide nexus," *Aggression and Violent Behavior* 44 (2019): 67–79.

12. D. E. Reidy et al., "Psychopathy traits and violent assault among men with and without history of arrest," *Journal of Interpersonal Violence* 34, no. 12 (2019): 2438–2457.

13. Adrian Raine and José Sanmartín, *Violence and psychopathy* (New York: Springer, 2001).

14. Robert Hare, "Psychopathy: A clinical construct whose time has come," *Criminal Justice and Behavior* 23, no. 1 (1996): 25–54; Stephen Porter, Leanne Brinke, and Kevin Wilson, "Crime profiles and conditional release performance of psychopathic and non-psychopathic sexual offenders," *Legal and Criminological Psychology* 14, no. 1 (2009): 109–118.

15. Dylan T. Gatner et al., "How much does that cost? Examining the economic costs of crime in North America attributable to people with psychopathic personality disorder," *Personality Disorders: Theory, Research, and Treatment* 14, no. 4 (2023): 391–400; Kiehl and Hoffman, "Criminal psychopath"; Dennis E. Reidy et al., "Why psychopathy matters," *Aggression and Violent Behavior* 24 (2015): 214–225.

16. S. D. Benning, N. Venables, and J. Hall, "Successful psychopathy," in *Handbook of psychopathy*, ed. C. Patrick (New York: The Guilford Press, 2018), 585–608; S. O. Lilienfeld, A. L. Watts, and S. F. Smith, "Successful psychopathy: A scientific status report," *Current Directions in Psychological Science* 24, no. 4 (2015): 298–303.

17. Paul Babiak and Robert Hare, *Snakes in suits: When psychopaths go to work* (New York: Regan Books, 2006); Paul Babiak, Craig S. Neumann, and Robert Hare, "Corporate psychopathy: Talking the walk," *Behavioral Sciences & the Law* 28, no. 2 (2010): 174–193; Kevin Dutton, *The wisdom of psychopaths: What saints, spies, and serial killers can teach us about success*, vol. 1 (New York: Scientific American/Farrar, Straus and Giroux, 2012); C. Mathieu et al., "A dark side of leadership: Corporate psychopathy and its influence on employee well-being and job satisfaction," *Personality and Individual Differences* 59 (2014): 83–88.

18. Babiak and Hare, *Snakes in suits*; Clive Roland Boddy, "Organisational psychopaths: A ten year update," *Management Decision* 53, no. 10 (2015): 2407–2432.

19. Scott O. Lilienfeld et al., "Correlates of psychopathic personality traits in everyday life: Results from a large community survey," original research, *Frontiers in*

Psychology 5 (2014): 740; Désiré Palmen, Jan Derksen, and Emile Kolthoff, "House of cards: Psychopathy in politics," *Public Integrity* 20, no. 5 (2018): 427–443.

20. Scott O. Lilienfeld et al., "Fearless dominance and the U.S. presidency: Implications of psychopathic personality traits for successful and unsuccessful political leadership," *Journal of Personality and Social Psychology* 103, no. 3 (2012): 489–505.

21. Robert Hare et al., "Psychopathy and crimes against humanity: A conceptual and empirical examination of human rights violators," *Journal of Criminal Justice* 81 (2022): 101901.

22. Carl B. Gacono, "Introduction," in *The clinical and forensic assessment of psychopathy: A practitioner's guide*, ed. C. Gacono (New York: Routledge, 2016), 3–13; Robert Hare, "Psychopathy, the PCL-R, and criminal justice: Some new findings and current issues," *Canadian Psychology-Psychologie Canadienne* 57, no. 1 (2016): 21–34.

23. Robert Hare, "Forty years aren't enough: Recollections, prognostications, and random musings," in *The Psychopath: Theory, Research, and Practice*, ed. Hugues Hervé and John C. Yuille (Mahwah, NJ: Lawrence Erlbaum Associates, 2007), 3–28.

24. Grant T. Harris and Marnie E. Rice, "Treatment of psychopathy: A review of empirical findings," in *Handbook of psychopathy*, ed. C. Patrick (New York: The Guilford Press, 2006), 555–572; W. H. Reid et al., eds., *Unmasking the psychopath: Antisocial personality and related syndromes* (New York: W. W. Norton & Company, 1986); Georg K. Stürup, *Treating the "untreatable"* (Baltimore: John Hopkins Press, 1968).

25. Robert Hare, C. Neumann, and A. Mokros, "The PCL-R assessment of psychopathy: Development, properties, debates, and new directions," in *Handbook of psychopathy*, ed. C. Patrick (New York: The Guilford Press, 2018), 39–79.

26. D. DeMatteo et al., "Investigating the role of the Psychopathy Checklist–Revised in United States case law," *Psychology Public Policy and Law* 20, no. 1 (2014): 96–107; Tess M. S. Neal and Thomas Grisso, "Assessment practices and expert judgment methods in forensic psychology and psychiatry: An international snapshot," *Criminal Justice and Behavior* 41, no. 12 (2014): 1406–1421; Jay P. Singh et al., "International perspectives on the practical application of violence risk assessment: A global survey of 44 countries," *International Journal of Forensic Mental Health* 13, no. 3 (2014): 193–206.

27. David DeMatteo and Mark E. Olver, "Use of the Psychopathy Checklist-Revised in legal contexts: Validity, reliability, admissibility, and evidentiary issues," *Journal of Personality Assessment* 104, no. 2 (2022): 234–251; Rasmus Rosenberg Larsen et al., "Are psychopathy assessments ethical? A view from forensic mental health," *Journal of Threat Assessment and Management* 9, no. 4 (2022): 260–286; Randy K. Otto and Kirk Heilbrun, "The practice of forensic psychology: A look toward the future in light of the past," *American Psychologist* 57, no. 1 (2002): 5–18.

28. DeMatteo and Olver, "Use of the Psychopathy Checklist-Revised"; Carl B. Gacono, ed., *The clinical and forensic assessment of psychopathy: A practitioner's guide*, 2nd ed. (New York: Routledge, 2016); Larsen et al., "Are psychopathy assessments ethical?"

29. DeMatteo and Olver, "Use of the Psychopathy Checklist-Revised"; Larsen et al., "Are psychopathy assessments ethical?"

30. P. Babiak et al., "Psychopathy: An important forensic concept for the 21st century," *FBI Law Enforcement Bulletin* 81, no. 7 (2012): 3–8; Robert Hare, "Psychopaths and their nature: Implications for the mental health and criminal justice system," in *Psychopathy: Antisocial, criminal, and violent behavior*, ed. T. Millon et al. (New York: Guilford Press, 1998), 188–212; Robert Hare, "Psychopathy: A clinical and forensic overview," *Psychiatric Clinics of North America* 29, no. 3 (2006): 709–724; Harris et al., "The construct of psychopathy"; J. Monahan, "Statement on book jacket," in *Handbook of psychopathy*, ed. C. Patrick (New York: Guilford Press, 2006); Gacono, "Suggestions for implementation and use of the Psychopathy Checklist in forensic and clinical practice"; Robert Hare, "Foreword," in *The clinical and forensic assessment of psychopathy: A practitioner's guide*, ed. C. Gacono (New York: Routledge, 2016), xvi-xviii.

31. Hare, "Psychopathy, the PCL-R, and criminal justice"; Kiehl and Hoffman, "Criminal psychopath"; DeLisi, "Psychopathy and crime are inextricably linked."

32. American Psychological Association (APA), "Evidence-based practice in psychology," *American Psychologist* 61, no. 4 (2006): 271–285; G. D. Glancy et al., "AAPL practice guideline for the forensic assessment," *Journal of the American Academy of Psychiatry and the Law* 43, no. 2 suppl. (2015): S3–53; G. D. Glancy and Michael Saini, "The confluence of evidence-based practice and Daubert within the fields of forensic psychiatry and the law," *Journal of the American Academy of Psychiatry and the Law Online* 37, no. 4 (2009): 438–441.

33. Hare, "Psychopathy: A clinical construct whose time has come"; Porter et al., "Psychopathy and aggression."

34. Robert Hare and Craig S. Neumann, "Psychopathy: Assessment and forensic implications," *Canadian Journal of Psychiatry* 54, no. 12 (2009): 791–802; G. T. Harris and M. E. Rice, "Psychopathy research at Oak Ridge: Skepticism overcome," in *The psychopath: Theory, research, and practice*, ed. H. Hervé and J. C. Yuille (New York: Routledge, 2007), 57–76.

35. Jacqueline B. Helfgott, *No remorse: Psychopathy and criminal justice* (Santa Barbara, CA: Praeger, 2019); James Blair, "Traits of empathy and anger: Implications for psychopathy and other disorders associated with aggression," *Philosophical Transactions of the Royal Society B: Biological Sciences* 373, no. 1744 (2018): 1–8; Hare, *Without conscience*.

36. De Brito et al., "Psychopathy"; Andrea L. Glenn and Adrian Raine, *Psychopathy: An introduction to biological findings and their implications* (New York: New York University Press, 2014); Matt DeLisi, *Psychopathy as unified theory of crime* (New York: Palgrave Macmillan, 2016).

37. James Blair, Derek Mitchell, and Karina Blair, *The psychopath: Emotion and the brain* (Malden, MA: Blackwell Publishing, 2005); Rachel Hamilton and Joseph Newman, "The response modulation hypothesis: Formulation, development, and implications for psychopathy," in *Handbook of Psychopathy*, ed. C. Patrick (New York: The Guilford Press, 2018), 80–93; Hare, "Psychopaths and their nature."

38. Hare, "Psychopathy: A clinical construct whose time has come"; De Brito et al., "Psychopathy"; Kiehl and Hoffman, "Criminal psychopath"; Reidy et al., "Why psychopathy matters."

39. M. DeLisi, ed., *Routledge international handbook of psychopathy and crime* (New York: Routledge, 2018); Gacono, *Clinical and forensic assessment of psychopathy*; Hugues Hervé and John C. Yuille, eds., *The psychopath: Theory, research, and practice* (Mahwah, NJ: Lawrence Erlbaum Associates, 2007); Kent A. Kiehl and Walter P. Sinnott-Armstrong, eds., *Handbook on psychopathy and law* (New York: Oxford University Press, 2013); Christopher J. Patrick, ed., *Handbook of psychopathy*, 2nd ed. (New York: The Guilford Press, 2018); A. R. Felthous and H. Sass, eds., *The Wiley international handbook on psychopathic disorders and the law* (Hoboken, NJ: John Wiley & Sons, 2020).

40. James Blair, D. Mitchell, and K. Blair, *The psychopath: Emotion and the brain* (Malden, MA: Blackwell Publishing, 2005); Hare, *Without conscience*; Helfgott, *No remorse: Psychopathy and criminal justice*; Glenn and Raine, *Psychopathy*; Kent A. Kiehl, *The psychopath whisperer: The science of those without conscience* (New York: Crown Publishers, 2014); Minna Lyons, *The Dark Triad of personality: Narcissism, Machiavellianism, and psychopathy in everyday life* (London: Academic Press, 2019); Nicholas Thomson, *Understanding psychopathy: The biopsychosocial perspective* (New York: Routledge, 2019); Essi Viding, *Psychopathy: A very short introduction* (New York: Oxford University Press, 2019).

Chapter 1

1. Robert Hare, Craig S. Neumann, and Thomas A. Widiger, "Psychopathy," in *The Oxford handbook of personality disorders*, ed. T. A. Widiger (New York: Oxford University Press, 2012), 478–504, 478.

2. Theodore Millon, Erik Simonsen, and Morten Birket-Smith, "Historical conceptions of psychopathy in the United States and Europe," in *Psychopathy: Antisocial, criminal, and violent behavior*, ed. T. Millon et al. (New York: Guilford Press, 1998), 3–31; Henning Sass and Alan R. Felthous, "The heterogeneous construct of psychopathy," in *Being amoral: Psychopathy and moral incapacity*, ed. T. Schramme, Psychopathy and Moral Incapacity (Cambridge, MA: MIT Press, 2014), 41–68; Henry Werlinder, "Psychopathy: A history of the concepts, analysis of the origin, and development of a family of concepts in psychopathology" (6 PhD, Uppsala Universitet, 1978).

3. Thomas A. Widiger, "Psychopathy and the DSM-IV psychopathology," in *The handbook of psychopathy*, ed. C. Patrick (New York: The Guildford Press, 2006),

156–171; Cristina Crego and Thomas A. Widiger, "Psychopathy and the DSM," *Journal of Personality* 83, no. 6 (2015): 665–677.

4. J. R. Oltmanns and T. A. Widiger, "Evaluating the assessment of the ICD-11 personality disorder diagnostic system," *Psychological Assessment* 31, no. 5 (2019): 674–684; Bo Bach and Roger Mulder, "Empirical foundation of the ICD-11 classification of personality disorders," in *Personality disorders and pathology: Integrating clinical assessment and practice in the DSM-5 and ICD-11 era*. (Washington, DC: American Psychological Association, 2022), 27–52.

5. Benjamin Rush, *An inquiry into the influence of physical causes upon the moral faculty: Delivered before the American Philosophical Society, held in Philadelphia on the twenty-seventh of February, 1786* (Philadelphia: Haswell, Barrington, and Haswell, 1839).

6. Rush, *An inquiry*, 4.

7. David Freeman Hawke, *Benjamin Rush: Revolutionary gadfly* (Indianapolis, IN: Bobbs-Merrill, 1971).

8. E. T. Carlson and M. M. Simpson, "Benjamin Rush's medical use of the moral faculty," *Bulletin of the History of Medicine* 39 (1965): 22–33.

9. Philippe Pinel, *Traite medico-philosophique sur l'alienation mentale ou la manie* (Paris: Richard (An IX), 1801).

10. Jean-Étienne D. Esquirol, *Des maladies mentales* (Paris: Chez J. B. Baillière, 1838).

11. E. T. Carlson and N. Dain, "The meaning of moral insanity," *Norman Bulletin on the History of Medicine* 36, no. 2 (1962): 130–140.

12. Sass and Felthous, "The heterogeneous construct of psychopathy."

13. B. Karpman, "The myth of the psychopathic personality," *The American Journal of Psychiatry* 104, no. 9 (1948): 523–534.

14. David Kennedy Henderson, *Psychopathic states* (New York: W. W. Norton & Company, 1947).

15. Silvano Arieti, "Psychopathic personality: Some views on its psychopathology and psychodynamics," *Comprehensive Psychiatry* 4, no. 5 (1963): 301–312.

16. Hervey M. Cleckley, *The mask of sanity: An attempt to clarify some issues about the so-called psychopathic personality*, 5th ed. (St. Louis: Mosby, 1988).

17. William Maxwell McCord and Joan McCord, *Psychopathy and delinquency* (New York: Grune & Stratton, 1956).

18. David T. Lykken, *The antisocial personalities* (Hillsdale, NJ: Lawrence Erlbaum Associates, 1995).

19. Lee N. Robins, *Deviant children grown up: A sociological and psychiatric study of sociopathic personality* (Oxford: Williams & Wilkins, 1966).

20. Cleckley, *Mask of sanity*, viii-xi.

21. S. O. Lilienfeld et al., "Hervey Cleckley (1903–1984): Contributions to the study of psychopathy," *Journal of Personality Disorders* 9, no. 6 (2018): 510–520.

22. Crego and Widiger, "Psychopathy and the DSM."

23. T. A. Widiger and Elizabeth M. Corbitt, "Antisocial Personality Disorder," in *The DSM-IV personality disorders*, ed. W. J. Livesley (New York: The Guilford Press, 1995), 103–126.

24. C. Crego and T. A. Widiger, "Psychopathy, DSM-5, and a caution," *Personality Disorders-Theory Research and Treatment* 5, no. 4 (2014): 335–347; APA, *Diagnostic and statistical manual of mental disorders, fifth edition, text revisions (DSM-5-TR)* (Washington, DC: American Psychiatric Publication, 2022); C. M. Strickland et al., "Characterizing psychopathy using DSM-5 personality traits," *Assessment* 20, no. 3 (2013): 327–338; Erin K. Fuller, Dylan T. Gatner, and Kevin S. Douglas, "Concurrent, convergent, and discriminant validity of the DSM-5 section III psychopathy specifier," *Assessment* 30, no. 6 (2023): 1790–1810.

25. David DeMatteo and John F. Edens, "The role and relevance of the Psychopathy Checklist-Revised in court: A case law survey of U.S. courts (1991–2004)," *Psychology, Public Policy, and Law* 12, no. 2 (2006): 214–241.

26. DeMatteo et al., "Investigating the role"; John F. Edens et al., "How reliable are Psychopathy Checklist-Revised scores in Canadian criminal trials? A case law review," *Psychological Assessment* 27, no. 2 (2015): 447–456.

27. Harris and Rice, "Psychopathy research at Oak Ridge"; James R. P. Ogloff, Stephen Wong, and Anthony Greenwood, "Treating criminal psychopaths in a therapeutic community program," *Behavioral Sciences & the Law* 8, no. 2 (1990): 181–190; Georg K. Stürup, "The management and treatment of psychopaths in a special institution in Denmark," *Proceedings of the Royal Society of Medicine* 41, no. 11 (1948): 765–768.

28. Aubrey Lewis, "Psychopathic personality: A most elusive category," *Psychological Medicine* 4, no. 2 (1974): 133–140.

29. Susanna Shapland, "Defining the elephant: A history of psychopathy, 1891–1959" (Birkbeck: University of London, 2020), https://eprints.bbk.ac.uk/id/eprint /40393/. Henry Werlinder, "Psychopathy: A history of the concepts, analysis of the origin, and development of a family of concepts in psychopathology" (Stockholm: Uppsala Universitet, 1978).

30. Hare, "Forty years aren't enough," 8.

31. Robert Hare, "The NATO advanced study institute on psychopathy, Alvor, Portugal, 1995," *Journal of Personality Disorders* 11, no. 3 (1997): 301–303, 301.

32. Robert Hare and D. Schalling, eds., *Psychopathic behaviour: Approaches to research* (Chichester, UK: Wiley, 1978).

33. Hare, "The NATO advanced study institute on psychopathy," 301.

34. See also Scott O. Lilienfeld, "Conceptual problems in the assessment of psychopathy," *Clinical Psychology Review* 14, no. 1 (1994): 17–38.

35. Karpman, "The myth of the psychopathic personality"; B. A. Thomas-Peter, "The Classification of Psychopathy—a Review of the Hare Vs Blackburn Debate," *Personality and Individual Differences* 13, no. 3 (1992): 337–342; James S. Wulach, "Diagnosing the DSM-III antisocial personality disorder," *Professional Psychology: Research and Practice* 14, no. 3 (1983): 330–340.

36. Ronald Blackburn, "On moral judgements and personality disorders: The myth of psychopathic personality revisited," *British Journal of Psychiatry* 153, no. 4 (1988): 505–512, 511.

37. De Brito et al., "Psychopathy"; Hare et al., "Psychopathy."

38. Hare, "Psychopathy, the PCL-R, and criminal justice."

39. DeMatteo et al., "Investigating the role."

40. J. L. Viljoen et al., "Psychopathy evidence in legal proceedings involving adolescent offenders," *Psychology Public Policy and Law* 16, no. 3 (2010): 254–283.

41. Viljoen et al., "Psychopathy evidence," 254.

42. C. D. Lloyd, H. J. Clark, and A. E. Forth, "Psychopathy, expert testimony, and indeterminate sentences: Exploring the relationship between Psychopathy Checklist-Revised testimony and trial outcome in Canada," *Legal and Criminological Psychology* 15, no. 2 (2010): 323–339.

43. Edens et al., "How reliable?"

44. Hare, "Psychopathy, the PCL-R, and criminal justice," 30.

45. DeMatteo and Olver, "Use of the Psychopathy Checklist-Revised," 14.

46. For similar examples, see Cleckley, *Mask of sanity*; Hare and Schalling, *Psychopathic behaviour*; McCord and McCord, *Psychopathy and delinquency*; J. Reid Meloy, *The psychopathic mind: Origins, dynamics, and treatment* (Northvale, NJ: Jason Aronson, 1988).

47. Kiehl and Sinnott-Armstrong, *Handbook on psychopathy and law.*

48. Gacono, *Clinical and forensic assessment of psychopathy.*

49. Patrick, *Handbook of psychopathy.*

50. DeLisi, *Routledge international handbook of psychopathy and crime.*

51. Curt R. Bartol and Anne M. Bartol, *Introduction to forensic psychology: Research and application* 6th ed. (Thousand Oaks, CA: Sage Publications, 2021).

52. Joanna Pozzulo, Craig Bennell, and Adelle Forth, *Forensic Psychology*, 6th ed. (Toronto: Pearson Canada, 2022).

53. Viding, *Psychopathy: A very short introduction*.

54. Lykken, *Antisocial personalities*.

55. Raine and Sanmartín, *Violence and psychopathy*.

56. Blair et al., *The psychopath: Emotion and the brain*.

57. Glenn and Raine, *Psychopathy*.

58. DeLisi, *Psychopathy as unified theory of crime*.

59. Helfgott, *No remorse*.

60. Hare, *Without conscience*.

61. Babiak and Hare, *Snakes in suits*.

62. Jon Ronson, *The psychopath test: A journey through the madness industry* (New York: Riverhead Books, 2011).

63. Dutton, *The wisdom of psychopaths*.

64. James H. Fallon, *The psychopath inside: A neuroscientist's personal journey into the dark side of the brain* (New York: Current, 2013).

65. Martha Stout, *The sociopath next door: The ruthless versus the rest of us* (New York: Broadway Books, 2005).

66. Kiehl, *Psychopath whisperer*.

67. Hare, "Foreword," xvii.

68. David J. Cooke et al., "Explicating the construct of psychopathy: Development and validation of a conceptual model, the Comprehensive Assessment of Psychopathic Personality (CAPP)," *International Journal of Forensic Mental Health* 11, no. 4 (2012): 242–252.

69. Michael R. Levenson, Kent A. Kiehl, and Cory M. Fitzpatrick, "Assessing psychopathic attributes in a noninstitutionalized population," *Journal of Personality and Social Psychology* 68, no. 1 (1995): 151–158.

70. A. K. Andershed et al., "Psychopathic traits in non-referred youths: A new assessment tool," in *Psychopaths: Current international perspectives*, ed. E. Blaauw and L. Sheridan (The Hague: Elsevier, 2002), 131–158.

71. S. O. Lilienfeld and M. R. Widows, *Psychopathic Personality Inventory—Revised: Professional manual* (Lutz: Psychological Assessment Resources, 2005).

72. Christopher J. Patrick, D. C. Fowles, and Robert F. Krueger, "Triarchic Conceptualization of Psychopathy: Developmental origins of disinhibition, boldness, and
meanness," *Development and Psychopathology* 21, no. 3 (2009): 913–938.

73. Robert Hare, *Hare Psychopathy Checklist–Revised*, 2nd ed. (Toronto: Multi Health
System, 2003).

74. Robert Hare, *Psychopathy Checklist–Revised (PCL–R)*, 1st ed. (Toronto, ON: Multi-
Health Systems, 1991).

75. Stephen D. Hart, Robert Hare, and Timothy J. Harpur, "The Psychopathy
Checklist—Revised (PCL-R)," in *Advances in psychological assessment*, ed. James C.
Rosen and Paul McReynolds (Boston: Springer US, 1992), 103–130.

76. For the most recent review of the PCL-R, see Hare et al., "The PCL-R assessment
of psychopathy"; A. Debowska, D. Boduszek, and R. Woodfield, "The PCL–R family
of psychopathy measures: Dimensionality and predictive utility of the PCL–R, PCL:
SV, PCL: YV, SRP–III, and SRP–SF," in *Routledge international handbook of psychopathy
and crime*, ed. Matt DeLisi (New York: Routledge, 2018), 225–240.

77. Christopher J. Patrick, "Back to the future: Cleckley as a guide to the next generation of psychopathy research," in *Handbook of psychopathy*, ed. Christopher J.
Patrick (New York: Guilford Press, 2006), 605–617; Robert Hare and Craig S. Neumann, "Psychopathy as a clinical and empirical construct," *Annual Review of Clinical
Psychology* 4 (2008): 217–246.

78. Hare et al., "The PCL-R assessment of psychopathy."

79. Robert Hare, "Diagnosis of antisocial personality disorder in two prison populations," *Am J Psychiatry* 140, no. 7 (1983): 887–890; Hare, "Research scale"; Robert Hare,
"Psychopathy and Antisocial Personality Disorder: A Case of Diagnostic Confusion,"
Psychiatric Times 13, no. 2 (1996): 39–40; Robert Hare and Stephen Hart, "Commentary on Antisocial Personality Disorder: The DSM-IV field trial," in *The DSM-IV Personality Disorders*, ed. John Livesley (New York: The Guilford Press, 1995), 127–134.

80. J. P. Verma, *Statistics and research methods in psychology with Excel* (Singapore:
Springer Nature, 2019).

81. Hare, *Hare Psychopathy Checklist–Revised*, 69.

82. See also D. J. Cooke and C. Michie, "An item response theory analysis of the
Hare Psychopathy Checklist—Revised," *Psychological Assessment* 9, no. 1 (1997):
3–14; B. Verschuere et al., "What features of psychopathy might be central? A network analysis of the Psychopathy Checklist-Revised (PCL-R) in three large samples,"
Journal of Abnormal Psychology 127, no. 1 (2018): 51–65.

83. Bruce Thompson and Larry G. Daniel, "Factor analytic evidence for the construct validity of scores: A Historical overview and some guidelines," *Educational and
Psychological Measurement* 56, no. 2 (1996): 197–208.

84. Martin Sellbom and Laura E. Drislane, "The classification of psychopathy," *Aggression and Violent Behavior* 59 (2021): 101473; Michael D. Maraun et al., "The dimensionality of the Hare Psychopathy Checklist-Revised, revisited: Its purported multidimensionality might well be artifactual," *Personality and Individual Differences* 138 (2019): 24–32.

85. Crego and Widiger, "Psychopathy, DSM-5, and a caution"; Crego and Widiger, "Psychopathy and the DSM"; J. Coid and S. Ullrich, "Antisocial personality disorder is on a continuum with psychopathy," *Comprehensive Psychiatry* 51, no. 4 (2010): 426–433; Dustin B. Wygant et al., "Examining the DSM-5 alternative personality disorder model operationalization of antisocial personality disorder and psychopathy in a male correctional sample," *Personality Disorders: Theory, Research, and Treatment* 7, no. 3 (2016): 229–239.

86. Thomas A. Widiger and Douglas B. Samuel, "Diagnostic categories or dimensions? A question for the Diagnostic and Statistical Manual of Mental Disorders–fifth edition," *Journal of Abnormal Psychology* 114, no. 4 (2005): 494–504.

87. A. Forth, D. Kosson, and Robert Hare, *The Psychopathy Checklist: Youth Version* (North Tonawanda, NY: Multi-Health Systems, 2003).

88. Stephen D. Hart, D. N. Cox, and Robert Hare, *The Hare PCL:SV—Psychopathy Checklist Screening Version* (Toronto: Multi-Health Systems, 1995).

89. D. L. Paulhus et al., *The Self-Report Psychopathy Scale—fourth edition (SRP 4)* (North Tonawanda, NY: Multi-Health Systems, 2016).

90. Hare et al., "The PCL-R assessment of psychopathy."

91. James V. Ray et al., "The relation between self-reported psychopathic traits and distorted response styles: A meta-analytic review," Clinical Psychological Testing 2224, *Personality Disorders: Theory, Research, and Treatment* 4, no. 1 (2013): 1–14; S. Lilienfeld and K. A. Fowler, "The self-report assessment of psychopathy: Problems, pitfalls, and promises," in *Handbook of psychopathy*, ed. C. Patrick (New York: The Guilford Press, 2006), 107–132.

92. Neal and Grisso, "Assessment practices and expert judgment."

93. Glancy et al., "AAPL practice guideline."

94. John F. Edens and Tiffany N. Truong, "Psychopathy evidence in legal proceedings," in *Psychopathy and criminal behavior: Current trends and challenges*, ed. Paulo Barbosa Marques, Mauro Paulino, and Laura Alho (Cambridge, MA: Academic Press, 2022), 241–272; Larsen et al., "Are psychopathy assessments ethical?"

95. Glancy et al., "AAPL practice guideline"; APA, "Evidence-based practice in psychology"; Leam A. Craig, Louise Dixon, and Theresa A. Gannon, eds., *What works in offender rehabilitation: An evidence-based approach to assessment and treatment* (Malden, MA: John Wiley & Sons, 2013); Rüdiger Müller-Isberner et al., "Implementation of

evidence-based practices in forensic mental health services," in *Handbook of forensic mental health services*, ed. Ronald Roesch and Alana N. Cook (New York: Routledge, 2017), 443–469.

96. Hare, "Psychopathy: A clinical construct whose time has come"; Porter et al., "Psychopathy and aggression"; Fox and DeLisi, "Psychopathic killers."

97. Harris and Rice, "Treatment of psychopathy"; Hare and Neumann, "Psychopathy: Assessment and forensic implications."

98. Blair, "Traits of empathy and anger: Implications for psychopathy and other disorders associated with aggression"; Hare, *Without conscience*; Helfgott, *No remorse*.

99. De Brito et al., "Psychopathy"; Glenn and Raine, *Psychopathy*.

100. Blair et al., *The psychopath*; DeLisi, *Psychopathy as unified theory of crime*.

101. Gacono, *Clinical and forensic assessment of psychopathy*; D. DeMatteo, John F. Edens, and A. Hart, "The use of measures of psychopathy in violence risk assessment," in *Handbook of violence risk assessment*, ed. R. K. Otto and K. S. Douglas (New York: Routledge, 2010), 19–40; DeMatteo and Olver, "Use of the Psychopathy Checklist-Revised"; Edens and Truong, "Psychopathy evidence in legal proceedings."

102. L. G. Aspinwall, T. R. Brown, and J. Tabery, "The double-edged sword: Does biomechanism increase or decrease judges' sentencing of psychopaths?," *Science* 337, no. 6096 (2012): 846–849; Katie Davey, "Psychopathy and sentencing: An investigative look into when the PCL-R is admitted into Canadian courtrooms and how a PCL-R score affects sentencing outcome" (MA University of Western Ontario, 2013); DeMatteo and Olver, "Use of the Psychopathy Checklist-Revised"; John F. Edens, L. S. Guy, and K. Fernandez, "Psychopathic traits predict attitudes toward a juvenile capital murderer," *Behavioral Sciences & the Law* 21, no. 6 (2003): 807–828; S. Jones and E. Cauffman, "Juvenile psychopathy and judicial decision making: An empirical analysis of an ethical dilemma," *Behavioral Sciences & the Law* 26, no. 2 (2008): 151–165; Lloyd et al., "Psychopathy, expert testimony"; John Monahan and Jennifer L. Skeem, "Risk assessment in criminal sentencing," *Annual Review of Clinical Psychology* 12 (2016): 489–513; Edens and Truong, "Psychopathy evidence in legal proceedings"; C. M. Berryessa and B. Wohlstetter, "The psychopathic "label" and effects on punishment outcomes: A meta-analysis," *Law and Human Behavior* 43, no. 1 (2019): 9–25; Mia A. Thomaidou, Alisha Patel, Sandy S. Xie, and Colleen M. Berryessa, "Machine learning analysis of a national sample of U.S. case law involving mental health evidence," *Journal of Criminal Justice* 94 (2024): 102266.

103. Berryessa and Wohlstetter, "The psychopathic 'label' and effects on punishment outcomes: A meta-analysis"; Neal and Grisso, "Assessment practices and expert judgment"; J. L. Viljoen et al., "Impact of risk assessment instruments on rates of pretrial detention, postconviction placements, and release: A systematic review and meta-analysis," *Law and Human Behavior* 43, no. 5 (2019): 397–420.

104. David DeMatteo et al., "Statement of concerned experts on the use of the Hare Psychopathy Checklist—Revised in capital sentencing to assess risk for institutional violence," *Psychology, Public Policy, and Law* 26, no. 2 (2020): 133–144; J. Cox et al., "Jury panel member perceptions of interpersonal-affective traits of psychopathy predict support for execution in a capital murder trial simulation," *Behavioral Sciences & the Law* 31, no. 4 (2013): 411–428; John F. Edens and Jennifer Cox, "Examining the prevalence, role and impact of evidence regarding antisocial personality, sociopathy and psychopathy in capital cases: A survey of defense team members," *Behavioral Sciences & the Law* 30, no. 3 (2012): 239–255; Edens et al., "Psychopathic traits predict attitudes toward a juvenile capital murderer"; John F. Edens et al., "The impact of mental health evidence on support for capital punishment: Are defendants labeled psychopathic considered more deserving of death?," *Behavioral Sciences & the Law* 23, no. 5 (2005): 603–625; S. E. Kelley et al., "Dangerous, depraved, and death-worthy: A meta-analysis of the correlates of perceived psychopathy in jury simulation studies," *Journal of Clinical Psychology* 75, no. 4 (2019): 627–643; Michael J. Saks et al., "The impact of neuroimages in the sentencing phase of capital trials," *Journal of Empirical Legal Studies* 11, no. 1 (2014): 105–131; Tiffany N. Truong, Shannon E. Kelley, and John F. Edens, "Does psychopathy influence juror decision-making in capital murder trials? 'The devil is in the (methodological) details,'" *Criminal Justice and Behavior* 48, no. 5 (2020): 690–707.

105. Davey, "Psychopathy and Sentencing: An Investigative look into when the PCL-R is admitted into Canadian courtrooms and how a PCL-R score affects sentencing outcome"; Edens et al., "How reliable"; Lloyd, "Psychopathy, expert testimony"; D. DeMatteo et al., "The role and reliability of the Psychopathy Checklist-Revised in U.S. sexually violent predator evaluations: A case law survey," *Law and Human Behavior* 38, no. 3 (2014): 248–255; I. Zinger and A. E. Forth, "Psychopathy and Canadian criminal proceedings: The potential for human rights abuses," *Canadian Journal of Criminology-Revue Canadienne De Criminologie* 40, no. 3 (1998): 237–276; DeMatteo et al., "Investigating the role."

106. M. T. Boccaccini et al., "Psychopathy Checklist-Revised use and reporting practices in Sexually Violent Predator evaluations," *Sex Abuse* 29, no. 6 (2017): 592–614; Rebecca L. Jackson and Derek T. Hess, "Evaluation for civil commitment of sex offenders: A survey of experts," *Sexual Abuse: A Journal of Research and Treatment* 19, no. 4 (2007): 425–448; Rachel E. Kahn et al., "Risk-tinted spectacles: What influences evaluator decision making in sexually violent persons examinations," Criminal Law & Adjudication 4230, *Psychology, Public Policy, and Law* 28, no. 2 (2022): 292–305; DeMatteo et al., "Role and reliability of the Psychopathy Checklist-Revised"; Marcus T. Boccaccini, Darrel B. Turner, and Daniel C. Murrie, "Do some evaluators report consistently higher or lower PCL-R scores than others?: Findings from a statewide sample of Sexually Violent Predator evaluations," *Psychology, Public Policy, and Law* 14, no. 4 (2008): 262–283; S. M. Kelley, G. Ambroziak, D. Thornton, and R. M. Barahal, "How do professionals assess sexual recidivism risk? An updated survey of practices," *Sex*

Abuse 32, no. 1 (2020): 3–29; Steven K. Erickson, Marc A. Martinez, and David DeMatteo, "The promise and pitfalls of the Psychopathy Checklist-Revised in sexual offense evaluations," *Journal of Forensic Psychology Research and Practice*, 1–28.

107. M. T. Boccaccini et al., "Describing, diagnosing, and naming psychopathy: How do youth psychopathy labels influence jurors?," *Behavioral Sciences & the Law* 26, no. 4 (2008): 487–510; Jennifer L. Rockett, Daniel C. Murrie, and Marcus T. Boccaccini, "Diagnostic labeling in juvenile justice settings: Do psychopathy and conduct disorder findings influence clinicians?," *Psychological Services* 4, no. 2 (2007): 107–122; D. C. Murrie et al., "Diagnostic labeling in juvenile court: How do descriptions of psychopathy and conduct disorder influence judges?," *Journal of Clinical Child and Adolescent Psychology* 36, no. 2 (2007): 228–241; Daniel C. Murrie, Dewey G. Cornell, and Wendy K. McCoy, "Psychopathy, conduct disorder, and stigma: Does diagnostic labeling influence juvenile probation officer recommendations?," *Law and Human Behavior* 29, no. 3 (2005): 323–342; John F. Edens and G. M. Vincent, "Juvenile psychopathy: A clinical construct in need of restraint?," *Journal of Forensic Psychology Practice* 8, no. 2 (2008): 186–197; Jones and Cauffman, "Juvenile psychopathy and judicial decision making: An empirical analysis of an ethical dilemma"; J. Petrila and Jennifer L. Skeem, "An introduction to the special issues on juvenile psychopathy and some reflections on the current debate—Juvenile psychopathy: The debate," *Behavioral Sciences & the Law* 21, no. 6 (2003): 689–694; Stephanie R. Penney and Marlene M. Moretti, "The transfer of juveniles to adult court in Canada and the United States: Confused agendas and compromised assessment procedures," *International Journal of Forensic Mental Health* 4, no. 1 (2005): 19–37; Rockett et al., "Diagnostic labeling in juvenile justice settings"; Jennifer L. Skeem, Elizabeth Scott, and Edward P. Mulvey, "Justice policy reform for high-risk juveniles: Using science to achieve large-scale crime reduction," *Annual Review of Clinical Psychology* 10 (2014): 709–739; Viljoen et al., "Psychopathy evidence"; Gina M. Vincent, Eva R. Kimonis, and Alisa Clark, "Juvenile psychopathy: Appropriate and inappropriate uses in legal proceedings," in *APA handbook of psychology and juvenile justice*, ed. Kirk Heilbrun, David DeMatteo, and Naomi Goldstein (Washington, DC: American Psychological Association, 2016), 197–232.

108. John F. Edens, J. Petrila, and S. E. Kelley, "Legal and ethical issues in the assessment and treatment of psychopathy," in *Handbook of Psychopathy*, ed. C. Patrick (New York: The Guilford Press, 2018), 732–751; Evelyn K. Haneveld et al., "Clinical appraisals of individual differences in treatment responsivity among patients with psychopathy: A consensual qualitative research study," *Criminal Justice and Behavior* 48, no. 8 (2020): 1031–1051; Rasmus Rosenberg Larsen, "Psychopathy treatment and the stigma of yesterday's research," *Kennedy Institute of Ethics Journal* 29, no. 3 (2019): 243–272; Mark E. Olver, "Can psychopathy be treated? What the research tells us," in *New frontiers in offender treatment: The translation of evidence-based practices to correctional settings*, ed. Elizabeth L. Jeglic and Cynthia Calkins (Cham: Springer International Publishing, 2018), 287–306; Devon Polaschek and Jennifer L. Skeem,

"Treatment of adults and juveniles with psychopathy," in *Handbook of Psychopathy*, ed. C. Patrick (New York: The Guilford Press, 2018), 710–731; Devon Polaschek, "Treatment of psychopathy," in *Routledge international handbook of psychopathy and crime*, ed. M. DeLisi (New York: Routledge, 2019), 610–634.

109. DeMatteo and Edens, "Role and relevance of the Psychopathy Checklist-Revised"; DeMatteo et al., "Role and reliability of the Psychopathy Checklist-Revised"; William James Denomme, Jamie Curno, and Adelle Forth, "Psychopathic traits, risk and protective factors, and attractiveness in forensic psychiatric patients: Their role in review board dispositions," *Journal of Forensic Psychology Research and Practice* 20, no. 3 (2020): 264–289; L. S. Guy et al., "Influence of the HCR-20, LS/CMI, and PCL-R on decisions about parole suitability among lifers," *Law and Human Behavior* 39, no. 3 (2015): 232–243; Christa McLeod, "Why California got it right: Assessing psychopaths before release," *Rutgers Law Record* 45 (2018): 145–169; Monahan and Skeem, "Risk assessment in criminal sentencing"; Shannon Marie Stewart, "Mental health problems in parole decisions: The re-conceptualization of mental health problems as risk factors" (PhD University of Ottawa, 2016); Robin Fitzgerald et al., "The practicalities of parole board decision-making," in *Parole on Probation: Parole Decision-Making, Public Opinion and Public Confidence*, ed. Robin Fitzgerald et al. (Cham: Springer International Publishing, 2023), 19–54; Gina M. Manguno-Mire et al., "Are release recommendations for NGRI acquittees informed by relevant data?," *Behavioral Sciences & the Law* 25, no. 1 (2007): 43–55.

110. Ryan M. Labrecque, "Security threat management in prison: Revalidation and revision of the Inmate Risk Assessment for Segregation Placement," *The Prison Journal* 102, no. 1 (2022): 47–63; Manguno-Mire et al., "Are release recommendations for NGRI acquittees informed by relevant data?"; Guy Bourgon et al., "Offender risk assessment practices vary across Canada," *Canadian Journal of Criminology and Criminal Justice* 60, no. 2 (2018): 167–205; Malin Pauli et al., "Gendered expressions of psychopathy: Correctional staffs' perceptions of the CAPP and CABP models," *International Journal of Forensic Mental Health* 17, no. 2 (2018): 97–110; S. Vidal and Jennifer L. Skeem, "Effect of psychopathy, abuse, and ethnicity on juvenile probation officers' decision-making and supervision strategies," *Law and Human Behavior* 31, no. 5 (2007): 479–498; Murrie et al., "Psychopathy, conduct disorder, and stigma"; R. C. Wagoner, C. A. Schubert, and E. P. Mulvey, "Probation intensity, self-reported offending, and psychopathy in juveniles on probation for serious offenses," *Journal of the American Academy of Psychiatry and the Law* 43, no. 2 (2015): 191–200.

111. DeMatteo et al., "Statement of concerned experts."

112. Harris et al., "Construct of psychopathy," 230–231. See also Babiak et al., "Psychopathy: An important forensic concept for the 21st century"; Hare, "Psychopathy: A clinical construct whose time has come"; Hare, "Psychopathy: A clinical and forensic overview"; Monahan, "Statement on book jacket"; Hare, "Foreword"; James F. Hemphill and Robert Hare, "Some misconceptions about the Hare PCL-R and risk

assessment: A reply to Gendreau, Goggin, and Smith," *Criminal Justice and Behavior* 31, no. 2 (2004): 203–243.

113. DeMatteo et al., "Investigating the role"; DeMatteo and Edens, "Role and relevance of the Psychopathy Checklist-Revised."

114. See also Viljoen et al., "Psychopathy evidence."

115. DeMatteo and Olver, "Use of the Psychopathy Checklist-Revised"; Hare, "Psychopathy, the PCL-R, and criminal justice."

116. Boccaccini et al., "Psychopathy Checklist-Revised use."

117. Jackson and Hess, "Evaluation for civil commitment of sex offenders."

118. A. R. Felthous and J. Ko, "Sexually Violent Predator law in the United States," *East Asian Archives of Psychiatry* 28, no. 4 (2018): 159–173.

119. Lloyd et al., "Psychopathy, expert testimony."

120. L. MacAulay, *High-risk offenders: A handbook for criminal justice professionals*, Solicitor General Canada (2001), 76, 131, https://www.publicsafety.gc.ca/cnt/rsrcs/pblctns/hghrsk-ffndrs-hndb/index-en.aspx.

121. Edens and Cox, "Examining the prevalence, role and impact of evidence regarding antisocial personality, sociopathy and psychopathy in capital cases: A survey of defense team members."

122. Murrie et al., "Diagnostic labeling in juvenile court"

123. See also Boccaccini et al., "Describing, diagnosing, and naming psychopathy"

124. Melissa R. Jonnson and Jodi L. Viljoen, "What are judges' views of risk assessments, and how do tools affect adolescent dispositions?," *Psychology, Public Policy, and Law* 27, no. 1 (2021): 112–123.

125. See also Tess M. S. Neal et al., "Psychological assessments in legal contexts: Are courts keeping "junk science" out of the courtroom?," *Psychological Science in the Public Interest* 20, no. 3 (2019): 135–164.

126. John F. Edens et al., "Bold, smart, dangerous and evil: Perceived correlates of core psychopathic traits among jury panel members," *Personality and Mental Health* 7, no. 2 (2013): 143–153.

127. See also Adrian Furnham, Yasmine Daoud, and Viren Swami, "'How to spot a psychopath.' Lay theories of psychopathy," journal article, *Social Psychiatry and Psychiatric Epidemiology* 44, no. 6 (2008): 464–472.

128. S. T. Smith et al., "'So, what is a psychopath?' Venireperson perceptions, beliefs, and attitudes about psychopathic personality," *Law and Human Behavior* 38, no. 5 (2014): 490–500.

129. See also Truong et al., "Does psychopathy influence?"

130. For a review of the literature, see Kelley et al., "Dangerous, depraved."

131. Singh et al., "International perspectives."

132. Viljoen et al., "Psychopathy evidence."

133. David Hill and Sabrina Demetrioff, "Clinical-forensic psychology in Canada: A survey of practitioner characteristics, attitudes, and psychological assessment practices," *Canadian Psychology/Psychologie Canadienne* 60, no. 1 (2019): 55–63.

134. Neal and Grisso, "Assessment practices and expert judgment."

135. For more surveys with comparable methods and results, see R. P. Archer et al., "A survey of psychological test use patterns among forensic psychologists," *Journal of Personality Assessment* 87, no. 1 (2006): 84–94; Stephen J. Lally, "What tests are acceptable for use in forensic evaluations? A survey of experts," *Professional Psychology: Research and Practice* 34, no. 5 (2003): 491–498; Claudia C. Hurducas et al., "Violence risk assessment tools: A systematic review of surveys," *International Journal of Forensic Mental Health* 13, no. 3 (2014): 181–192.

136. Monahan and Skeem, "Risk assessment in criminal sentencing"; Glancy et al., "AAPL practice guideline"; Gacono, "Suggestions for implementation."

137. Hill and Demetrioff, "Clinical-forensic psychology in Canada"; Singh et al., "International perspectives."

138. Rich Kluckow and Zhen Zeng, *Correctional populations in the United States, 2020—statistical tables*, Bureau of Justice Statistics (2022), https://bjs.ojp.gov/library/publications/correctional-populations-united-states-2020-statistical-tables.

139. Otto and Heilbrun, "Practice of forensic psychology."

140. Kiehl and Hoffman, "Criminal psychopath."

141. Hare, "Foreword."

Chapter 2

1. M. DeLisi and Bryanna Fox, "Psychopathy is integral to understanding homicide and violence," in *Psychopathy and criminal behavior: Current trends and challenges*, ed. Paulo B. Marques, Mauro Paulino, and Laura Alho (San Diego: Academic Press, 2022), 357–367.

2. Hare, "Psychopathy: A clinical construct whose time has come."

3. Hare, *Without conscience.*

4. Kiehl and Hoffman, "Criminal psychopath."

5. Reidy et al., "Why psychopathy matters," 216.

6. Reidy and Bogen, "Public health considerations in psychopathy"; Gatner et al., "How much does that cost?"; Kiehl and Hoffman, "The criminal psychopath."

7. Jay P. Singh, "Predictive validity performance indicators in violence risk assessment: A methodological primer," *Behavioral Sciences & the Law* 31, no. 1 (2013): 8–22; Jay P. Singh and Seena Fazel, "Forensic risk assessment: A metareview," *Criminal Justice and Behavior* 37, no. 9 (2010): 965–988.

8. Rasmus Rosenberg Larsen, Jarkko Jalava, and Stephanie Griffiths, "Are Psychopathy Checklist (PCL) psychopaths dangerous, untreatable, and without conscience? A systematic review of the empirical evidence," *Psychology, Public Policy, and Law* 26, no. 3 (2020): 297–311; Steven M. Gillespie, Andrew Jones, and Carlo Garofalo, "Psychopathy and dangerousness: An umbrella review and meta-analysis," *Clinical Psychology Review* 100 (2023): 102240.

9. J. Cohen, *Statistical power analysis for the behavioral sciences*, 2nd ed. (Mahwah, NJ: Lawrence Erlbaum Associates, 1988); Daniel Lakens, *Improving your statistical inferences* (Github2023), https://lakens.github.io/statistical_inferences/.

10. James F. Hemphill and Robert Hare, "Psychopathy Checklist factor scores and recidivism," *Issues in Criminological & Legal Psychology* 24 (1995): 68–73; Paul Gendreau, Tracy Little, and Claire Goggin, "A meta-analysis of the predictors of adult offender recidivism: What works!," *Criminology* 34, no. 4 (1996): 575–608; R. T. Salekin, R. Rogers, and K. W. Sewell, "A review and meta-analysis of the psychopathy checklist and psychopathy checklist-revised: Predictive validity of dangerousness," *Clinical Psychology-Science and Practice* 3, no. 3 (1996): 203–215; James F. Hemphill, Robert Hare, and Stephen Wong, "Psychopathy and recidivism: A review," *Legal and Criminological Psychology* 3, no. 1 (1998): 139–170; Paul Gendreau, Claire Goggin, and Paula Smith, "Is the PCL-R really the "unparalleled" measure of offender risk? A lesson in knowledge cumulation," *Criminal Justice and Behavior* 29, no. 4 (2002): 397–426; G. D. Walters, "Predicting criminal justice outcomes with the psychopathy checklist and lifestyle criminality screening form: A meta-analytic comparison," *Behavioral Sciences & the Law* 21, no. 1 (2003): 89–102; G. D. Walters, "Predicting institutional adjustment and recidivism with the Psychopathy Checklist factor scores: A meta-analysis," *Law and Human Behavior* 27, no. 5 (2003): 541–558; John F. Edens, J. S. Campbell, and J. M. Weir, "Youth psychopathy and criminal recidivism: A meta-analysis of the psychopathy checklist measures," *Law and Human Behavior* 31, no. 1 (2006): 53–75; A.M.R. Leistico et al., "A large-scale meta-analysis relating the hare measures of psychopathy to antisocial conduct," *Law and Human Behavior* 32, no. 1 (2008): 28–45; Mark E. Olver, Keira C. Stockdale, and J. Stephen Wormith, "Risk assessment with young offenders: A meta-analysis of three assessment measures," *Criminal Justice and Behavior* 36, no. 4 (2009): 329–353; Jessica J. Asscher et al., "The relationship between juvenile psychopathic traits, delinquency

and (violent) recidivism: A meta-analysis," *Journal of Child Psychology and Psychiatry* 52, no. 11 (2011): 1134–1143; J. P. Singh, M. Grann, and S. Fazel, "A comparative study of violence risk assessment tools: A systematic review and metaregression analysis of 68 studies involving 25,980 participants," *Clinical Psychology Review* 31, no. 3 (2011): 499–513; Mary Ann Campbell, Sheila French, and Paul Gendreau, "The prediction of violence in adult offenders: A meta-analytic comparison of instruments and methods of assessment," *Criminal Justice and Behavior* 36, no. 6 (2009): 567–590; Patrick J. Kennealy et al., "Do core interpersonal and affective traits of PCL-R psychopathy interact with antisocial behavior and disinhibition to predict violence?," *Psychological Assessment* 22, no. 3 (2010): 569–580; M. Yang, Stephen C. Wong, and J. Coid, "The efficacy of violence prediction: A meta-analytic comparison of nine risk assessment tools," *Psychological Bulletin* 136, no. 5 (2010): 740–767; S. W. Hawes, M. T. Boccaccini, and D. C. Murrie, "Psychopathy and the combination of psychopathy and sexual deviance as predictors of sexual recidivism: Meta-analytic findings using the Psychopathy Checklist-Revised," *Psychological Assessment* 25, no. 1 (2013): 233–243; Julie Blais, Elizabeth Solodukhin, and Adelle E. Forth, "A meta-analysis exploring the relationship between psychopathy and instrumental versus reactive violence," *Criminal Justice and Behavior* 41, no. 7 (2014): 797–821; Andreas Mokros, Knut Vohs, and Elmar Habermeyer, "Psychopathy and violent reoffending in german-speaking countries: A meta-Analysis," *European Journal of Psychological Assessment* 30, no. 2 (2014): 117–129; L. S. Guy et al., "Does psychopathy predict institutional misconduct among adults? A meta-analytic investigation," *Journal of Consulting and Clinical Psychology* 73, no. 6 (2005): 1056–1064; Neil Hogan and Liam Ennis, "Assessing risk for forensic psychiatric inpatient violence: A meta-analysis," *Open Access Journal of Forensic Psychology* 2 (2010): 137–147; Mark E. Olver et al., "Reliability and validity of the Psychopathy Checklist-Revised in the assessment of risk for institutional violence: A cautionary note on DeMatteo et al. (2020)," *Psychology, Public Policy, and Law* 26, no. 4 (2020): 490–510.

11. David C. Funder and Daniel J. Ozer, "Evaluating effect size in psychological research: Sense and nonsense," *Advances in Methods and Practices in Psychological Science* 2, no. 2 (2019): 156–168; Ralph Rosnow and Robert Rosenthal, "Effect sizes for experimenting psychologists," *Canadian Journal of Experimental Psychology-Revue Canadienne De Psychologie Experimentale* 57, no. 3 (2003): 221–237.

12. A tool for visualizing and interpreting Cohen's *d*, see "Interpreting Cohen's *d* effect size: An interactive visualization," 2023, accessed April 19, 2023, https://rpsychologist.com/cohend/.

13. Cohen, *Statistical power analysis for the behavioral sciences.*

14. Steven M. Gillespie, Andrew Jones, and Carlo Garofalo, "Psychopathy and dangerousness: An umbrella review and meta-analysis," *Clinical Psychology Review* 100 (2023): 102240.

15. J. P. Ioannidis, "Why most discovered true associations are inflated," *Epidemiology* 19, no. 5 (2008): 640–648; Julian P. T. Higgins et al., eds., *Cochrane handbook for systematic reviews of interventions* (West Sussex: Wiley-Blackwell, 2019).

16. Petar Milin and Olga Hadžić, "Moderating and mediating variables in psychological research," in *International encyclopedia of statistical science*, ed. Miodrag Lovric (Berlin: Springer, 2011), 849–852.

17. Yang et al., "The efficacy of violence prediction."

18. Seena Fazel et al., "The predictive performance of criminal risk assessment tools used at sentencing: Systematic review of validation studies," *Journal of Criminal Justice* 81 (2022): 101902.

19. Here are some examples: R. C. Serin and N. L. Amos, "The role of psychopathy in the assessment of dangerousness," *International Journal of Law and Psychiatry* 18, no. 2 (1995): 231–238; L. A. Sewall and Mark E. Olver, "Psychopathy and treatment outcome: Results from a sexual violence reduction program," *Personality Disorders: Theory, Research, and Treatment* 10, no. 1 (2019): 59–69; S. M. Shepherd, R. E. Campbell, and J.R.P. Ogloff, "Psychopathy, Antisocial Personality Disorder, and Reconviction in an Australian Sample of Forensic Patients," *International Journal of Offender Therapy and Comparative Criminology* 62, no. 3 (2018): 609–628; Z. Walsh, M. T. Swogger, and D. S. Kosson, "Psychopathy, IQ, and violence in European American and African American county jail inmates," *Journal of Consulting and Clinical Psychology* 72, no. 6 (2004): 1165–1169; G. T. Harris, M. E. Rice, and C. A. Cormier, "Psychopathy and violent recidivism," *Law and Human Behavior* 15, no. 6 (1991): 625–637; C. Huchzermeier et al., "Psychopathy checklist score predicts negative events during the sentences of prisoners with hare psychopathy: A prospective study at a German prison," *Canadian Journal of Psychiatry-Revue Canadienne De Psychiatrie* 51, no. 11 (2006): 692–697.

20. Angelika M. Stefan and Felix D. Schönbrodt, "Big little lies: A compendium and simulation of p-hacking strategies," *Royal Society Open Science* 10, no. 2 (2023): 220346.

21. Gendreau et al., "A lesson in knowledge cumulation."

22. Jennifer L. Skeem and David J. Cooke, "Is criminal behavior a central component of psychopathy? Conceptual directions for resolving the debate," *Psychological Assessment* 22, no. 2 (2010): 433–445; J. P. Camp et al., "Psychopathic predators? Getting specific about the relation between psychopathy and violence," *Journal of Consulting and Clinical Psychology* 81, no. 3 (2013): 467–480; Jennifer L. Skeem et al., "Psychopathic personality: Bridging the gap between scientific evidence and public policy," *Psychological Science in the Public Interest* 12, no. 3 (2011): 95–162; R. Serin, S. Brown, and A. Wolf, "The clinical use of the Hare Psychopathy Checklist–Revised (PCL-R) in contemporary risk assessment," in *The clinical and forensic assessment of psychopathy: A practitioner's guide*, ed. C. Gacono (New York: Routledge, 2016), 293–310; G. D. Walters, "The trouble with psychopathy as a general theory of crime," *International Journal of Offender Therapy and Comparative Criminology* 48, no. 2 (2004): 133–148.

23. Leistico et al., "A large-scale meta-analysis," 40.

24. Kluckow and Zeng, *Correctional populations in the United States, 2020—statistical tables*; Wendy Sawyer, *The gender divide: Tracking women's state prison growth* (Prison Policy Initiative, 2018), https://www.prisonpolicy.org/reports/women_overtime.html; Wendy Sawyer and Peter Wagner, *Mass incarceration: The whole pie 2023*, Prison Policy Initiative (2023), https://www.prisonpolicy.org/reports/pie2023.html.

Chapter 3

1. Hare, *Hare Psychopathy Checklist–Revised*, 158.

2. Donald A. Andrews and James Bonta, *The psychology of criminal conduct*, 6th ed. (New York: Routledge, 2017); Daniel M. Blonigen, "Explaining the relationship between age and crime: Contributions from the developmental literature on personality," *Clinical Psychology Review* 30, no. 1 (2010): 89–100; Shadd Maruna, *Making good: How ex-convicts reform and rebuild their lives* (Washington, DC: American Psychological Association, 2001).

3. Robert Hare, P. Black, and Z. Walsh, "The Psychopathy Checklist-Revised: Forensic applications and limitations," in *Forensic uses of clinical assessment instruments*, ed. R. P. Archer and E.M.A. Wheeler (New York: Routledge, 2013), 230–265, 244–245.

4. Carlson and Simpson, "Benjamin Rush's medical use of the moral faculty"; Rush, *An inquiry*.

5. David W. Jones, *Disordered personalities and crime: An analysis of the history of moral insanity* (New York: Routledge, 2016); Nicole Hahn Rafter, *The criminal brain: Understanding biological theories of crime*, 1st ed. (New York: New York University Press, 2008).

6. Hare, "Psychopaths and their nature"; McCord and McCord, *Psychopathy and delinquency*.

7. Cleckley, *Mask of sanity*.

8. Cleckley, *Mask of sanity*, 433.

9. Cleckley, *Mask of sanity*, 417.

10. Ronald Blackburn, *The psychology of criminal conduct: Theory, research and practice* (Oxford: John Wiley & Sons, 1993); Reid et al., *Unmasking the psychopath*; R. T. Salekin, "Psychopathy and therapeutic pessimism: Clinical lore or clinical reality?," *Clinical Psychology Review* 22, no. 1 (2002): 79–112; Stürup, *Treating the "untreatable."*

11. P. Suedfeld and P. B. Landon, "Approaches to treatment," in *Psychopathic behavior: Approaches to research*, ed. Robert Hare and D. Schalling (Chichester, UK: Wiley, 1978), 347–376; Harris and Rice, "Treatment of psychopathy: A review of empirical findings"; Harris and Rice, "Psychopathy research at Oak Ridge: Skepticism overcome"; Carl B. Gacono et al., "Treating conduct disorder, antisocial, and

psychopathic personalities," in *Treating adult and juvenile offenders with special needs,* ed. José B. Ashford, Bruce Dennis Sales, and William H. Reid (Washington, DC: American Psychological Association, 2001), 99–129.

12. Francis T. Cullen, "Rehabilitation: Beyond nothing works," *Crime and Justice* 42, no. 1 (2013): 299–376.

13. Hare, *Without conscience,* 194.

14. Hare, *Hare Psychopathy Checklist–Revised,* 158.

15. Hare, *Hare Psychopathy Checklist–Revised,* 158.

16. Hare, *Hare Psychopathy Checklist–Revised,* 159.

17. Harris and Rice, "Psychopathy research at Oak Ridge: Skepticism overcome"; Hare, "Psychopaths and their nature."

18. Polaschek, "Treatment of psychopathy"; Craig et al., *What works in offender rehabilitation.*

19. A. R. Baskin-Sommers, J. J. Curtin, and J. P. Newman, "Altering the cognitive-affective dysfunctions of psychopathic and externalizing offender subtypes with cognitive remediation," *Clinical Psychological Science* 3, no. 1 (2015): 45–57.

20. Anthony W. Bateman, John Gunderson, and Roger Mulder, "Treatment of personality disorder," *The Lancet* 385, no. 9969 (2015): 735–743; De Brito et al., "Psychopathy."

21. Aaron T. Beck, "Cognitive therapy: Nature and relation to behavior therapy," *Behavior Therapy* 1, no. 2 (1970): 184–200.

22. For other treatment methods, see Geraldine Akerman and Richard Shuker, eds., *Global perspectives on interventions in forensic therapeutic communities* (New York: Routledge, 2020); Trudy van der Stouwe et al., "The effectiveness of Multisystemic Therapy (MST): A meta-analysis," *Clinical Psychology Review* 34, no. 6 (2014): 468–481.

23. Daniel David, Ioana Cristea, and Stefan G. Hofmann, "Why Cognitive Behavioral Therapy is the current gold standard of psychotherapy," Opinion, *Frontiers in Psychiatry* 9 (2018): 4.

24. Raymond Chip Tafrate and Damon Mitchell, eds., *Forensic CBT: A handbook for clinical practice* (Malden, MA: Wiley Blackwell, 2014); Damon Mitchell, Raymond Chip Tafrate, and Tom Hogan, "Cognitive Behavioral Therapy in forensic treatment," in *New frontiers in offender treatment: The translation of evidence-based practices to correctional settings,* ed. Elizabeth L. Jeglic and Cynthia Calkins (Cham: Springer International Publishing, 2018), 57–84.

25. Andrews and Bonta, *The psychology of criminal conduct;* Devon Polaschek, ed., *Treatment programmes for high risk offenders* (New York: Routledge, 2016).

26. Keith S. Dobson and David J. A. Dozois, eds., *Handbook of cognitive-behavioral therapies*, 4th ed. (New York: The Guilford Press, 2019); Tafrate and Mitchell, *Forensic CBT: A handbook for clinical practice*.

27. Karen D'Silva, Conor Duggan, and Lucy McCarthy, "Does treatment really make psychopaths worse? A review of the evidence," *Journal of Personality Disorders* 18, no. 2 (2004): 163–177; R. T. Salekin, C. Worley, and R. D. Grimes, "Treatment of psychopathy: A review and brief introduction to the mental model approach for psychopathy," *Behavioral Sciences & the Law* 28, no. 2 (2010): 235–266.

28. M. F. Caldwell and G. J. Van Rybroek, "Reducing violence in serious juvenile offenders using intensive treatment," *International Journal of Law Psychiatry* 28, no. 6 (2005): 622–636.

29. Jennifer L. Skeem, J. Monahan, and E. P. Mulvey, "Psychopathy, treatment involvement, and subsequent violence among civil psychiatric patients," *Law and Human Behavior* 26, no. 6 (2002): 577–603.

30. Andrews and Bonta, *The psychology of criminal conduct*; Polaschek, *Treatment programmes for high risk offenders*.

31. Vicente Garrido, Cristina Esteban, and C. Molero, "The effectiveness in the treatment of psychopathy: A meta-analysis," *Issues in Criminological & Legal Psychology* 24 (1995): 57–59; Gacono et al., "Treating conduct disorder, antisocial, and psychopathic personalities"; Salekin, "Psychopathy and therapeutic pessimism"; D'Silva et al., "Does treatment really make psychopaths worse?"; Harris and Rice, "Treatment of psychopathy: A review of empirical findings"; D. M. Doren and P. M. Yates, "Effectiveness of sex offender treatment for psychopathic sexual offenders," *International Journal of Offender Therapy and Comparative Criminology* 52, no. 2 (2008): 234–245; Jeffrey Abracen, Jan Looman, and Calvin M. Langton, "Treatment of sexual offenders with psychopathic traits: Recent research developments and clinical implications," *Trauma, Violence, & Abuse* 9, no. 3 (2008): 144–166; Salekin et al., "Treatment of psychopathy"; Chaturaka Rodrigo, Senaka Rajapakse, and Gamini Jayananda, "The 'antisocial' person: An insight in to biology, classification and current evidence on treatment," *Annals of General Psychiatry* 9 (2010): 31; Julia Shaw and Stephen Porter, "Forever a psychopath? Psychopathy and the criminal career trajectory," in *Psychopathy and law: A practitioner's guide*, ed. Helinä Häkkänen-Nyholm and Jan-Olof Nyholm (Hoboken, NJ: Wiley-Blackwell, 2012), 201–221; Dennis E. Reidy, Megan C. Kearns, and Sarah DeGue, "Reducing psychopathic violence: A review of the treatment literature," *Aggression and Violent Behavior* 18, no. 5 (2013): 527–538; Chasity Bailey et al., "Exploring treatment options for an allegedly "untreatable" disorder, psychopathy: An integrative literature review," in *Psychopathy: Risk factors, behavioral symptoms and treatment options*, ed. Michael Fitzgerald, Psychiatry—theory, applications and treatments (Hauppauge, NY: Nova Science Publishers, 2015), 203–219; Barry Rosenfeld et al., "Mental health treatment of criminal offenders," in *APA handbook of forensic psychology, Vol. 1: Individual and*

situational influences in criminal and civil contexts, ed. Brian L. Cutler and Patricia A. Zapf (Washington, DC: American Psychological Association,, 2015), 159–190; C. de Ruiter, F. Chakhssi, and D. Bernstein, "Treating the untreatable psychopath," in *The clinical and forensic assessment of psychopathy: A practitioner's guide*, ed. C. Gacono (New York: Routledge, 2016), 288–402; Lisa K. Hecht, Robert D. Latzman, and Scott O. Lilienfeld, "The psychological treatment of psychopathy: Theory and research," in *Evidence-based psychotherapy: The state of the science and practice*, ed. Daniel David, Steven Jay Lynn, and Guy H. Montgomery (Hoboken, NJ: Wiley-Blackwell, 2018), 271–298; Polaschek and Skeem, "Treatment of adults and juveniles"; Olver, "Can psychopathy be treated? What the research tells us"; Polaschek, "Treatment of psychopathy"; Mark E. Olver, "Treatment of psychopathic offenders: A review of research, past, and current practice," in *Psychopathy and criminal behavior*, ed. Paulo Barbosa Marques, Mauro Paulino, and Laura Alho (Cambridge, MA: Academic Press, 2022), 469–481; Larsen et al., "Are Psychopathy Checklist (PCL) psychopaths dangerous?"

32. Devon Polaschek and E. C. Ross, "Do early therapeutic alliance, motivation, and stages of change predict therapy change for high-risk, psychopathic violent prisoners?," *Criminal Behaviour and Mental Health* 20, no. 2 (2010): 100–111; Julia A. Yesberg and Devon Polaschek, "How does offender rehabilitation actually work? Exploring mechanisms of change in high-risk treated parolees," *International Journal of Offender Therapy and Comparative Criminology* 63, no. 15–16 (2019): 2672–2692.

33. Andrews and Bonta, *The psychology of criminal conduct.*

34. See also Craig et al., *What works in offender rehabilitation.*

35. Olver, "Can psychopathy be treated?"; Polaschek and Skeem, "Treatment of adults and juveniles"; Polaschek, "Treatment of psychopathy"; S. C. P. Wong and Mark E. Olver, "Risk reduction treatment of psychopathy and applications to mentally disordered offenders," *CNS Spectrums* 20, no. 3 (2015): 303–310.

36. Danielle R. DeSorcy, Mark E. Olver, and J. Stephen Wormith, "Working alliance and psychopathy: Linkages to treatment outcome in a sample of treated sexual offenders," *Journal of Interpersonal Violence* 35, no. 7–8 (2020): 1739–1760.

37. See also Evelyn Klein Haneveld et al., "Treatment responsiveness of replicated psychopathy profiles," *Law and Human Behavior* 42, no. 5 (2018): 484–495; Ogloff et al., "Treating criminal psychopaths"; Mark E. Olver and S. C. Wong, "Therapeutic responses of psychopathic sexual offenders: Treatment attrition, therapeutic change, and long-term recidivism," *Journal of Consulting and Clinical Psychology* 77, no. 2 (2009): 328–336; Mark E. Olver and S. Wong, "Predictors of sex offender treatment dropout: Psychopathy, sex offender risk, and responsivity implications," *Psychology Crime & Law* 17, no. 5 (2011): 457–471; Polaschek and Ross, "Do early therapeutic alliance, motivation, and stages of change predict therapy change for high-risk, psychopathic violent prisoners?".

38. Polaschek, "Treatment of psychopathy," 617.

39. Julie Hobson, John Shine, and Russell Roberts, "How do psychopaths behave in a prison therapeutic community?," *Psychology, Crime & Law* 6, no. 2 (2000): 139–154; Gareth Hughes et al., "First-stage evaluation of a treatment programme for personality disordered offenders," *Journal of Forensic Psychiatry* 8, no. 3 (1997): 515–527; H. J. Richards, J. O. Casey, and S. W. Lucente, "Psychopathy and treatment response in incarcerated female substance abusers," *Criminal Justice and Behavior* 30, no. 2 (2003): 251–276; M. C. Seto and H. E. Barbaree, "Psychopathy, treatment behavior, and sex offender recidivism," *Journal of Interpersonal Violence* 14, no. 12 (1999): 1235–1248.

40. Skeem et al., "Psychopathy, treatment involvement."

41. Skeem et al., "Psychopathy, treatment involvement"; M. Hildebrand, C. de Ruiter, and V. de Vogel, "Psychopathy and sexual deviance in treated rapists: Association with sexual and nonsexual recidivism," *Sexual Abuse: A Journal of Research and Treatment* 16, no. 1 (2004): 1–24; C. M. Langton et al., "Sex offenders' response to treatment and its association with recidivism as a function of psychopathy," *Sexual Abuse: A Journal of Research and Treatment* 18, no. 1 (2006): 99–120; Caldwell and Van Rybroek, "Reducing violence in serious juvenile offenders using intensive treatment"; M. Caldwell et al., "Treatment response of adolescent offenders with psychopathy features: A 2-year follow-up," *Criminal Justice and Behavior* 33, no. 5 (2006): 571–596; Michael F. Caldwell, Michael Vitacco, and Gregory J. Van Rybroek, "Are violent delinquents worth treating? A cost–benefit analysis," *Journal of Research in Crime and Delinquency* 43, no. 2 (2006): 148–168; J. Looman et al., "Psychopathy, treatment change, and recidivism in high-risk, high-need sexual offenders," *Journal of Interpersonal Violence* 20, no. 5 (2005): 549–568; Michael F. Caldwell et al., "Evidence of treatment progress and therapeutic outcomes among adolescents with psychopathic features," *Criminal Justice and Behavior* 34, no. 5 (2007): 573–587; Olver and Wong, "Therapeutic responses of psychopathic sexual offenders"; M. F. Caldwell, "Treatment-related changes in behavioral outcomes of psychopathy facets in adolescent offenders," *Law and Human Behavior* 35, no. 4 (2011): 275–287; Jeffrey Abracen et al., "Recidivism among treated sexual offenders and comparison subjects: Recent outcome data from the Regional Treatment Centre (Ontario) high-intensity Sex Offender Treatment Programme," *Journal of Sexual Aggression* 17, no. 2 (2011): 142–152; Stephen C. P. Wong et al., "The effectiveness of violence reduction treatment for psychopathic offenders: Empirical evidence and a treatment model," *International Journal of Forensic Mental Health* 11, no. 4 (2012): 336–349; M. F. Caldwell et al., "Treatment-related changes in psychopathy features and behavior in adolescent offenders," *Criminal Justice and Behavior* 39, no. 2 (2012): 144–155; Mark E. Olver, Kathy Lewis, and Stephen C. P. Wong, "Risk reduction treatment of high-risk psychopathic offenders: The relationship of psychopathy and treatment change to violent recidivism," *Personality Disorders: Theory, Research, and Treatment* 4, no. 2 (2013): 160–167; Grant N. Burt, Mark E. Olver, and Stephen C. Wong, "Investigating characteristics of the nonrecidivating psychopathic offender," *Criminal Justice and Behavior*

43, no. 12 (2016): 1741–1760; Sewall and Olver, "Psychopathy and treatment outcome: Results from a sexual violence reduction program"; Erika Y. Rojas and Mark E. Olver, "Juvenile psychopathy and community treatment response in youth adjudicated for sexual offenses," *International Journal of Offender Therapy and Comparative Criminology* 66, no. 15 (2021): 1575–1602; Mark E. Olver and Emily K. Riemer, "High-psychopathy men with a history of sexual offending have protective factors too: But are these risk relevant and can they change in treatment?," *Journal of Consulting and Clinical Psychology* 89, no. 5 (2021): 406–420; T. E. Daly, "Why are psychopaths difficult to treat? Testing the two-component model for the treatment of PCL psychopaths" (PhD Victoria University of Wellington, 2017).

42. J. Monahan et al., *Rethinking risk assessment: The MacArthur study of mental disorder and violence* (New York: Oxford University Press, 2001).

43. Olver and Wong, "Therapeutic responses of psychopathic sexual offenders."

44. Caldwell et al., "Are violent delinquents worth treating?"

45. Polaschek and Skeem, "Treatment of adults and juveniles," 725.

46. Farid Chakhssi, Corine de Ruiter, and David Bernstein, "Change during forensic treatment in psychopathic versus nonpsychopathic offenders," *Journal of Forensic Psychiatry & Psychology* 21, no. 5 (2010): 660–682; DeSorcy et al., "Working alliance and psychopathy"; Richards et al., "Psychopathy and treatment response in incarcerated female substance abusers"; S. C. Wong, A. Gordon, and D. Gu, "Assessment and treatment of violence-prone forensic clients: An integrated approach," *British Journal of Psychiatry Suppl* 49 (2007): s66–74.

47. Mark E. Olver, "Treatment of psychopathic offenders: A review of research, past, and current practice," in *Psychopathy and criminal behavior: Current trends and challenges*, ed. Paulo B. Marques, Mauro Paulino, and Laura Alho (Cambridge, MA: Academic Press, 2022), 469–481, 479.

48. Harris and Rice, "Treatment of psychopathy," 560.

49. Thomas C. Chalmers et al., "A method for assessing the quality of a randomized control trial," *Controlled Clinical Trials* 2, no. 1 (1981): 31–49.

50. Clive R. Hollin, "Evaluating offending behaviour programmes: Does only randomization glister?," *Criminology & Criminal Justice* 8, no. 1 (2008): 89–106; Glenn Shean, "Limitations of randomized control designs in psychotherapy research," *Advances in Psychiatry* 2014 (2014): 561452.

51. For a similar criticism of Harris and Rice's view, see Polaschek and Skeem, "Treatment of adults and juveniles"; Polaschek, "Treatment of psychopathy"; Salekin et al., "Treatment of psychopathy: A review and brief introduction to the mental model approach for psychopathy."

52. Marnie E. Rice, Grant T. Harris, and Catherine A. Cormier, "An evaluation of a maximum security therapeutic community for psychopaths and other mentally disordered offenders," *Law and Human Behavior* 16, no. 4 (1992): 399–412.

53. Seto and Barbaree, "Psychopathy, treatment behavior."

54. H. E. Barbaree, "Psychopathy, treatment behavior, and recidivism: An extended follow-up of Seto and Barbaree," *Journal of Interpersonal Violence* 20, no. 9 (2005): 1115–1131.

55. Hughes et al., "First-stage evaluation."

56. Rice et al., "Maximum security therapeutic community," 408.

57. Richard Weisman, "Reflections on the Oak Ridge experiment with mentally disordered offenders, 1965–1968," *International Journal of Law and Psychiatry* 18, no. 3 (1995): 265–290; J. Gunn, "Misplaced enthusiasm with neglect of human rights: Beneficence is not enough," *Criminal Behaviour and Mental Health* 31, no. 3 (2021): 156–161.

58. Rice et al., "Maximum security therapeutic community," 401, 408.

59. Elliott T. Barker and M. F. Buck, "LSD in a coercive milieu therapy program," *Canadian Psychiatric Association Journal* 22, no. 6 (1977): 311–314.

60. Elliott T. Barker and Alan J. McLaughlin, "The total encounter capsule," *Canadian Psychiatric Association Journal* 22, no. 7 (1977): 355–160.

61. Gunn, "Misplaced enthusiasm"; Weisman, "Reflections on the Oak Ridge experiment"; Ronson, *Psychopath test.*

62. Elliott T. Barker and M. H. Mason, "Buber behind bars," *Canadian Psychiatric Association Journal* 13, no. 1 (1968): 61–72; Barker and Buck, "LSD in a coercive milieu therapy program"; Barker and McLaughlin, "The total encounter capsule"; Elliott Barker, "The Penetanguishene program: A personal review," in *Therapeutic communities in corrections*, ed. Hans Toch (New York: Praeger, 1980), 73–81.

63. Barker v. Barker, 3397 ONSC CV-00-199551 (Ontario Superior Court of Justice 2017); Barker v. Barker, 3746 ONSC (Ontario Superior Court of Justice 2020).

64. Sean Fine, "Doctors tortured patients at Ontario mental-health centre, judge rules," *The Globe and Mail*, June 7, 2017, https://www.theglobeandmail.com/news/national/doctors-at-ontario-mental-health-facility-tortured-patients-court-finds/article35246519/.

65. R. R. Larsen, E. Ades, Y. Shroff, S. McLaren, S. Griffiths, and J. Jalava, "Unreliable and unethical, yet highly influential: A critical rebuttal of Rice et al.'s 1992-study suggesting adverse treatment effects in psychopathic patients," https://osf.io/ktp4r/.

66. Olver, "Can psychopathy be treated?"; Reidy et al., "Reducing psychopathic violence."

67. Devon Polaschek and T. E. Daly, "Treatment and psychopathy in forensic settings," *Aggression and Violent Behavior* 18, no. 5 (2013): 592–603, 595.

68. Harris and Rice, "Treatment of psychopathy," 556.

69. Weisman, "Reflections on the Oak Ridge experiment."

70. Stephen J. Hucker et al., *Oak Ridge: A review and an alternative*, Ontario Ministry of Health (1984).

71. Gunn, "Misplaced enthusiasm."

72. Harris and Rice, "Psychopathy research at Oak Ridge: Skepticism overcome."

73. Hecht et al., "The psychological treatment of psychopathy," 290.

74. Polaschek and Skeem, "Treatment of adults and juveniles," 725.

75. Maeve Moosburner et al., "Is psychopathy a dynamic risk factor? An empirical investigation of changes in psychopathic personality traits over the course of correctional treatment," *Criminal Justice and Behavior* 51, no. 2 (2023): 230–246; Alyssa Roberson and Michael J. Vitacco, "Psychopathy in correctional settings: Considerations for developing and implementing a treatment program," *Journal of Correctional Health Care* 29, no. 3 (2023): 232–238; Olver and Riemer, "High-psychopathy men with a history of sexual offending have protective factors too." B. Mohajerin and R. C. Howard, "Effects of two treatments on interpersonal, affective, and lifestyle features of psychopathy and Emotion Dysregulation," *Personality and Mental Health* 18, no. 1 (2024): 43–59.

76. Saul M. Kassin, Itiel E. Dror, and Jeff Kukucka, "The forensic confirmation bias: Problems, perspectives, and proposed solutions," *Journal of Applied Research in Memory and Cognition* 2, no. 1 (2013): 42–52; Chris Chambers, *The seven deadly sins of psychology: A manifesto for reforming the culture of scientific practice* (Princeton, NJ: Princeton University Press, 2017).

77. Hare, *Hare Psychopathy Checklist–Revised*, 158.

78. Barbaree, "Psychopathy, treatment behavior, and recidivism," 1117.

79. Seto and Barbaree, "Psychopathy, treatment behavior."

Chapter 4

1. Cleckley, *Mask of sanity*, 40.

2. Rush, *An inquiry*.

3. Sass and Felthous, "Heterogeneous construct of psychopathy"; Millon et al., "Historical conceptions of psychopathy."

4. Cleckley, *Mask of sanity*.

5. Hare, *Without conscience*, 129.

6. E. Viding, E. McCrory, and A. Seara-Cardoso, "Psychopathy," *Current Biology* 24, no. 18 (2014): R871–R74; James Blair, "Empathic dysfunction in psychopathic individuals," in *Empathy in mental illness*, ed. T. Farrow and P. Woodruff (New York: Cambridge University Press, 2007), 3–16.

7. Hare, "Psychopaths and their nature"; Helfgott, *No remorse*; Joshua W. Buckholtz and Kent A. Kiehl, "Inside the mind of a psychopath," *Scientific American Mind*, September 1, 2010, https://www.scientificamerican.com/article/inside-the-mind-of-a-psychopath/.

8. R. J. R. Blair, "Emotion-based learning systems and the development of morality," *Cognition* 167 (2017): 38–45; Cleckley, *Mask of sanity*; Lykken, *Antisocial personalities*.

9. Hare, *Hare Psychopathy Checklist–Revised*.

10. Hare, *Hare Psychopathy Checklist–Revised*, 5.

11. S. Greenland et al., "Statistical tests, p values, confidence intervals, and power: A guide to misinterpretations," *European Journal of Epidemiology* 31, no. 4 (2016): 337–350; Lakens, *Improving your statistical inferences*.

12. Greenland et al., "Statistical tests, p values, confidence intervals, and power: A guide to misinterpretations"; S. Goodman, "A dirty dozen: Twelve p-value misconceptions," *Seminars in Hematology* 45, no. 3 (2008): 135–140.

13. Cohen, *Statistical power analysis for the behavioral sciences*.

14. Barry Cohen, *Explaining psychological statistics*, 4th ed. (Hoboken, NJ: John Wiley & Sons, 2013); Lakens, *Improving your statistical inferences*.

15. Shannon Spaulding, "Cognitive empathy," in *The Routledge handbook of philosophy of empathy*, Routledge handbooks in philosophy. (New York: Routledge, 2017), 13–21.

16. Heidi L. Maibom, "Affective empathy," in *The Routledge handbook of philosophy of empathy*, Routledge handbooks in philosophy (New York: Routledge/Taylor & Francis Group, 2017), 22–32.

17. For more discussions on empathy subtypes, see C. Daniel Batson, "These things called empathy: Eight related but distinct phenomena," in *The social neuroscience of empathy*, ed. J. Decety and W. Ickes (Cambridge, MA: MIT Press, 2009), 3–15; Morgan D. Stosic et al., "What is your empathy scale not measuring? The convergent, discriminant, and predictive validity of five empathy scales," *Journal of Social Psychology* 162, no. 1 (2022): 7–25.

18. R. R. Larsen, Sonya McLaren, Jarkko Jalava, and Stephanie Griffiths, "Do psychopathic persons lack empathy? An exploratory systematic review of empathy

assessment and emotion recognition studies in psychopathy checklist samples," *Psychology, Public Policy, and Law* (2024).

19. Mark H. Davis, "Measuring individual differences in empathy: Evidence for a multidimensional approach," *Journal of Personality and Social Psychology* 44, no. 1 (1983): 113–126; Mark H. Davis, *Empathy: A social psychological approach* (Boulder, CO: Westview Press, 1996).

20. For an overview of different kinds of empathy assessment tools, see Larsen et al., "Do psychopathic persons lack empathy?".

21. A. Szollosi and C. Donkin, "Arrested Theory Development: The misguided distinction between exploratory and confirmatory research," *Perspectives on Psychological Science* 16, no. 4 (2021): 717–724.

22. Mark Rubin, "Do p values lose their meaning in exploratory analyses? It depends how you define the familywise error rate," *Review of General Psychology* 21, no. 3 (2017): 269–275.

23. Lakens, *Improving your statistical inferences.*

24. Vivian Ta and W. Ickes, "Empathic accuracy," in *The Routledge handbook of philosophy of empathy*, ed. Heidi Maibom (London: Routledge, 2017); T. Ruffman et al., "A meta-analytic review of emotion recognition and aging: Implications for neuropsychological models of aging," *Neuroscience and Biobehavioral Reviews* 32, no. 4 (2008): 863–881; M. N. Dalili et al., "Meta-analysis of emotion recognition deficits in major depressive disorder," *Psychological Medicine* 45, no. 6 (2015): 1135–1144.

25. Edouard Machery and Stephen Stich, "The moral/conventional distinction," in *The Stanford Encyclopedia of Philosophy* ed. Edward Zalta (Metaphysics Research Lab, Stanford University, 2022).

26. Machery and Stich, "The moral/conventional distinction."

27. James Blair, "A cognitive developmental approach to morality: Investigating the psychopath," *Cognition* 57, no. 1 (1995): 1–29.

28. Blair, "Cognitive developmental approach," 23.

29. Jarkko Jalava and Stephanie Griffiths, "Philosophers on psychopaths: A cautionary tale in interdisciplinarity," *Philosophy, Psychiatry, & Psychology* 24, no. 1 (2017): 1–12; David Sackris, "The disunity of moral judgment: Implications for the study of psychopathy," *Philosophical Psychology* (2022): 1–26.

30. James Blair et al., "Is the psychopath 'morally insane'?," *Personality and Individual Differences* 19, no. 5 (1995): 741–752.

31. M. C. Dolan and R. S. Fullam, "Moral/conventional transgression distinction and psychopathy in conduct disordered adolescent offenders," *Personality and Individual Differences* 49, no. 8 (2010): 995–1000.

32. E. Aharoni, W. Sinnott-Armstrong, and K. A. Kiehl, "Can psychopathic offenders discern moral wrongs? A new look at the moral/conventional distinction," *Journal of Abnormal Psychology* 121, no. 2 (2012): 484–497.

33. Eyal Aharoni, Walter Sinnott-Armstrong, and Kent A. Kiehl, "What's wrong? Moral understanding in psychopathic offenders," *Journal of Research in Personality* 53 (2014): 175–181.

34. Julia Marshall, Ashley L. Watts, and Scott O. Lilienfeld, "Do psychopathic individuals possess a misaligned moral compass? A meta-analytic examination of psychopathy's relations with moral judgment," *Personality Disorders: Theory, Research, and Treatment* 9, no. 1 (2018): 40–50; J. S. Borg and W. Sinnott-Armstrong, "Do psychopaths make moral judgments?," in *Handbook on psychopathy and law*, ed. K. A. Kiehl and W. Sinnott-Armstrong (New York: Oxford University Press, 2013), 107–128; Jalava and Griffiths, "Philosophers on Psychopaths."

35. Joshua Greene and Jonathan Haidt, "How (and where) does moral judgment work?," *Trends in Cognitive Sciences* 6, no. 12 (2002): 517–523.

36. Joshua D. Greene et al., "An fMRI Investigation of emotional engagement in moral judgment," *Science* 293, no. 5537 (2001): 2105–2108.

37. M. Cima, F. Tonnaer, and M. Hauser, "Psychopaths know right from wrong but don't care," *Social Cognitive and Affective Neuroscience* 5, no. 1 (2010): 59–67.

38. M. Koenigs et al., "Utilitarian moral judgment in psychopathy," *Social Cognitive and Affective Neuroscience* 7, no. 6 (2012): 708–714.

39. A. L. Glenn et al., "Increased DLPFC activity during moral decision-making in psychopathy," *Molecular Psychiatry* 14, no. 10 (2009): 909–911.

40. J. Pujol et al., "Breakdown in the brain network subserving moral judgment in criminal psychopathy," *Social Cognitive and Affective Neuroscience* 7, no. 8 (2012): 917–923.

41. N. C. Hauser et al., "Rational, emotional, or both? Subcomponents of psychopathy predict opposing moral decisions," *Behavioral Sciences & the Law* 39, no. 5 (2021): 541–566.

42. Aisling O'Kane, Diane Fawcett, and Ronald Blackburn, "Psychopathy and moral reasoning: Comparison of two classifications," *Personality and Individual Differences* 20, no. 4 (1996): 505–514.

43. James Rest et al., "Judging the important issues in moral dilemmas: An objective measure of development," *Developmental Psychology* 10, no. 4 (1974): 491–501.

44. Carla L. Harenski et al., "Aberrant neural processing of moral violations in criminal psychopaths," *Journal of Abnormal Psychology* 119, no. 4 (2010): 863–874.

45. Harenski et al., "Aberrant neural processing of moral violations," 867.

46. Carla L. Harenski, A. Harenski, and Kent A. Kiehl, "Neural processing of moral violations among incarcerated adolescents with psychopathic traits," *Developmental Cognitive Neuroscience* 10 (2014): 181–189.

47. Eyal Aharoni, Olga Antonenko, and Kent A. Kiehl, "Disparities in the moral intuitions of criminal offenders: The role of psychopathy," *Journal of Research in Personality* 45, no. 3 (2011): 322–327.

48. Jesse Graham et al., "Moral foundations theory: The pragmatic validity of moral pluralism," in *Advances in experimental social psychology*, ed. Patricia Devine and Ashby Plant (Cambridge, MA: Academic Press, 2013), 55–130.

49. Maya A. Irvin-Vitela et al., "Reduced endorsement of specific moral foundations in incarcerated adult women with elevated psychopathic traits," *Personality and individual Differences* 181 (2021): 110998.

50. A. L. Glenn et al., "Are all types of morality compromised in psychopathy?," *Journal of Personality Disorder* 23, no. 4 (2009): 384–398.

51. J. Graham, J. Haidt, and B. A. Nosek, "Liberals and conservatives rely on different sets of moral foundations," *Journal of Personality and Social Psychology* 96, no. 5 (2009): 1029–1046.

52. S. J. Fede et al., "Distinct neuronal patterns of positive and negative moral processing in psychopathy," *Cognitive, Affective, & Behavioral Neuroscience* 16, no. 6 (2016): 1074–1085.

53. Ankur Joshi et al., "Likert scale: Explored and explained," *British Journal of Applied Science & Technology* 7 (2015): 396–403.

54. Sandra Baez et al., "Increased moral condemnation of accidental harm in institutionalized adolescents," *Scientific Reports* 8, no. 1 (2018): 11609; Indrajeet Patil and Bastien Trémolière, "Reasoning supports forgiving accidental harms," *Scientific Reports* 11 (2021): 14418; Liane Young and Rebecca Saxe, "When ignorance is no excuse: Different roles for intent across moral domains," *Cognition* 120, no. 2 (2011): 202–214.

55. Paul Robinson and Robert Kurzban, "Concordance & conflict in intuitions of justice," *Public Law and Legal Theory Research Paper Series*, no. 06–38 (2007): 1829–1907.

56. K. J. Yoder et al., "Neural networks underlying implicit and explicit moral evaluations in psychopathy," *Translational Psychiatry* 5 (2015): e625.

57. G. Weizmann-Henelius, E. Sailas, V. Viemerö, and M. Eronen, "Violent women, blame attribution, crime, and personality," *Psychopathology* 35, no. 6 (2002): 355–361.

58. Andrew Spice et al., "Remorse, psychopathology, and psychopathy among adolescent offenders," *Law and Human Behavior* 39, no. 5 (2015): 451–462.

59. Blair, "Empathic dysfunction in psychopathic individuals," 9.

60. Christopher J. Patrick, *Handbook of psychopathy* (New York: Guilford Press, 2006); Hervé and Yuille, *Psychopath: Theory, research, and practice.*

61. Larsen et al., "Do psychopathic persons lack empathy?"

62. Viding, *Psychopathy: A very short introduction*, 2.

63. Helfgott, *No remorse*, 2.

64. De Brito et al., "Psychopathy," 1, 4.

Chapter 5

1. De Brito et al., "Psychopathy."

2. Rafter, *Criminal brain*; Sass and Felthous, "The heterogeneous construct of psychopathy."

3. Rush, *An inquiry.*

4. Marvin E. Wolfgang, "Pioneers in criminology: Cesare Lombroso (1835–1909)," *Journal of Criminal Law, Criminology, and Police Science* 52, no. 4 (1961): 361–391.

5. Soraya de Chadarevian, *Heredity under the microscope: Chromosomes and the study of the human genome* (Chicago: University of Chicago Press, 2020); Theodore R. Sarbin and Jeffrey E. Miller, "Demonism revisited: The XYY chromosomal anomaly," *Issues in Criminology* 5, no. 2 (1970): 195–207; Robert Hare, "Some empirical studies of psychopathy," *Canadas Mental Health* 18, no. 1 (1970): 4–9.

6. Glenn and Raine, *Psychopathy*; Hare, "Some empirical studies of psychopathy."

7. Stephanie Griffiths et al., "Genetic correlates of PCL-R psychopathy: A systematic review," *Aggression and Violent Behavior* 66 (2022): 101765.

8. Ronald Blackburn, "Other theoretical models of psychopathy," in *Handbook of Psychopathy*, ed. Christopher J. Patrick (New York: Guilford Press, 2006), 35–57; Inti A. Brazil and Maaike Cima, "Contemporary approaches to psychopathy," in *The handbook of forensic psychopathology and treatment*, ed. Maaike Cima (New York: Routledge, 2016), 206–226; Blair et al., *Psychopath*; Glenn and Raine, *Psychopathy.*

9. Cleckley, *Mask of sanity.*

10. B. Karpman, "Psychopathy in the scheme of human typology," *Journal of Nervous and Mental Disease* 103, no. 3 (1946): 276–288.

11. Lykken, *Antisocial personalities.*

12. Hare, "Some empirical studies of psychopathy."

13. Werlinder, "Psychopathy"; Sass and Felthous, "The heterogeneous construct of psychopathy."

14. L. Pessoa, *The entangled brain: How perception, cognition, and emotion are woven together* (Cambridge, MA: The MIT Press, 2022); Christiana Westlin et al., "Improving the study of brain-behavior relationships by revisiting basic assumptions," *Trends in Cognitive Sciences* 27, no. 3 (2023): 246–257.

15. M. Lanczik and G. Keil, "Carl Wernicke's localization theory and its significance for the development of scientific psychiatry," *History of Psychiatry* 2, no. 6 (1991): 171–180; Leonard L. LaPointe, *Paul Broca and the origins of language in the brain* (San Diego: Plural Publishing, 2013).

16. Cleckley, *Mask of sanity*, 12–14, 286, 379, 403.

17. Russell A. Poldrack, "Can cognitive processes be inferred from neuroimaging data?," *Trends in Cognitive Sciences* 10, no. 2 (2006): 59–63; Shaun Gallagher, "Phenomenology and experimental design: Toward a phenomenologically enlightened experimental science," *Journal of Consciousness Studies* 10, no. 9–10 (2003): 85–99.

18. Felix Schirmann, "'The wondrous eyes of a new technology': A history of the early electroencephalography (EEG) of psychopathy, delinquency, and immorality," Hypothesis and Theory, *Frontiers in Human Neuroscience* 8 (2014): 232.

19. Cleckley, *Mask of sanity*, 413.

20. Cleckley, *Mask of sanity*, 415.

21. Glenn and Raine, *Psychopathy*, 196.

22. Kiehl and Hoffman, "Criminal psychopath," 357.

23. De Brito et al., "Psychopathy."

24. Blackburn, "On moral judgements and personality disorders."

25. N. C. Andreasen, "The validation of psychiatric diagnosis: New models and approaches," *The American journal of psychiatry* 152, no. 2 (1995): 161–162; S. O. Lilienfeld, S. F. Smith, and A. L. Watts, "Issues in diagnosis: Conceptual issues and controversies," in *Psychopathology: History, diagnosis, and empirical foundation*, ed. W. Edward Craighead, David J. Miklowitz, and Linda W. Craighead (Hoboken, NJ: John Wiley & Sons, 2013), 1–35.

26. Rebecca Umbach, Colleen M. Berryessa, and Adrian Raine, "Brain imaging research on psychopathy: Implications for punishment, prediction, and treatment in youth and adults," *Journal of Criminal Justice* 43, no. 4 (2015): 295–306; Annalise Perricone, Arielle Baskin-Sommers, and Woo-kyoung Ahn, "The effect of neuroscientific evidence on sentencing depends on how one conceives of reasons for incarceration," *PLOS ONE* 17, no. 11 (2022): e0276237; Shichun Ling and Adrian Raine, "The neuroscience of psychopathy and forensic implications," *Psychology, Crime & Law* 24, no. 3 (2018): 296–312.

27. Glenn and Raine, *Psychopathy*, 13.

28. Kiehl and Hoffman, "Criminal psychopath," 357.

29. Matt Carter et al., *Guide to research techniques in neuroscience* (London: Academic Press, 2022).

30. Suzana Herculano-Houzel, "The remarkable, yet not extraordinary, human brain as a scaled-up primate brain and its associated cost," *Proceedings of the National Academy of Sciences* 109, no. supplement 1 (2012): 10661–10668.

31. Ranganatha Sitaram et al., "Closed-loop brain training: The science of neurofeedback," *Nature Reviews Neuroscience* 18, no. 2 (2017): 86–100; Zen J. Lau et al., "Brain entropy, fractal dimensions and predictability: A review of complexity measures for EEG in healthy and neuropsychiatric populations," *European Journal of Neuroscience* 56, no. 7 (2022): 5047–5069.

32. Abby P. Clark et al., "Psychopathy and neurodynamic brain functioning: A review of EEG research," *Neuroscience & Biobehavioral Reviews* 103 (2019): 352–373.

33. Schirmann, "History of the early electroencephalography"; Paul E. Mullen, "On building arguments on shifting sands," *Philosophy, Psychiatry, & Psychology* 14, no. 2 (2007): 143–147; Yuri G. Pavlov et al., "#EEGManyLabs: Investigating the replicability of influential EEG experiments," *Cortex* 144 (2021): 213–229; Lau et al., "Brain entropy."

34. For a detailed introduction to MRI technology, see R. A. Poldrack, J. A. Mumford, and T. E. Nichols, *Handbook of functional MRI data analysis* (New York: Cambridge University Press, 2011); Peter A. Bandettini, *fMRI*, The MIT Press Essential Knowledge series, (Cambridge, MA: MIT Press, 2020).

35. J. Ashburner and K. J. Friston, "Voxel based morphometry," in *Encyclopedia of neuroscience*, ed. Larry R. Squire (Oxford: Academic Press, 2009), 471–477.

36. Poldrack et al., *Handbook of functional MRI data analysis*.

37. Adina Roskies, "Are neuroimages like photographs of the brain?," *Philosophy of Science* 74, no. 5 (2007): 860–872; Colin Klein, "Images are not the evidence in neuroimaging," *The British Journal for the Philosophy of Science* 61, no. 2 (2010): 265–278.

38. A. Shapson-Coe, M. Januszewski, D. R. Berger, A. Pope, Y. Wu, T. Blakely, R. L. Schalek, et al., "A petavoxel fragment of human cerebral cortex reconstructed at nanoscale resolution," *Science* 384, no. 6696 (2024): eadk4858.

39. A. Raine et al., "Reduced prefrontal gray matter volume and reduced autonomic activity in antisocial personality disorder," *Archives of General Psychiatry* 57, no. 2 (2000): 119–127.

40. For the most recent systematic reviews, see Mika Johanson et al., "A systematic literature review of neuroimaging of psychopathic traits," Neuropsychology & Neurology, *Frontiers in Psychiatry* 10 (2020): 1027; Stephanie Y. Griffiths and Jarkko V. Jalava, "A comprehensive neuroimaging review of PCL-R defined psychopathy," *Aggression and Violent Behavior* 36 (2017): 60–75; Timm B. Poeppl et al., "A view behind the mask of sanity: Meta-analysis of aberrant brain activity in psychopaths," Behavior Disorders & Antisocial Behavior 3230, *Molecular Psychiatry* 24, no. 3 (2019): 463–470; Philip Deming and Michael Koenigs, "Functional neural correlates of psychopathy: A meta-analysis of MRI data," *Translational psychiatry* 10, no. 1 (2020): 133.

41. M. Koenigs et al., "Investigating the neural correlates of psychopathy: A critical review," *Molecular Psychiatry* 16, no. 8 (2011): 792–799.

42. Griffiths and Jalava, "A comprehensive neuroimaging review."

43. Jarkko Jalava, Stephanie Griffiths, and Rasmus Rosenberg Larsen, "How to keep unreproducible neuroimaging evidence out of court: A case study in fMRI and psychopathy," *Psychology, Public Policy, and Law* 29, no. 1 (2023): 1–18.

44. James Blair, "The amygdala and ventromedial prefrontal cortex in morality and psychopathy," *Trends in Cognitive Sciences* 11, no. 9 (2007): 387–392; Philip Deming, Mickela Heilicher, and Michael Koenigs, "How reliable are amygdala findings in psychopathy? A systematic review of MRI studies," *Neuroscience & Biobehavioral Reviews* 142 (2022): 104875.

45. Michael Koenigs, "The role of prefrontal cortex in psychopathy," *Reviews in the Neurosciences* 23, no. 3 (2012): 253–262; Deming, Philip, Stephanie Griffiths, Jarkko Jalava, Michael Koenigs, and Rasmus Rosenberg Larsen. "Psychopathy and Medial Frontal Cortex: A Systematic Review Reveals Predominantly Null Relationships." *Neuroscience & Biobehavioral Reviews* 167 (2024): 105904. https://doi.org/10.1016/j .neubiorev.2024.105904.

46. Deming, Heilicher, and Koenigs, "How reliable are amygdala findings?"

47. Deming et al., "Psychopathy and medial frontal cortex."

48. Elsa Ermer et al., "Aberrant paralimbic gray matter in incarcerated male adolescents with psychopathic traits," *Journal of the American Academy of Child and Adolescent Psychiatry* 52, no. 1 (2013): 94–103.e3; Gina M. Vincent et al., "Callous-unemotional traits modulate brain drug craving response in high-risk young offenders," Criminal Behavior & Juvenile Delinquency 3236, *Journal of Abnormal Child Psychology* 46, no. 5 (2018): 993–1009; A. A. Marsh et al., "Reduced amygdala response to fearful expressions in children and adolescents with callous-unemotional traits and disruptive behavior disorders," *American Journal of Psychiatry* 165, no. 6 (2008): 712–720; Vaughn R. Steele et al., "Machine learning of structural magnetic resonance imaging predicts psychopathic traits in adolescent offenders," *NeuroImage* 145, B (2017): 265–273.

49. Scott Marek et al., "Reproducible brain-wide association studies require thousands of individuals," *Nature* 603 (2022): 654–660.

50. Kevin R. Murphy, Brett Myors, and Allen Wolach, *Statistical power analysis: A simple and general model for traditional and modern hypothesis tests*, 4th ed. (New York: Routledge, 2014); Alex Reinhart, *Statistics done wrong: The woefully complete guide* (San Francisco: No Starch Press, 2015).

51. Deming et al., "Psychopathy and medial frontal cortex."

52. Deming et al., "Psychopathy and medial frontal cortex."

53. Daniel R. Weinberger and Eugenia Radulescu, "Structural magnetic resonance imaging all over again," *JAMA Psychiatry* 78, no. 1 (2021): 11–12.

54. Milin and Hadžić, "Moderating and mediating variables in psychological research."

55. Marek et al., "Reproducible brain-wide association studies"; A. Eklund, T. E. Nichols, and H. Knutsson, "Cluster failure: Why fMRI inferences for spatial extent have inflated false-positive rates," *Proceedings of the National Academy of Sciences of the United States of America* 113, no. 28 (2016): 7900–7905; Benjamin O. Turner et al., "Small sample sizes reduce the replicability of task-based fMRI studies," *Communications Biology* 1, no. 1 (2018): 62.

56. Russell A. Poldrack, "Region of interest analysis for fMRI," *Social Cognitive and Affective Neuroscience* 2, no. 1 (2007): 67–70.

57. Marek et al., "Reproducible brain-wide association studies."

58. State v. Brian Dugan, No. #05 CF 3491 (Circuit Court of Du Page County for the Eighteenth Judicial Circuit of Illinois 2009); Julia R. Lushing, Lyn M. Gaudet, and Kent A. Kiehl, "Brain imaging in psychopathy," in *The clinical and forensic assessment of psychopathy: A practitioner's guide*, 2nd ed., ed. C. Gacono, Personality and clinical psychology series. (New York: Routledge, 2016), 32–53; V. Hughes, "Science in court: head case," *Nature* 464, no. 7287 (2010): 340–342.

59. Darby Aono, Gideon Yaffe, and Hedy Kober, "Neuroscientific evidence in the courtroom: A review," *Cognitive Research: Principles and Implications* 4, no. 1 (2019): 40. Aspinwall et al., "The double-edged sword"; Jalave et al., "How to keep unreproducible neuroimaging evidence out of court"; Lushing et al., "Brain imaging in psychopathy."

60. Deborah Denno, "The myth of the double-edged sword: An empirical study of neuroscience evidence in criminal cases," *Boston College Law Review.* 56 (2015): 493–551; Ling and Raine, "The neuroscience of psychopathy and forensic implications"; Sally Satel and Scott O. Lilienfeld, *Brainwashed: The seductive appeal of mindless neuroscience* (New York: Basic Books, 2013); Jarkko Jalava and Stephanie Griffiths, "Psychopathy: Neurohype and its consequences," in *Psychopathy: Its uses, validity and status*, ed. Luca Malatesti, John McMillan, and Predrag Šustar (Cham: Springer International Publishing, 2022), 79–98.

61. Jarkko Jalava et al., "Is the psychopathic brain an artifact of coding bias? A systematic review," Systematic Review, *Frontiers in Psychology* 12 (2021): 654336.

62. Gendreau et al., "A lesson in knowledge cumulation."

63. Salekin, "Psychopathy and therapeutic pessimism."

64. Borg and Sinnott-Armstrong, "Do psychopaths make moral judgments?"

65. R. H. Fletcher and B. Black, "'Spin' in scientific writing: Scientific mischief and legal jeopardy," *Medical Law* 26, no. 3 (2007): 511–525; Clément Lazarus et al., "Classification and prevalence of spin in abstracts of non-randomized studies evaluating an intervention," *BMC Medical Research Methodology* 15, no. 1 (2015): 85.

66. John et al., "Measuring the prevalence."

67. Chambers, *The seven deadly sins of psychology*; Scott O. Lilienfeld and Irwin D. Waldman, eds., *Psychological science under scrutiny: Recent challenges and proposed solutions* (Hoboken, NJ: Wiley Blackwell, 2017).

68. Griffiths and Jalava, "A comprehensive neuroimaging review."

69. John Seabrook, "Suffering souls: The search for the roots of psychopathy," *New Yorker*, November 2, 2008, https://www.newyorker.com/magazine/2008/11/10/suffering-souls; Judith Ohikuare, "Life as a nonviolent psychopath," *The Atlantic*, 01.21.2014, 2014, http://www.theatlantic.com/health/archive/2014/01/life-as-a-non violent-psychopath/282271/; Greg Miller, "What it's like to spend 20 years listening to psychopaths for science," interview with Kent Kiehl, *Wired*, 04.17.2014 2014.

70. Kiehl, *Psychopath whisperer*; Hare, *Without conscience*; Babiak and Hare, *Snakes in suits*; Fallon, *Psychopath inside*; Stout, *Sociopath next door*.

71. Jeremy Torrie, "The psychopath next door," (Canada: Bandwidth Digital Releasing, 2014), CBC Canada (TV).

72. Koenigs et al., "Investigating the neural correlates of psychopathy."

73. Kiehl, *Psychopath whisperer*, 262.

Chapter 6

1. Robert Hare, "Psychopathy, affect, and behavior," in *Psychopathy: Theory, research, and implications for society*, ed. D. J. Cooke, A. E. Forth, and Robert Hare (Dordrecht: Spinger, Kluwer, 1998), 105–137, 105.

2. Smith, "On construct validity"; D. Borsboom, A. Cramer, and A. Kalis, "Brain disorders? Not really . . . Why network structures block reductionism in psychopathology research," *Behavioral and Brain Sciences* 42 (2019): e2; Lee Anna Clark et al., "Three approaches to understanding and classifying mental disorder: ICD-11, DSM-5, and

the National Institute of Mental Health's Research Domain Criteria (RDoC)," *Psychological Science in the Public Interest* 18, no. 2 (2017): 72–145; Paul E. Meehl, "Cliometric metatheory: II. Criteria scientists use in theory appraisal and why it is rational to do so," *Psychological Reports* 91, no. 2 (2002): 339–404.

3. Robert I. Sutton and Barry M. Staw, "What theory is not," *Administrative Science Quarterly* 40, no. 3 (1995): 371–384.

4. Carl G. Hempel and Paul Oppenheim, "Studies in the logic of explanation," *Philosophy of Science* 15, no. 2 (1948): 135–175.

5. Jerry A. Coyne, *Why evolution is true* (New York: Viking, 2009); Neil Shubin, *Your inner fish: A journey into the 3.5-billion-year history of the human body* (New York: Vintage Books, 2009).

6. Lewis, "Psychopathic personality."

7. Hare and Schalling, *Psychopathic behaviour.*

8. Hare, "Forty years aren't enough."

9. Patrick et al., "Triarchic Conceptualization of Psychopathy"; Cooke et al., "Explicating the construct of psychopathy."

100. Cleckley, *Mask of sanity*, 370.

11. Blackburn, "Other theoretical models of psychopathy."

12. J. Reid Meloy and Andrew Shiva, "A psychoanalytic view of the psychopath," in *International handbook on psychopathic disorders and the law, Vol. 1*, ed. A. R. Felthous and H. Sass (Hoboken, NJ: John Wiley & Sons, 2007), 335–346; Robert Mitchell Lindner, *Rebel without a cause: The hypnoanalysis of a criminal psychopath* (New York: Grune & Stratton, 1944).

13. Linda Mealey, "The sociobiology of sociopathy: An integrated evolutionary model," *Behavioral and Brain Sciences* 18, no. 3 (1995): 523–541.

14. Brazil and Cima, "Contemporary approaches to psychopathy."

15. James Blair, Derek Mitchell, and Karina Blair, *The psychopath: Emotion and the brain* (Malden: Blackwell Publishing, 2005).

16. Hamilton and Newman, "The response modulation hypothesis."

17. R.K.B. Hamilton, K. H. Racer, and J. P. Newman, "Impaired integration in psychopathy: A unified theory of psychopathic dysfunction," *Psychological Review* 122, no. 4 (2015): 770–791.

18. Hamilton et al., "Impaired integration in psychopathy"; D. Fowles and L. Dindo, "A dual-deficit model of psychopathy," in *Handbook of psychopathy*, ed. C. Patrick (New York: The Guildford Press, 2006), 14–34.

19. C. S. Neumann, Robert Hare, and J. P. Newman, "The super-ordinate nature of the psychopathy checklist-revised," *Journal of Personality Disorders* 21, no. 2 (2007): 102–117; Christopher J. Patrick, "Psychopathy: Current knowledge and future directions," *Annual Review of Clinical Psychology* 18, no. 1 (2022): 387–415; De Brito et al., "Psychopathy."

20. Pessoa, Luiz. "On the Relationship between Emotion and Cognition." *Nature Reviews Neuroscience* 9, no. 2 (2008): 148–58.

21. Rush, *An inquiry*; Benjamin Rush, *Medical inquiries and observations upon the diseases of the mind* (Philadelphia: Kimber & Richardson, 1812).

22. Karpman, "Psychopathy in the scheme of human typology."

23. Henderson, *Psychopathic states.*

24. Cleckley, *Mask of sanity.*

25. Arieti, "Psychopathic personality."

26. Lykken, *Antisocial personalities.*

27. Blair et al., *The psychopath.*

28. Christopher J. Patrick, "Getting to the heart of psychopathy," in *The psychopath: Theory, research, and practice*, ed. H. Herve and J. C. Yuille (Mahwah, NJ: Lawrence Erlbaum Associates, 2007), 207–252.

29. Sass and Felthous, "The heterogeneous construct of psychopathy"; Brazil and Cima, "Contemporary approaches to psychopathy"; Lilienfeld et al., "Hervey Cleckley (1903–1984)."

30. Lisa Feldman Barrett, *How emotions are made: The secret life of the brain* (Boston: Houghton Mifflin Harcourt, 2017).

31. Nico H. Frijda et al., "The complexity of intensity: Issues concerning the structure of emotion intensity," in *Emotion and social behavior*, ed. M. S. Clark, Review of personality and social psychology, No. 13. (Thousand Oaks, CA: Sage Publications, 1992), 60–89.

32. Jonathan Posner, James A. Russell, and Bradley S. Peterson, "The circumplex model of affect: An integrative approach to affective neuroscience, cognitive development, and psychopathology," *Development and psychopathology* 17, no. 3 (2005): 715–734; Barrett, *How emotions are made.*

33. Kevin Mulligan and Klaus R. Scherer, "Toward a working definition of emotion," *Emotion Review* 4, no. 4 (2012): 345–357; Barrett, *How emotions are made.*

34. Cleckley, *Mask of sanity*, 373–374.

35. Hare, *Without conscience*, 52.

36. Cleckley, *Mask of sanity*, 373.

37. John H. Johns and Herbert C. Quay, "The effect of social reward on verbal conditioning in psychopathic and neurotic military offenders," *Journal of Consulting Psychology* 26, no. 3 (1962): 217–220, 217.

38. Patrick, "Back to the future"; C. Crego and T. A. Widiger, "Cleckley's psychopaths: Revisited," *Journal of Abnormal Psychology* 125, no. 1 (2016): 75–87; Lilienfeld et al., "Hervey Cleckley (1903–1984)."

39. Rasmus Rosenberg Larsen, "Psychopathy as moral blindness: A qualifying exploration of the blindness-analogy in psychopathy theory and research," *Philosophical Explorations* 23, no. 3 (2020): 214–233.

40. Larsen, "Psychopathy as moral blindness"; Rasmus Rosenberg Larsen, "Are psychopaths moral-psychologically impaired? Reassessing emotion-theoretical explanations," *Mind & Language* 37, no. 2 (2020): 177–193.

41. Cleckley, *Mask of sanity*, 348, 64; Hare, *Without conscience*, 38–40.

42. Lykken, *Antisocial personalities*.

43. Blair et al., *The psychopath*; Blair, "Emotion-based learning systems."

44. R. R. Larsen, Sonya McLaren, Jarkko Jalava, and Stephanie Griffiths, "Do psychopathic persons lack empathy? An exploratory systematic review of empathy assessment and emotion recognition studies in psychopathy checklist samples," *Psychology, Public Policy, and Law* (2024).

45. Klaus R. Scherer, "What are emotions? And how can they be measured?," *Social Science Information* 44, no. 4 (2005): 695–729.

46. Elizabeth A. Phelps and Joseph E. LeDoux, "Contributions of the amygdala to emotion processing: From animal models to human behavior," *Neuron* 48, no. 2 (2005): 175–187.

47. Ralph Adolphs and David J. Anderson, *The neuroscience of emotion: A new synthesis* (Princeton, NJ: Princeton University Press, 2018).

48. Deming et al., "How reliable are amygdala findings?"

49. Kent A. Kiehl et al., "Limbic abnormalities in affective processing by criminal psychopaths as revealed by functional magnetic resonance imaging," Neuropsychology & Neurology 2520, *Biological Psychiatry* 50, no. 9 (2001): 677–684.

50. Niels Birbaumer et al., "Deficient fear conditioning in psychopathy: A functional magnetic resonance imaging study," *Archives of General Psychiatry* 62, no. 7 (2005): 799–805.

51. Glenn et al., "Neural correlates."

52. Griffiths and Jalava, "A comprehensive neuroimaging review."

53. Jalava et al., "Coding Bias in Psychopathy Reviews."

54. Griffiths and Jalava, "A comprehensive neuroimaging review," 70.

55. Christopher J. Patrick, Bruce N. Cuthbert, and Peter J. Lang, "Emotion in the criminal psychopath: Fear image processing," *Journal of Abnormal Psychology* 103, no. 3 (1994): 523–534; A. L. Hansen et al., "Facets of psychopathy, heart rate variability and cognitive function," *Journal of Personality Disorders* 21, no. 5 (2007): 568–582; S. C. Ling et al., "The mediating role of emotional intelligence on the autonomic functioning—Psychopathy relationship," *Biological Psychology* 136 (2018): 136–143; Lisa A. Rosenberger et al., "Fairness norm violations in anti-social psychopathic offenders in a repeated trust game," *Translational psychiatry* 9, no. 1 (2019): 266; Deming et al., "Autonomic and facial responses"; Allison J. Lake et al., "Evidence for unique threat-processing mechanisms in psychopathic and anxious individuals," *Cognitive, Affective, & Behavioral Neuroscience* 11, no. 4 (2011): 451–462.

56. Brian D. Earp and David Trafimow, "Replication, falsification, and the crisis of confidence in social psychology," *Frontiers in Psychology* 6 (2015): 621; Meehl, "Cliometric metatheory: II."

57. A. Raine, "Effect of early environment on electrodermal and cognitive correlates of schizotypy and psychopathy in criminals," *International Journal of Psychophysiology* 4, no. 4 (1987): 277–287; C. J. Patrick and W. G. Iacono, "Psychopathy, threat, and polygraph test accuracy," *Journal of Applied Psychology* 74, no. 2 (1989): 347–355; James R. Ogloff and Stephen Wong, "Electrodermal and cardiovascular evidence of a coping response in psychopaths," *Criminal Justice and Behavior* 17, no. 2 (1990): 231–245; Peter A. Arnett et al., "Autonomic responsivity during passive avoidance in incarcerated psychopaths," *Personality and Individual Differences* 14, no. 1 (1993): 173–184; C. J. Patrick, M. M. Bradley, and P. J. Lang, "Emotion in the criminal psychopath: Startle reflex modulation," *Journal of Abnormal Psychology* 102, no. 1 (1993): 82–92; James Blair et al., "The psychopathic individual: A lack of responsiveness to distress cues?," *Psychophysiology* 34, no. 2 (1997): 192–198; G. K. Levenston et al., "The psychopath as observer: Emotion and attention in picture processing," Article, *Journal of Abnormal Psychology* 109, no. 3 (2000): 373–385; S. C. Herpertz et al., "Emotion in criminal offenders with psychopathy and borderline personality disorder," *Archives of General Psychiatry* 58, no. 8 (2001): 737–745; H. Flor et al., "Aversive pavlovian conditioning in psychopaths: Peripheral and central correlates," Article, *Psychophysiology* 39, no. 4 (2002): 505–518; S. K. Sutton, J. E. Vitale, and J. P. Newman, "Emotion among women with psychopathy during picture perception," *Journal of Abnormal Psychology* 111, no. 4 (2002): 610–619; M. C. Pastor et al., "Startle reflex modulation, affective ratings and autonomic reactivity in incarcerated Spanish psychopaths," conference paper, *Psychophysiology* 40, no. 6 (2003): 934–938; E. J. Vanman et al., "Modification of the startle reflex in a community sample: Do one or two dimensions of psychopathy underlie emotional processing?," *Personality and Individual Differences* 35, no. 8 (2003):

2007–2021; Edelyn Verona et al., "Psychopathy and physiological response to emotionally evocative sounds," *Journal of Abnormal Psychology* 113, no. 1 (2004): 99–108; P. Serafim Ade et al., "Cardiac response and anxiety levels in psychopathic murderers," *Braz J Psychiatry* 31, no. 3 (2009): 214–218; Joseph P. Newman et al., "Attention moderates the fearlessness of psychopathic offenders," *Biological Psychiatry* 67, no. 1 (2010): 66–70; U. Vaidyanathan et al., "Clarifying the role of defensive reactivity deficits in psychopathy and antisocial personality using startle reflex methodology," *Journal of Abnormal Psychology* 120, no. 1 (2011): 253–258; A. R. Baskin-Sommers et al., "Evaluating the generalizability of a fear deficit in psychopathic African American offenders," *Journal of Abnormal Psychology* 120, no. 1 (2011): 71–78; A. R. Baskin-Sommers, J. J. Curtin, and J. P. Newman, "Specifying the attentional selection that moderates the fearlessness of psychopathic offenders," *Psychological Science* 22, no. 2 (2011): 226–234; A. R. Baskin-Sommers, J. J. Curtin, and J. P. Newman, "Emotion-modulated startle in psychopathy: Clarifying familiar effects," *Journal of Abnormal Psychology* 122, no. 2 (2013): 458–468; Yu Gao, Adrian Raine, and Robert A. Schug, "Somatic aphasia: Mismatch of body sensations with autonomic stress reactivity in psychopathy," Physiological Processes 2540, *Biological Psychology* 90, no. 3 (2012): 228–233; Yvonne Rothemund et al., "Fear conditioning in psychopaths: Event-related potentials and peripheral measures," *Biological Psychology* 90, no. 1 (2012): 50–59; N. Sadeh and E. Verona, "Visual complexity attenuates emotional processing in psychopathy: Implications for fear-potentiated startle deficits," *Cognitive Affective & Behavioral Neuroscience* 12, no. 2 (2012): 346–360; M. E. Anton et al., "Differential effects of psychopathy and antisocial personality disorder symptoms on cognitive and fear processing in female offenders," *Cognitive Affective & Behavioral Neuroscience* 12, no. 4 (2012): 761–776; Ralf Veit et al., "Deficient fear conditioning in psychopathy as a function of interpersonal and affective disturbances," *Frontiers in Human Neuroscience* 7 (2013): 706; Edelyn Verona, Konrad Bresin, and Christopher J. Patrick, "Revisiting psychopathy in women: Cleckley/Hare conceptions and affective response," Personality Disorders 3217, *Journal of Abnormal Psychology* 122, no. 4 (2013): 1088–1093; M. M. Loomans, J.H.M. Tulen, and H.J.C. van Marle, "The startle paradigm in a forensic psychiatric setting: Elucidating psychopathy," *Criminal Behaviour and Mental Health* 25, no. 1 (2015): 42–53; Fanny Degouis et al., "How do people with antisocial personality disorder with or without psychopathic personality disorder activate and regulate emotions? Neurovegetative responses during an autobiographical task," *Journal of Experimental Psychopathology* 14, no. 4 (2023): 20438087231210477.

58. Joseph P. Simmons, Leif D. Nelson, and Uri Simonsohn, "False-positive psychology: Undisclosed flexibility in data collection and analysis allows presenting anything as significant," *Psychological Science* 22, no. 11 (2011): 1359–1366; Katherine S. Button et al., "Preventing the ends from justifying the means: Withholding results to address publication bias in peer-review," *BMC Psychology* 4, no. 1 (2016): 59.

59. Ogloff and Wong, "Electrodermal and cardiovascular evidence."

60. Rothemund et al., "Fear conditioning in psychopaths."

61. Frances K. McSweeney and Eric S. Murphy, eds., *The Wiley Blackwell handbook of operant and classical conditioning* (Hoboken, NJ: Wiley Blackwell, 2014).

62. Robert M. West, "Best practice in statistics: The use of log transformation," *Annals of Clinical Biochemistry* 59, no. 3 (2022): 162–165.

63. Grant L. Iverson, "Z Scores," in *Encyclopedia of Clinical Neuropsychology*, ed. Jeffrey S. Kreutzer, John DeLuca, and Bruce Caplan (New York: Springer New York, 2011), 2739–2740.

64. C. Feng et al., "Log-transformation and its implications for data analysis," *Shanghai Archives of Psychiatry* 26, no. 2 (2014): 105–109; Stefan and Schönbrodt, "Big little lies"; M. Elson et al., "Press CRTT to measure aggressive behavior: the unstandardized use of the competitive reaction time task in aggression research," *Psychological Assessment* 26, no. 2 (2014): 419–432.

65. Stefan and Schönbrodt, "Big little lies"; William T. O'Donohue, Akihiko Masuda, and S. Lilienfeld, eds., *Avoiding questionable research practices in applied psychology* (Cham: Springer Nature, 2022).

66. M. K. Sigurdson, K. L. Sainani, and J.P.A. Ioannidis, "Homeopathy can offer empirical insights on treatment effects in a null field," *Journal of Clinical Epidemiology* 155 (2023): 64–72; John P. A. Ioannidis, "Why most published research findings are false," *PLOS Medicine* 2, no. 8 (2005): e1004085.

67. Chambers, *The seven deadly sins of psychology*; Klaus Fiedler and Norbert Schwarz, "Questionable research practices revisited," *Social Psychological and Personality Science* 7, no. 1 (2015): 45–52; Stefan and Schönbrodt, "Big little lies."

68. Scott O. Lilienfeld, "Can psychology become a science?," *Personality and Individual Differences* 49, no. 4 (2010): 281–288; Simmons, Nelson, and Simonsohn, "False-positive psychology"; John, Loewenstein, and Prelec, "Measuring the prevalence."

69. Ethan E. Gorenstein and Joseph P. Newman, "Disinhibitory psychopathology: A new perspective and a model for research," *Psychological Review* 87 (1980): 301–315.

70. Hamilton and Newman, "The response modulation hypothesis."

71. Hamilton et al., "Impaired integration in psychopathy."

72. C. Mark Patterson and Joseph P. Newman, "Reflectivity and learning from aversive events: Toward a psychological mechanism for the syndromes of disinhibition," *Psychological Review* 100, no. 4 (1993): 716–736.

73. Baskin-Sommers et al., "Specifying the attentional selection."

74. Caroline Moul, Simon Killcross, and Mark R. Dadds, "A model of differential amygdala activation in psychopathy," *Psychological Review* 119, no. 4 (2012): 789–806.

75. For other proposals, see Hamilton and Newman, "Response modulation hypothesis."

76. Patterson and Newman, "Reflectivity and learning from aversive events."

77. Patterson and Newman, "Reflectivity and learning from aversive events"; Steven M. Gillespie et al., "Psychopathy and response inhibition: A meta-analysis of go/no-go and stop signal task performance," *Neuroscience & Biobehavioral Reviews* 142 (2022): 104868.

78. Matthias Burghart, Sergej Schmidt, and Daniela Mier, "Executive functions in psychopathy: A meta-analysis of inhibition, planning, shifting, and working memory performance," *Psychological Medicine* (2024), 1–15.

79. S. F. Smith and S. O. Lilienfeld, "The response modulation hypothesis of psychopathy: A meta-analytic and narrative analysis," *Psychological Bulletin* 141, no. 6 (2015): 1145–1177.

80. M. Baliousis et al., "Executive function, attention, and memory deficits in antisocial personality disorder and psychopathy," *Psychiatry Research* 278 (2019): 151–161; C. Delfin et al., "Examining associations between psychopathic traits and executive functions in incarcerated violent offenders," *Front Psychiatry* 9 (2018): 310; Malin Pauli et al., "Assessing the relevance of self-reported ADHD symptoms and cognitive functioning for psychopathy using the PCL-R and the TriPM," *Journal of Forensic Psychiatry & Psychology* 30, no. 4 (2019): 642–657; V. Pera-Guardiola et al., "Modulatory effects of psychopathy on Wisconsin Card Sorting Test performance in male offenders with antisocial personality disorder," *Psychiatry Research* 235 (2016): 43–48; V. R. Steele et al., "Error-related processing in adult males with elevated psychopathic traits," *Personality Disorders* 7, no. 1 (2016): 80–90; F. Tonnaer, M. Cima, and A. Arntz, "Modeling impulsivity in forensic patients: A three-dimensional model of impulsivity," *American Journal of Psychology* 129, no. 4 (2016): 429–441.

81. D. J. Ozer, "Correlation and the coefficient of determination," *Psychological Bulletin* 97, no. 2 (1985): 307–315; Funder and Ozer, "Evaluating Effect Size in Psychological Research."

82. S. Duval and R. Tweedie, "Trim and fill: A simple funnel-plot-based method of testing and adjusting for publication bias in meta-analysis," *Biometrics* 56, no. 2 (2000): 455–463.

83. Smith and Lilienfeld, "The response modulation hypothesis," 1168.

84. Burghart et al., "Executive functions in psychopathy," 10.

85. Leah Wright, Jonathan Lipszyc, Annie Dupuis, Sathees Waran Thayapararajah, and Russell Schachar, "Response inhibition and psychopathology: A meta-analysis of go/no-go task performance," *Journal of Abnormal Psychology* 123, no. 2 (2014): 429–439.

86. M. Spaniol and H. Danielsson, "A meta-analysis of the executive function components inhibition, shifting, and attention in intellectual disabilities," *Journal of Intellectual Disability Research* 66, no. 1–2 (2022): 9–31.

87. D. Senkowski, Theresa Ziegler, Mervyn Singh, Andreas Heinz, Jason He, Tim Silk, and Robert C. Lorenz, "Assessing inhibitory control deficits in adult ADHD: A systematic review and meta-analysis of the stop-signal task," *Neuropsychology Review* 34, no. 2 (2023): 548–567.

88. Evangelia Argyriou, Christopher B. Davison, and Tayla T. C. Lee, "Response inhibition and internet gaming disorder: A meta-analysis," *Addictive Behaviors* 71 (2017): 54–60.

89. Paul E. Meehl, "Appraising and amending theories: The strategy of Lakatosian defense and two principles that warrant it," *Psychological Inquiry* 1, no. 2 (1990): 108–141.

90. Earp and Trafimow, "Replication, falsification, and psychology."

91. Kristina D. Hiatt and Joseph P. Newman, "Understanding psychopathy: The cognitive side," in *Handbook of psychopathy*, ed. C. Patrick (New York: The Guilford Press, 2006), 334–352; Hamilton et al., "Impaired integration in psychopathy"; M. Brook, C. L. Brieman, and D. S. Kosson, "Emotion processing in Psychopathy Checklist-assessed psychopathy: A review of the literature," *Clinical Psychology Review* 33, no. 8 (2013): 979–995.

92. David S. Kosson, M. J. Vitacco, M. T. Swogger, and B. L. Steuerwald, "Emotional experiences of the psychopath," in *The clinical and forensic assessment of psychopathy: A practitioner's guide*, ed. Carl B. Gacono (New York: Routledge, 2016), 73–95, 73.

93. Paul E. Meehl, "Why summaries of research on psychological theories are often uninterpretable," *Psychological Reports* 66, no. 1 (1990): 195–244, 196.

Chapter 7

1. American Psychological Association, *Ethical principles of psychologists and code of conduct* (Washington, DC: American Psychological Association, 2017), 2.

2. APA, "Evidence-based practice in psychology"; Robbie Busch and Sharon McCarthy, "The emergence of evidence-based practice in psychology," in *The Palgrave encyclopedia of critical perspectives on mental health*, ed. Jessica Nina Lester and Michelle O'Reilly (Cham: Springer International Publishing, 2020), 1–15.

3. DeMatteo et al., "Statement of concerned experts."

4. David DeMatteo et al., "Death is different: Reply to Olver et al. (2020)," *Psychology, Public Policy, and Law* 26, no. 4 (2020): 511–518; Olver et al., "Reliability and validity of the Psychopathy Checklist-Revised"; Robert Hare et al., "The PCL–R and

capital sentencing: A commentary on "Death is different" DeMatteo et al. (2020a)," *Psychology, Public Policy, and Law* 26, no. 4 (2020): 519–522.

5. Polaschek and Skeem, "Treatment of adults and juveniles"; Larsen, "Psychopathy treatment and the stigma of yesterday's research."

6. Salekin, "Psychopathy and therapeutic pessimism"; Skeem et al., "Psychopathy, treatment involvement"; John F. Edens and John Petrila, "Legal and ethical issues in the assessment and treatment of psychopathy," in *Handbook of psychopathy*, ed. Christopher J. Patrick (New York: The Guilford Press, 2006), 573–588.

7. John F. Edens et al., "Assessment of "juvenile psychopathy" and its association with violence: A critical review," *Behavioral Sciences & the Law* 19, no. 1 (2001): 53–80; D. Seagrave and T. Grisso, "Adolescent development and the measurement of juvenile psychopathy," *Law and Human Behavior* 26, no. 2 (2002): 219–239; Penney and Moretti, "Transfer of juveniles."

8. Brett O. Gardner, Marcus T. Boccaccini, and Daniel C. Murrie, "Which PCL-R scores best predict forensic clinicians' opinions of offender risk?," *Criminal Justice and Behavior* 45, no. 9 (2018): 1404–1419; M. E. Keesler and D. DeMatteo, "How media exposure relates to laypersons' understanding of psychopathy," *Journal of Forensic Science* 62, no. 6 (2017): 1522–1533; Kelley et al., "Dangerous, depraved"; Truong et al., "Does psychopathy influence?"

9. See also Edens et al., "Legal and ethical issues"; S. Lilienfeld, "Afterword: Key unresolved questions," in *Psychopathy and criminal behavior: Current trends and challenges*, ed. Paulo B. Marques, Mauro Paulino, and Laura Alho (San Diego: Academic Press, 2021), 483–489; Skeem et al., "Psychopathic personality"; Stephen D. Hart, "Culture and violence risk assessment: The case of Ewert v. Canada," *Journal of Threat Assessment and Management* 3, no. 2 (2016): 76–96; Brian K. Steverson, *The ethics of employment screening for psychopathy* (Lanham, MD: Rowman & Littlefield, 2020); D. Fowles, "Current scientific views of psychopathy," *Psychological Science in the Public Interest* 12, no. 3 (2011): 93–94; Zinger and Forth, "Psychopathy and Canadian criminal proceedings."

10. For a deeper discussion, see Skeem et al., "Psychopathic personality"; DeMatteo and Olver, "Use of the Psychopathy Checklist-Revised"; Edens and Truong, "Psychopathy evidence in legal proceedings"; and Larsen et al., "Are psychopathy assessments ethical?"

11. Glancy et al., "AAPL practice guideline"; Neal and Grisso, "Assessment practices and expert judgment"; Gary B. Melton et al., *Psychological evaluations for the courts: A handbook for mental health professionals and lawyers*, 4th ed. (New York: The Guilford Press, 2018).

12. Sidney Bloch and Stephen Green, eds., *Psychiatric ethics*, 5th ed. (New York: Oxford University Press, 2021); Shane S. Bush, Mary Connell, and Robert L. Denney,

Ethical practice in forensic psychology: A guide for mental health professionals, 2nd ed. (Washington, DC: American Psychological Association, 2020).

13. Russ Shafer-Landau, *The fundamentals of ethics* (Oxford: Oxford University Press, 2020).

14. Tom L. Beauchamp, "The nature of applied ethics," in *A companion to applied ethics*, ed. R. G. Frey and Christopher Heath Wellman (Malden, MA: Blackwell, 2005), 1–16; David Luban, "Professional ethics," in *A companion to applied ethics*, ed. R.G. Frey and C. Heath Wellman (Malden, MA: Blackwell Publisher, 2005), 583–596.

15. S. R. Cruess and R. L. Cruess, "Professionalism: A contract between medicine and society," *Canadian Medical Association Journal* 162, no. 5 (2000): 668–669; W. M. Sullivan, "Medicine under threat: Professionalism and professional identity," *Canadian Medical Association Journal* 162, no. 5 (2000): 673–675.

16. Gerald P. Koocher and Patricia Keith-Spiegel, *Ethics in psychology and the mental health professions: Standards and cases*, 4th ed. (New York: Oxford University Press, 2016); Robert Sadoff, ed., *Ethical issues in forensic psychiatry: Minimizing harm* (Oxford: Wiley-Blackwell, 2010).

17. T. L. Beauchamp and J. F. Childress, *Principles of biomedical ethics* (New York: Oxford University Press, 2001).

18. W. M. Sullivan, *Work and integrity: The crisis and promise of professionalism in North America* (Stanford, CA: Jossey-Bass, 2005); W. Schupmann and J. D. Moreno, "Belmont in context," *Perspectives in Biology and Medicine* 63, no. 2 (2020): 220–239; Albert R. Jonson, *The birth of bioethics* (New York: Oxford University Press, 1998).

19. For a discussion, see Beauchamp and Childress, *Principles of biomedical ethics*.

20. Tom Beauchamp and James Childress, *"Principles of Biomedical Ethics*: Marking its fortieth anniversary," *The American Journal of Bioethics* 19, no. 11 (2019): 9–12; Koocher and Keith-Spiegel, *Ethics in psychology*; Sadoff, *Ethical issues*.

21. The American Medical Association, "AMA principles of medical ethics," 2021, accessed May 10, 2023, https://code-medical-ethics.ama-assn.org; American Psychiatric Association, *Opinions of the Ethics Committee on the Principles of Medical Ethics: With annotations especially applicable to psychiatry* (Washington, DC: American Psychiatric Association, 2022).

22. AMA, "AMA principles of medical ethics"; APA, *Ethical principles*.

23. APA, *Ethical principles*.

24. APA, *Opinions of the Ethics Committee*; APA, *Ethical principles*; AMA, "AMA principles of medical ethics."

25. For further discussions, see Larsen et al., "Are psychopathy assessments ethical?"; Edens et al., "Legal and ethical issues"; J. Dvoskin et al., "Psychopathic personality in

early childhood: A critical comment on Lopez-Romero et al. (2021)," *Journal of Personality Disorders* 36, no. 3 (2022): 249–253; W. H. Martens, "The problem with Robert Hare's Psychopathy Checklist: Incorrect conclusions, high risk of misuse, and lack of reliability," *Medical Law* 27, no. 2 (2008): 449–462; Norbert Leygraf and Klaus Elsner, "Risks of diagnosing psychopathic disorders," in *The Wiley International Handbook on Psychopathic Disorders and the Law*, ed. A. R. Felthous and H. Sass (Hoboken, NJ: John Wiley & Sons, 2020), 145–158; Fowles, "Current scientific views of psychopathy"; Skeem et al., "Psychopathic personality"; Zinger and Forth, "Psychopathy and Canadian criminal proceedings"; Wulach, "Diagnosing the DSM-III antisocial personality disorder."

26. AMA, "AMA principles of medical ethics."

27. APA, *Ethical principles*, 15.

28. Hare, *Hare Psychopathy Checklist–Revised*, 15.

29. Polaschek and Skeem, "Treatment of adults and juveniles."

30. Wong et al., "The effectiveness of violence reduction treatment"; Roberson and Vitacco, "Psychopathy in correctional settings"; S. Wong and Robert Hare, *Guidelines for a Psychopathy Treatment Program* (Toronto: Multi-Health Systems, 2005).

31. DeMatteo and Olver, "Use of the Psychopathy Checklist-Revised"; Edens et al., "Legal and ethical issues"; Edens and Truong, "Psychopathy evidence in legal proceedings"; David R. Lyon, James R. Ogloff, and Stephane M. Shepherd, "Legal and ethical issues in the assessment of psychopathy," in *The clinical and forensic assessment of psychopathy: A practitioner's guide*, ed. C. Gacono (New York: Routledge, 2016), 193–216.

32. AMA, "AMA principles of medical ethics."

33. APA, *Ethical principles*.

34. AMA, "AMA principles of medical ethics."

35. APA, *Ethical principles*, 6.

36. R. R. Faden and T. L. Beauchamp, *A history and theory of informed consent* (New York: Oxford University Press, 1986).

37. Tom L. Beauchamp, "Informed consent: Its history, meaning, and present challenges," *Cambridge Quarterly of Healthcare Ethics* 20, no. 4 (2011): 515–523; Tom L. Beauchamp, "Autonomy and consent," in *The ethics of consent: Theory and practice*, ed. Franklin Miller and Alan Alan Wertheimer (New York: Oxford University Press, 2009), 55–78.

38. AMA, "AMA principles of medical ethics."

39. APA, *Ethical principles*, 7.

40. Michael S. Purcell, Jennifer A. Chandler, and J. Paul Fedoroff, "The use of phallometric evidence in Canadian criminal law," *Journal of the American Academy of Psychiatry and the Law Online* 43, no. 2 (2015): 141–153; D. Bourget and J. Bradford, "Current ethics dilemmas in the assessment and treatment of sex offenders," in *Ethics challenges in forensic psychiatry and psychology practice*, ed. E. E. Griffith (New York: Columbia University Press, 2018), 141–153.

41. American Psychological Association, "Specialty guidelines for forensic psychology," *American Psychological Association* 68, no. 1 (2013): 7–19, 8.

42. Tom L. Beauchamp, *Standing on principles: Collected essays* (New York: Oxford University Press, 2010), 41.

43. AMA, "AMA principles of medical ethics."

44. APA, *Ethical principles*, 6.

45. APA, "Evidence-based practice in psychology"; Busch and McCarthy, "The emergence of evidence-based practice in psychology"; Gregory E. Gray, *Concise guide to evidence-based psychiatry* (Arlington, VA: American Psychiatric Publishing, 2004).

46. Kenneth Goodman, *Ethics and evidence-based medicine: Fallibility and responsibility in clinical science* (Cambridge: Cambridge University Press, 2002), 129.

47. APA, *Ethical principles*, 13.

48. Müller-Isberner et al., "Implementation of evidence-based practices"; Glancy and Saini, "Confluence of evidence-based practice"; APA, "Evidence-based practice in psychology"; Goodman, *Ethics and evidence-based medicine*.

49. Ian James Kidd, Lucienne Spencer, and Havi Carel, "Epistemic injustice in psychiatric research and practice," *Philosophical Psychology* (2022): 1–29.

50. Larsen et al., "Are psychopathy assessments ethical?"; Devon Polaschek, "Criminal justice responses to psychopathy," in *The complexity of psychopathy*, ed. Jennifer E. Vitale (Cham: Springer Nature, 2022), 571–610; Rasmus Rosenberg Larsen, "Psychopathy treatment and the stigma of yesterday's research," *Kennedy Institute of Ethics Journal* 29, no. 3 (2019): 243–272; Edens et al., "Legal and ethical"; DeMatteo et al., "Death is different"; Edens and Truong, "Psychopathy evidence in legal proceedings"; Steverson, *The ethics of employment screening for psychopathy*; T. Douglas et al., "Risk assessment tools in criminal justice and forensic psychiatry: The need for better data," *European Psychiatry* 42 (2017): 134–137.

51. For a more detailed discussion of various counterarguments, see Larsen et al., "Are psychopathy assessments ethical?"

52. APA, "Specialty guidelines for forensic psychology"; Glancy et al., "AAPL practice guideline"; AMA, "AMA principles of medical ethics"; D. Glancy Graham, Chatterjee Sumeeta, and Miller Daniel, "Ethics, empathy, and detached concern in

forensic psychiatry," *Journal of the American Academy of Psychiatry and the Law Online* 49, no. 2 (2021): 1–8.

53. AMA, "AMA principles of medical ethics," chapter 9, opinion 9.7.3.

54. American Psychological Association, "New APA policy bans psychologist participation in national security interrogations," *The Monitor* 46, no. 8 (2015): 8; Kenneth Pope, "The Hoffman Report and the American Psychological Association: Meeting the challenge of change," in *Ethics in psychotherapy and counseling: A practical guide*, ed. Kenneth S. Pope and Melba J. T. Vasquez (Hoboken, NJ: Wiley, 2016), 361–369; Roy J. Eidelson, *Doing harm: How the world's largest psychological association lost its way in the War on Terror* (Montreal: McGill-Queen's University Press, 2023).

55. AMA, "AMA principles of medical ethics."

56. APA, *Ethical principles*, 4.

57. John Monahan, "Tarasoff at thirty: How developments in science and policy shape the common law," *University of Cincinnati Law Review* 75 (2007).

58. R. E. Upshur, "Principles for the justification of public health intervention," *Canadian Journal of Public Health* 93, no. 2 (2002): 101–103.

59. M. G. Bloche, "Psychiatry, capital punishment, and the purposes of medicine," *International Journal of Law and Psychiatry* 16, no. 3–4 (1993): 301–357; Thomas Grisso, "Reply to Schafer: Doing harm ethically," *Journal of the American Academy of Psychiatry and the Law* 29, no. 4 (2001): 457–460; Piyal Sen et al., "Ethical dilemmas in forensic psychiatry: Two illustrative cases," *Journal of Medical Ethics* 33, no. 6 (2007): 337–341; Upshur, "Principles"; R. M. Veatch and J. B. Pitt, "The myth of presumed consent: Ethical problems in new organ procurement strategies," *Transplantation Proceedings* 27, no. 2 (1995): 1888–1892.

60. G. Spitale, "COVID-19 and the ethics of quarantine: A lesson from the Eyam plague," *Med Health Care Philos* 23, no. 4 (2020): 603–609; Upshur, "Principles."

61. P. S. Appelbaum, "A theory of ethics for forensic psychiatry," *Journal of the American Academy of Psychiatry and the Law* 25, no. 3 (1997): 233–247.

62. Appelbaum, "Theory of ethics," 237.

63. APA, *Opinions of the Ethics Committee*; APA, "Specialty guidelines for forensic psychology"; "American Academy of Psychiatry and the Law: Ethics guidelines for the practice of forensic psychiatry," American Academy of Psychiatry and the Law, 2005, https://aapl.org/ethics-guidelines.

64. Grisso, "Reply to Schafer: Doing harm ethically," 459.

65. Bloch and Green, *Psychiatric ethics*; Koocher and Keith-Spiegel, *Ethics in psychology*; Sadoff, *Ethical issues*.

66. Paul S. Appelbaum, "Ethics and forensic psychiatry: Translating principles into practice," *Journal of the American Academy of Psychiatry and the Law Online* 36, no. 2 (2008): 195–200.

67. Bloch and Green, *Psychiatric ethics*; Koocher and Keith-Spiegel, *Ethics in psychology*.

68. Yang et al., "Efficacy of violence prediction"; Fazel et al., "Predictive performance"; Maya G. T. Ogonah et al., "Violence risk assessment instruments in forensic psychiatric populations: A systematic review and meta-analysis," *The Lancet Psychiatry* 10, no. 10 (2023): 780–789.

Chapter 8

1. L. F. Barrett, "Zombie ideas," *Observer* 32, no. 8 (2019), https://www.psychologi calscience.org/observer/zombie-ideas.

2. Alexander Bird, "Scientific progress," in *The Oxford handbook of philosophy of science*, ed. Paul Humphreys (New York: Oxford University Press, 2016), 544–563; John P. A. Ioannidis, "Why science is not necessarily self-correcting," *Perspectives on Psychological Science* 7, no. 6 (2012): 645–654.

3. De Brito et al., "Psychopathy"; J. E. Vitale, ed., *The complexity of psychopathy* (Cham: Springer Nature, 2022); Glenn and Raine, *Psychopathy*; Patrick, "Psychopathy"; Patrick, *Handbook of psychopathy*.

4. David E. J. Linden, "The challenges and promise of neuroimaging in psychiatry," *Neuron* 73, no. 1 (2012): 8–22; D. R. Weinberger and E. Radulescu, "Finding the elusive psychiatric "lesion" with 21st-century neuroanatomy: A note of caution," *American Journal of Psychiatry* 173, no. 1 (2016): 27–33; Bernard Christophe, "Brain's best kept secret: Degeneracy," *eNeuro* 10, no. 11 (2023): 1–5.

5. Stephen D. Hart and Alana N. Cook, "Current issues in the assessment and diagnosis of Psychopathy (Psychopathic Personality Disorder)," *Neuropsychiatry* 2, no. 6 (2012): 497–508; Skeem and Cooke, "Is criminal behavior a central component of psychopathy?"; Jennifer L. Skeem and David J. Cooke, "One measure does not a construct make: Directions toward reinvigorating psychopathy research—Reply to Hare and Neumann (2010)," *Psychological Assessment* 22, no. 2 (2010): 455–459; Robert Hare and Craig S. Neumann, "The role of antisociality in the psychopathy construct: Comment on Skeem and Cooke (2010)," *Psychological Assessment* 22, no. 2 (2010): 446–454.

6. A. Nunes, T. Trappenberg, and M. Alda, "We need an operational framework for heterogeneity in psychiatric research," *Journal of Psychiatry & Neuroscience* 45, no. 1 (2020): 3–6.

7. Patrick, "Psychopathy"; Deming et al., "Psychopathy and medial frontal cortex"; Deming et al., "How reliable are amygdala findings?"

8. Barrett, "Zombie ideas"; Brainard Guy Peters and Maximilian Lennart Nagel, *Zombie ideas: Why failed policy ideas persist* (Cambridge: Cambridge University Press, 2020); Paul Krugman, *Arguing with zombies: Economics, politics and the fight for a better future* (New York: W. W. Norton, 2020).

9. Steven D. Hales, "Thinking tools: You can prove a negative," *Think* 4, no. 10 (2005): 109–112.

10. Douglas N. Walton, *Arguments from ignorance* (University Park, PA: The Pennsylvania State University Press, 1996).

11. M. Pigliucci, *Nonsense on stilts: How to tell science from bunk* (Chicago: University of Chicago Press, 2010); Meehl, "Cliometric metatheory: II."

12. M. Pigliucci and M. Boudry, eds., *Philosophy of pseudoscience: Reconsidering the demarcation problem* (Chicago: University of Chicago Press, 2013).

13. Igor Douven, "Abduction," in *The Stanford encyclopedia of philosophy*, ed. Edward Zalta (2021). https://plato.stanford.edu/archives/sum2021/entries/abduction; Dellsén, Finnur, *Abductive reasoning in science* (Cambridge: Cambridge University Press, 2024).

14. Atocha Aliseda, *Abductive reasoning: Logical investigations into discovery and explanations* (Dordrecht: Springer, 2006).

15. Adolfas Mackonis, "Inference to the best explanation, coherence and other explanatory virtues," *Synthese* 190, no. 6 (2013): 975–995.

16. Sass and Felthous, "The heterogeneous construct of psychopathy."

17. Lewis, "Psychopathic personality."

18. Hare, "Forty years aren't enough."

19. Lilienfeld et al., "Hervey Cleckley (1903–1984)."

20. Cleckley, *Mask of sanity*, 16.

21. Lilienfeld et al., "Hervey Cleckley (1903–1984)."

22. Reid et al., *Unmasking the psychopath*.

23. For a clear example of this view from that period, see Roy W. Persons and Carol E. Persons, "Some experimental support of psychopathic theory: A critique," *Psychological Reports* 16, no. 3 (1965): 745–749.

24. Hare, "Forty years aren't enough."

25. DeMatteo et al., "Role and reliability of the psychopathy"; Hare and Hart, "Commentary on antisocial personality disorder"; Crego and Widiger, "Psychopathy and the DSM."

26. Blair, "Cognitive developmental approach."

27. Serin and Amos, "Role of psychopathy."

28. Rice et al., "Maximum security therapeutic community."

29. Blackburn, "On moral judgements and personality disorders."

30. R. Blackburn and J. W. Coid, "Psychopathy and the dimensions of personality disorder in violent offenders," *Personality and Individual Differences* 25, no. 1 (1998): 129–145; O'Kane et al., "Psychopathy and moral reasoning"; Ronald Blackburn, "Psychopathy as a personality construct," in *Handbook of personology and psychopathoogy*, ed. S. Strack (New York: Wiley, 2005), 271–291.

31. Blackburn, *Psychology of criminal conduct.*

32. Hart et al., "The Psychopathy Checklist—Revised (PCL-R)"; Widiger and Corbitt, "Antisocial personality disorder"; Crego and Widiger, "Psychopathy and the DSM."

33. Meloy, *Psychopathic mind.*

34. Hare, *Without conscience.*

35. E. Hickey, *Serial murderers and their victims* (Boston: Cengage, 2015).

36. Hickey, *Serial murderers and their victims.*

37. Sara Sun Beale, "Still tough on crime? Prospects for restorative justice in the United States interdisciplinary perspectives on restorative justice," *Utah Law Review* 413, no. 1 (2003): 413–438.

38. Nathan J. Robinson, *Superpredator: Bill Clinton's use and abuse of black America* (W. Sommerville, MA: Current Affairs Press, 2016).

39. Cleckley, *Mask of sanity*, 372–373.

40. H. Merskey, "The manufacture of personalities: The production of multiple personality disorder," *British Journal of Psychiatry* 160 (1992): 327–340.

41. Corbett H. Thigpen and Hervey M. Cleckley, *The three faces of Eve* (New York: McGraw-Hill, 1957).

42. Evelyn Lancaster, *The final face of Eve* (New York: McGraw-Hill, 1958); Chris Costner Sizemore and Elen Sain Pittilo, *I'm Eve* (New York: Doubleday, 1977); Chris Costner Sizemore, *Mind of my own: The woman who was known as "Eve" tells the story of her triumph over Multiple Personality Disorder* (New York: William Morrow, 1989).

43. Philip J. Corr and Gerald Matthews, eds., *The Cambridge handbook of personality psychology*, 2nd ed. (New York: Cambridge University Press, 2020).

44. Pigliucci, *Nonsense on stilts.*

45. For a deeper discussion, see Peter Lipton, "Inference to the best explanation," in *A companion to the philosophy of science*, ed. W. H. Newton-Smith (Hoboken, NJ:

Blackwell, 2000), 184–193; Meehl, "Cliometric metatheory: II"; Paul R. Thagard, "The best explanation: Criteria for theory choice," *Journal of Philosophy* 75, no. 2 (1978): 76–92.

46. Dorothy Walsh, "Occam's razor: A principle of intellectual elegance," *American Philosophical Quarterly* 16, no. 3 (1979): 241–244.

47. Meehl, "Cliometric metatheory: II." 363.

48. Lilienfeld, "Afterword," 484.

49. Lilienfeld, "Afterword," 486.

50. See also David P. Farrington and Henriette Bergstrøm, "The development of psychopathy through the lifespan and its relation to offending," in *Psychopathy and criminal behavior: Current trends and challenges*, ed. Paulo Barbosa Marques, Mauro Paulino, and Laura Alho (Academic Press, 2021), 105–125; Joshua D. Miller and Donald R. Lynam, "Psychopathy and personality: Advances and debates," *Journal of Personality* 83, no. 6 (2015): 585–592.

51. S. O. Lilienfeld et al., "Personality disorders as emergent interpersonal syndromes: Psychopathic personality as a case example," *Journal of Personality Disorders* 33, no. 5 (2019): 577–622.

52. Jesse J. Prinz, "Against empathy," *The Southern Journal of Philosophy* 49 (2011): 214–233; Paul Bloom, *Against empathy: The case for rational compassion* (New York: Ecco, HarperCollins, 2016).

53. Jonathan Haidt, *The righteous mind: Why good people are divided by politics and religion* (New York: Pantheon Books, 2012); Alan Page Fiske and Tage Shakti Rai, *Virtuous violence: Hurting and killing to create, sustain, end, and honor social relationships* (Cambridge: Cambridge University Press, 2014).

54. Blair et al., *The psychopath.*

55. Cleckley, *Mask of sanity*, 50.

56. Cleckley, *Mask of sanity*, 109.

57. Cleckley, *Mask of sanity*, 166.

58. Cleckley, *Mask of sanity*, 69.

59. Cleckley, *Mask of sanity*, 59.

60. Cleckley, *Mask of sanity*, 159.

61. See also Hickey, *Serial murderers and their victims.*

62. For more information about the Ted Bundy-case, see Stephen Michaud and Hugh Aynesworth, *The only living witness: The true story of serial sex killer Ted Bundy* (Laguna,

TX: Authorlink Press, 1983); E. Hickey et al., "Deviance at its darkest: Serial murder and psychopathy," in *The Handbook of Psychopathy*, ed. C. Patrick (New York: The Guilford Press, 2018), 570–584; Hickey, *Serial murderers and their victims*.

63. Kiehl and Hoffman, "Criminal psychopath."

64. Hickey, *Serial murderers and their victims*.

65. Sigurdson et al., "Homeopathy can offer empirical insights on treatment effects in a null field"; S. Reisman, M. Balboul, and T. Jones, "P-curve accurately rejects evidence for homeopathic ultramolecular dilutions," *PeerJ* 7 (2019): e6318.

66. D. J. Bem, "Feeling the future: Experimental evidence for anomalous retroactive influences on cognition and affect," *Journal of Personality and Social Psychology* 100, no. 3 (2011): 407–425.

67. E. J. Wagenmakers et al., "Why psychologists must change the way they analyze their data: The case of psi: comment on Bem (2011)," *Journal of Personality and Social Psychology* 100, no. 3 (2011): 426–432; D. S. Schwarzkopf, "We should have seen this coming," *Frontiers in Human Neuroscience* 8 (2014): 332.

68. Jarkko Jalava, Michael Maraun, and Stephanie Griffiths, *The myth of the born criminal: Psychopathy, neurobiology, and the creation of the modern degenerate* (Toronto: University of Toronto Press, 2015).

69. Lilienfeld, "Afterword"; Lilienfeld et al., "Personality disorders."

70. Lykken, "Study of anxiety in the sociopathic personality."

71. Persons and Persons, "Some experimental support of psychopathic theory."

72. David T. Lykken, "Black cats, red herrings, and horses of another color," *Psychological Reports* 18, no. 2 (1966): 621–622.

73. Marcus R. Munafò et al., "A manifesto for reproducible science," *Nature Human Behaviour* 1, no. 1 (2017): 0021; Ioannidis, "Why most published research findings are false"; Leif D. Nelson, Joseph Simmons, and Uri Simonsohn, "Psychology's renaissance," *Annual Review of Psychology* 69, no. 1 (2018): 511–534.

Conclusion

1. C. J. Hopwood et al., "The time has come for dimensional personality disorder diagnosis," *Personality and Mental Health* 12, no. 1 (2018): 82–86; Bach and Mulder, "Empirical foundation"; Clark et al., "Three approaches to understanding."

2. Stefan and Schönbrodt, "Big little lies"; O'Donohue et al., *Avoiding questionable research practices in applied psychology*.

3. B. A. Nosek et al., "Promoting an open research culture: Author guidelines for journals could help to promote transparency, openness, and reproducibility," *Science* 348, no. 6242 (2015): 1422–1425.

4. John et al., "Measuring the prevalence."

5. Scott E. Maxwell, Michael Y. Lau, and George S. Howard, "Is psychology suffering from a replication crisis? What does "failure to replicate" really mean?," *American Psychologist* 70, no. 6 (2015): 487–498; Patrick E. Shrout and Joseph L. Rodgers, "Psychology, science, and knowledge construction: Broadening perspectives from the replication crisis," *Annual Review of Psychology* 69 (2018): 487–510.

6. Bruno Verschuere et al., "A plea for preregistration in personality disorders research: The case of psychopathy," Personality Disorders 3217, *Journal of Personality Disorders* 35, no. 2 (2021): 161–176.

7. Jalava et al., *The myth of the born criminal*; Lilienfeld, "Afterword."

8. Wulach, "Diagnosing the DSM-III antisocial personality disorder."

9. Barbaree, "Psychopathy, treatment behavior, and recidivism"; Gendreau et al., "A lesson in knowledge cumulation"; Edens et al., "Assessment of 'juvenile psychopathy.'"

10. John F. Edens et al., "'A psychopath by any other name?': Juror perceptions of the DSM-5 'Limited Prosocial Emotions' specifier," *Journal of Personality Disorders* 31, no. 1 (2017): 90–109.

11. Gianluca Sesso and Annarita Milone, "Conduct Disorder, empathy, and callous-unemotional traits," in *Handbook of anger, aggression, and violence*, ed. Colin Martin, Victor R. Preedy, and Vinood B. Patel (Cham: Springer International Publishing, 2022), 1–26.

12. Adrian Furnham, Steven C. Richards, and Delroy L. Paulhus, "The Dark Triad of personality: A 10 year review," *Social and Personality Psychology Compass* 7, no. 3 (2013): 199–216.

13. Lilienfeld et al., "Successful psychopathy: A scientific status report"; Patrick, "Psychopathy."

14. Maruna, *Making good*; Per-Olof H. Wikström, "Explaining crime and criminal careers: the DEA model of situational action theory," *Journal of Developmental and Life-Course Criminology* 6, no. 2 (2020): 188–203.

15. R. Rogers, K. Y. Tazi, and E. Y. Drogin, "Forensic assessment instruments: Their reliability and applicability to criminal forensic issues," *Behavioral Sciences & the Law* 41, no. 5 (2023): 415–431; Kirk Heilbrun, Richard Rogers, and Randy Otto, "Forensic assessment: Current status and future directions," in *Taking psychology and law into the*

twenty-first century, ed. J. R. Ogloff, Perspectives in law & psychology, vol. 14. (New York: Kluwer Academic, 2002), 119–146; Edens and Truong, "Psychopathy evidence in legal proceedings."

16. Corr and Matthews, *The Cambridge handbook of personality psychology*; Kenneth H. Craik, Robert Hogan, and Raymond N. Wolfe, eds., *Fifty years of personality psychology* (New York: Springer, 1993).

17. Rush, *An inquiry*, 4.

18. Fiske and Rai, *Virtuous Violence*; Smith, *Less than human*; Albert Bandura, *Moral disengagement: How people do harm and live with themselves* (New York: Worth Publishers, 2016).

Index

Publisher contact:
The MIT Press
Massachusetts Institute of Technology
77 Massachusetts Avenue, Cambridge, MA 02139
mitpress.mit.edu

EU Authorised Representative:
Easy Access System Europe, Mustamäe tee 50, 10621 Tallinn, Estonia
gpsr.requests@easproject.com

Printed by Integrated Books International, United States of America